Technologies, Applications and Assessments for Proper Sustainable Forest Operations (SFO)

Technologies, Applications and Assessments for Proper Sustainable Forest Operations (SFO)

Editors

Rachele Venanzi
Janine Schweier
Rodolfo Picchio

MDPI • Basel • Beijing • Wuhan • Barcelona • Belgrade • Manchester • Tokyo • Cluj • Tianjin

Editors
Rachele Venanzi
Tuscia University
Italy

Janine Schweier
Swiss Federal Institute for Forest
Switzerland

Rodolfo Picchio
Tuscia University
Italy

Editorial Office
MDPI
St. Alban-Anlage 66
4052 Basel, Switzerland

This is a reprint of articles from the Special Issue published online in the open access journal *Sustainability* (ISSN 2071-1050) (available at: https://www.mdpi.com/journal/sustainability/special_issues/SFO).

For citation purposes, cite each article independently as indicated on the article page online and as indicated below:

LastName, A.A.; LastName, B.B.; LastName, C.C. Article Title. *Journal Name* **Year**, *Volume Number*, Page Range.

ISBN 978-3-0365-4977-4 (Hbk)
ISBN 978-3-0365-4978-1 (PDF)

Cover image courtesy of Rachele Venanzi

Contents

About the Editors

Rachele Venanzi

Rachele Venanzi (Researcher) obtained the Environmental Science and the Forest Science degrees at Tuscia University in Viterbo and a PhD degree in Forest Mechanization at Padua University. Since 2013, she has both worked as a researcher in Forestry and Forest Logging and Wood Technologies at Tuscia University, and has been part of the 'Forest Utilization Work Group' at Tuscia University (DAFNE). Her fields of expertise include sustainable forest operations; reduced impact logging; QBS-ar index; sustainable forest management; and forest mechanization. She has worked as the Italian representative in an EU COST Action on silviculture. Furthermore, she has co-authored 81 peer-reviewed papers in international journals indexed by WoS and Scopus, 6 technical books, and over 24 papers in technical magazines. She also works on the editorial boards of international scientific journals covering forestry and forest engineering subjects as Guest Editor. She is a member of the Italian Society of Silviculture and Forest Ecology.

Janine Schweier

Janine Schweier (Researcher) obtained a Forest Science degree and a doctoral degree at the Faculty of Environment and Natural Resources of the University of Freiburg, Germany. From 2008 to 2019, she worked as a researcher in Forestry at the same University with a focus on Life-Cycle Assessment and Sustainability Impact Assessment of forest–wood chains. In 2019, she moved to the Swiss Federal Institute for Forest, Snow, and Landscape Research (WSL) where she is head of the research group Sustainable Forestry. Her research focus includes sustainable forest operations, forest ecosystem services, forest economy in a changing environment, and forest digitalization. She has co-authored more than 40 peer-reviewed papers in international journals indexed by WoS and Scopus and contributed to many technical magazines, guidelines, and book chapters. She is editorial board member of the *International Journal of Forest Engineering* and the *Croatian Journal of Forest Engineering*, member of the expert committees "silviculture and forest technique" and "forest machines" of the German Center for Forest Work and Technology (KWF) and board member of the Swiss Wood Innovation Network.

Rodolfo Picchio

Rodolfo Picchio (Professor) obtained a Forest Science degree and a PhD degree in Forest Mechanization at Tuscia University in Viterbo. From 2004 to 2014, he worked as a university researcher, and since 2014 has been an Associate Professor in Forest Logging and Wood Technologies. Since 2004, he has led the 'Forest Utilization Work Group' at Tuscia University (DAFNE). From 2015 to 2017, he served as the president of the master's degree of Forest Science "Conservation and Restoration of the Forest Environment and Soil Defense" at Tuscia University (DAFNE). He is on the board of the PhD course "Engineering for Energy and Environment" at Tuscia University. His fields of expertise include forest logging, forest mechanization, precision forestry, wood technologies, and the environmental impacts of wood harvesting operations. He has worked as the Italian representative in two EU COST Actions, on silviculture and biomass. Furthermore, he has co-authored 136 peer-reviewed papers in international journals indexed by WoS and Scopus, 18 technical books, and over 160 papers in technical magazines. He also works on the editorial boards of five international scientific journals covering forestry and forest engineering subjects. He is a member of the Italian Society of Silviculture and Forest Ecology.

Preface to "Technologies, Applications and Assessments for Proper Sustainable Forest Operations (SFO)"

This Special Issue focused on the "Technologies, Applications, and Assessments for Proper Sustainable Forest Operations (SFOs)". One of the main topics was to promote knowledge for future relations between forest logging, environmental protection, and management of forests in order to provide timber at reasonable costs and other ecosystem services, such as recreation, conservation, and biodiversity. Only efficient planning and management of forest operations and the implementation of sustainable supply chains will offer high social and environmental benefits and provide various ecosystem services in the long term. These aspects can be guaranteed only through sustainable forest management in synergy with SFO, tools essential for proper environmental protection, and they are mandatory in order to maintain forests and their multiple functions. In particular, forest operations are interesting but delicate issues to be analyzed and evaluated in order to achieve real sustainability.

Authors were invited to contribute to this Special Issue with original papers including experimental studies, monitoring approaches, and models on silviculture, logging activities, and forest operations engineering, in order to promote knowledge and future strategies for effective and efficient SFOs.

This Special Issue contains 12 original papers, some of them reporting the outcomes of long-time experiences, reviewed by international experts in forestry and forest operations engineering fields.

This Special Issue aims to increase the knowledge concerning the SFOs, with practical implications at scientific and technical levels.

We are pleased to share these works with the scientific community, forest engineers, private owners, and public managing authorities, in the hopes that this edition will provide a cognitive base to improve SFOs and reduce arguments between different interests and opinions.

As Guest Editors, we want to express our gratitude to MDPI, who agreed to publish this book; to the staff of *Sustainability* and MPDI, for their kindness and professional support; to the reviewers, for improving original texts and clearing little slips; and to the authors, for providing their papers with ethical and scientific rigor.

Guest Editor contributions: The three coeditors equally contributed to organizing the Special Issue, to the editorial work, and to writing this editorial.

Funding: No external funding was obtained for editing this Special Issue and for writing the preface.

Conflicts of Interest: The authors declare no conflicts of interest.

This book collects a representative sample of the most recent papers on the subject, which come from many different countries and cover a variety of subjects, confirming the wide scope covered by the Sustainable Forest Operations topic.

This work was supported by the Italian Ministry for education, University and Research (MIUR) for financial support (Law 232/2016, Italian University Departments of excellence 2023–2027) project "Digitali, Intelligenti, Verdi e Sostenibili (D.I.Ver.So)—UNITUS-DAFNE WP3 and the Agritech National Research Center receiving funding from the European Union Next-GenerationEU (PIANO NAZIONALE DI RIPRESA E RESILIENZA (PNRR)—MISSIONE 4 COMPONENTE 2, INVESTIMENTO 1.4—D.D. 1032 17/06/2022, CN00000022). These papers reflect only the authors' views and opinions, neither the European Union nor the European Commission can be considered responsible for them.

Rachele Venanzi, Janine Schweier, and Rodolfo Picchio
Editors

sustainability

MDPI

Article

Soil Recovery Assessment after Timber Harvesting Based on the Sustainable Forest Operation (SFO) Perspective in Iranian Temperate Forests

Hadi Sohrabi [1], Meghdad Jourgholami [1], Mohammad Jafari [2], Naghi Shabanian [3], Rachele Venanzi [4,5,*], Farzam Tavankar [6] and Rodolfo Picchio [4]

1 Department of Forestry and Forest Economics, Faculty of Natural Resources, University of Tehran, Karaj 999067, Alborz, Iran; hadi.sohrabi@ut.ac.ir (H.S.); mjgholami@ut.ac.ir (M.J.)
2 Department of Reclamation of Arid and Mountainous Regions, Faculty of Natural Resources, University of Tehran, Karaj 999067, Alborz, Iran; jafary@ut.ac.ir
3 Department of Forestry, Faculty of Natural Resources, University of Kurdistan, Khanagah Campus, Sanandaj 66177-1-5175, Iran; n.shabanian@uok.ac.ir
4 Department of Agricultural and Forest Sciences, University of Tuscia, 01100 Viterbo, Italy; r.picchio@unitus.it
5 Department of Land, Environment, Agriculture and Forestry, Università degli Studi di Padova, 35020 Legnaro, Padova, Italy
6 Department of Forestry, Khalkhal Branch, Islamic Azad University, Khalkhal 56817-31367, Iran; farzam.iaukh@gmail.com
* Correspondence: venanzi@unitus.it; Tel.: +39-0761-357-400; Fax: +39-0761-357-250

Received: 12 March 2020; Accepted: 1 April 2020; Published: 3 April 2020

Abstract: Minimizing the impact of timber harvesting on forest stands and soils is one of the main goals of sustainable forest operation (SFO). Thus, it is necessary to make an accurate assessment of forest operations on soil that is based on the SFO perspective. The present study was conducted according to SFO principles to investigate the time required for the natural recovery of soil after disturbance by skidding operations in some Iranian forests. The physical, chemical, and biological properties of soil found in abandoned skid trails from different time periods were compared with undisturbed forest soils. The soil bulk density, the penetration resistance, and the microporosity of a 25-year-old skid trail were 8.4–27.4% and 50.4% greater, and the total porosity, macroporosity, and soil moisture were 1.9–17.1% and 4.6% lower than the undisturbed area. In a 25-year-old skid trail, the values of pH, Electrical conductivity (EC), C, N, available P, K, Ca, and Mg, earthworm density, and biomass were lower than in the undisturbed area, and the C/N ratio value was higher than in the undisturbed area. High traffic intensity and slope classes of 20–30% in a three-year-old skid trail had the greatest impact on soil properties. In order to have sustainable timber production, SFO should be developed and soil recovery time should be reduced through post-harvest management operation.

Keywords: skidding operation; soil properties; sustainable forest operation; traffic intensity; soil recovery

1. Introduction

One of the main goals of sustainable forest operation (SFO) is to comply with the forest operations ecology [1–3]. Environmental issues and concerns have been increasing as quickly as the development of the mechanization of forest operations across most of the world over the last 50 years [4]. From an SFO perspective, forest operational planning must consider all possible factors affecting environmental impacts, as well as their interactions. Soil plays a vital role in forest ecosystems by providing nutrients, water, and energy flow, which leads to forest productivity and biodiversity conservation [5–7]. Due to their specific characteristics, forest soils are highly susceptible to disturbances caused by skidding operations, the effects of which may persist for years after skidding [8–10]. Log skidding in forest stands impacts the soil that, at the same time, is used as a growth substrate. However, due to soil disturbance, skidding operations lead to soil compaction, soil layer displacement, and the deformation of soil structure and texture [11–13]. Skidding operations generally require a wide and dense network of skid trails for proper access at the forest stand level. Also, finding the optimum space between the forest road to minimize the total cost of timber extraction and road construction plays a key role in planning sustainable forest operations [14]. In this case, a large area of forest soil is affected by the machinery traffic, with possible related changes or disturbances. In general, it may be possible to note an increasing trend in soil bulk density (BD) values [10,15–18], as well as shear and penetration resistance (PR) values [7,8,11,19,20]. At the same time, it may be possible to note a decreasing trend for the presence of macropores [21], hydraulic conductivity of water, air permeability, air–water and gas exchange [22], organic matter [5,20,23], the presence and quality of soil microorganisms [24,25], and carbon sequestration capacity [26]. The lack of recovery of soil properties, especially bulk density and penetration resistance, can also impact the regeneration and decrease of growth seedlings, as reported by Picchio et al. [9] and Sohrabi et al. [27].

Machinery traffic could also lead to variations in some soil features, but without a clear trend, including soil pH [20,28]; microbial biomass, due to inadequate respiration [29–31]; and nutrient availability, due to changes in mineralization process [20,29]. Variations in these soil features, depending on their degree and intensity, can influence plant growth [23,32–34].

The influence of important factors on soil property changes after skidding operation are traffic intensity (the number of passes), slope gradients, and skidding direction. Numerous studies have shown that high traffic intensity (more than 15 passes) and high slope gradient (more than 20%) in mountainous areas cause soil disturbance to a greater extent and with greater intensity [35–37].

The recovery of physical, chemical, and biological soil properties is a topic that has been extensively discussed but has not always been clearly stated due to the difficulties in assessing multiple complex variables. The processes of soil swelling and shrinkage, freezing and thawing, wetness and drought [38,39], root–soil interaction [40,41], fauna activities [39,41,42], precipitation, and the height of the stands are the main factors that can accelerate the recovery process of disturbed soils [43,44]. In addition, soil properties (structure, texture, thickness, and depth), soil moisture and litter layer, terrain slope, soil initial compaction, harvesting systems, and the type of machinery used in skidding operations also affect the soil recovery rate [8,45–48].

The recovery of the physical, chemical, and biological properties of compacted soils can take years and even decades without the use of improvement treatments [12,17,49–51]. Ezzati et al. [50] and Jaafari et al. [36] have shown that soil properties did not recover during a long-term period (20 years) after skidding operation. Von Wilpert and Schäffer [52] reached a similar conclusion; whereas Mohieddinne et al. [48] found that sandy neutral soils recovered in less than 20 years due to soil biological activities. According to some studies, the time required to recover the biological properties of soil is less than that of the physical and chemical properties [5,11]. For example, Mariani et al. [53] showed that soil biomass recovered on the forest floor between three and seven years after harvesting in Canada. Furthermore, Macedo et al. [54] reported that soil C and N were significantly recovered 13 years after harvesting. Picchio et al. [8] and Venanzi et al. [20] have shown that soil recovery after reduced impact logging activities can occur within 6–10 years, and this can be considered fairly fast. Accordingly, Hope [55] has shown that many of the changes that were visible in the first year after the harvesting operation disappeared in the 10 years following the operation, and the effects of machinery traffic on the chemical properties of the soil were significantly recovered, while the shortest time to recover physical properties such as bulk density was reported to be five years.

Three especially prominent approaches in the sustainability of forest operations are: environmentally sound forest harvesting, reduced-impact logging, and forest operations ecology [3]. Forest operations ecology applies the principles of industrial ecology to forest operations systems. It aims to develop and deploy environmentally sound forest operations technologies, to use resources efficiently, to minimize the overall production of waste and emissions, and to minimize the impacts on the structures and functions of environmental spheres (atmosphere, biosphere, hydrosphere, and lithosphere) [1]. Characterizing the effects of skidding operations on the physical, chemical, and biological properties of soil is one important step in protecting the forest ecosystem and forest soil fertility [56]. The aeration status and soil porosity, the quality and quantity of organic matter, the microbial biomass, and the soil nutrients are considered criteria for the evaluation of soil health and biological and chemical activity. Therefore, there is a need to investigate the changes in the physical, chemical, and biological properties of soil. Awareness of the time required for the recovery of disturbed soils is essential to sustainable forest management. We hypothesize that soil properties could not be recovered after skidding operations over a long-term period (20 years) under natural conditions. The main aim of this study is (1) to evaluate the time required for the recovery of the physical, chemical, and biological properties of soil; and (2) to assess the long-term effect of the slope of the skid trail, machine traffic, and soil depth on the recovery process of the soil environment after ground-based skidding operations in the Hyrcanian forests of Iran.

2. Materials and Methods

2.1. Site Description

This study was conducted in seven different forest compartments located in the Namkhaneh and Gorazbon districts of the Kheyrud forest, part of the Hyrcanian forest in northern Iran. The research area lies between 51°36′50″ E and 51°38′21″ E longitude and 36°34′21″ N and 36°33′34 5″ N latitude, at altitude of 1000–1232 m above sea level. The study area has no dry season and has a humid climate with an average annual rainfall of 1146 mm and an average annual temperature of 8.55 °C. The soil texture of the study site ranges from silt loam to loamy (Table 1). The soils are mainly brown forest (Alfisols) with good drainage and, in terms of geology, belong to the Jurassic period. This area is predominantly covered by deciduous trees such as oriental beech (*Fagus orientalis* Lipsky), hornbeam (*Carpinus betulus* L.), velvet maple (*Acer velutinum* Boiss.), Cappadocian maple (*Acer cappadocicum* Gled), large-leaved lime tree (*Tilia platyphyllos* Scop.), chestnut-leaved oak (*Quercus castaneifolia* C.A.M.), mountain elm (*Ulmus glabra* Huds.), and Caucasian alder (*Alnus subcordata* C.A. Mey). The silvicultural treatment applied in the study area was a combination of single-tree selection and group selection, resulting in uneven-aged stands. More general information of the experiments is given in Table 1. Motor-manual felling and processing (i.e., using chainsaws) were carried out at the felling site. The skidding machinery (Timberjack 450 C) was equipped with winch cables and was used to extract logs with a length of 5–15 m from the forest area to the landings. The Timberjack skidder is an articulated four-wheel drive vehicle with a weight of 10.3 tons (total weight with equipment) and an engine power of 177 hp. The front and rear axles in this machine are equipped with 775 × 813 mm tires with an average ground pressure of 220 kPa and a ground clearance of approximately 0.6 m, with an overall width of 3.1 m.

Table 1. Location and specifications of the study area.

Age of Skid Trail (years)	Replication	District (No. of Compartments)	Skid Trail Length (m)	Skid Trail Density (m ha^{-1})	Elevation (m)	Soil Texture
3	1st	Gorazbon (C. 315)	850	78.6	1186	Clay
	2nd	Gorazbon (C. 316)	1000	71.3	1214	Clay
	3rd	Gorazbon (C. 316)	900	71.3	1191	Silty clay loam
10	1st	Gorazbon (C. 318)	930	65.8	1186	Clay loam
	2nd	Gorazbon (C. 317)	1050	84.8	1200	Silty clay loam
	3rd	Gorazbon (C. 317)	860	84.8	1090	Clay loam
20	1st	Namkhaneh (C. 221)	1000	53.8	1120	Silt loam
	2nd	Namkhaneh (C. 220)	985	51.3	1010	Clay loam
	3rd	Namkhaneh (C. 220)	800	51.3	1117	Clay

Table 1. *Cont.*

Age of Skid Trail (years)	Replication	District (No. of Compartments)	Skid Trail Length (m)	Skid Trail Density (m ha^{-1})	Elevation (m)	Soil Texture
	1st	Namkhaneh (C. 218)	1040	74.5	1050	Clay
25	2nd	Namkhaneh (C. 218)	980	74.5	1140	Silty clay loam
	3rd	Namkhaneh (C. 221)	950	53.8	1020	Clay

2.2. Experimental Design

In this study, the long-term natural recovery of soil physical parameters (dry bulk density, total porosity, macroporosity, microporosity, penetration resistance, and soil moisture), chemical parameters (soil pH and Electrical conductivity (EC), C (%), N (%), C/N ratio, available P, K, Ca, and Mg), and biological parameters (earthworm density and biomass) on the surface soil layer of the skid trail was quantified at different levels of slope, traffic intensity, and soil depth and compared to the undisturbed area. Four abandoned skid trails with a downslope skidding direction that encompassed a wide range of longitudinal slope gradients (regardless of the lateral slope, area height, and soil texture and structure) in three replications at the forest level were selected for the study. The trails ranged from three years, through 10 years and 20 years, to 25 years since forest harvesting (Figure 1). A three-year-old skid trail instead of one-year-old skid trail was chosen because the study was conducted after the introduction of a prohibition of forest harvesting in the Hyrcanian forest, and the last trail on which the skidding operation was conducted was three years old. The average load volume and number of logs for each skidding cycle by the Timberjack skidder were 3.8 m^3 and 2, respectively. The motivation behind this research was to gain a retrospective view of the research conducted by Ezzati et al. [50] and Sohrabi et al. [27] in the Hyrcanian forest, who claimed that the recovery of soil physical properties can be lengthy. However, this research did not investigate the recovery of the chemical and biological properties of the soil in responses to soil compaction over the time, which is one of the most interesting ideas developed in the current research. On all the studied skid trails, the skidder only traveled on the skid trails at the time of skidding, and these skid trails were not used for timber extraction in later years. The entrances of skid trails were blocked by embankments after the timber extraction was completed. Felled trees were thick and high (dbh > 60 cm; height > 20 m), and were scattered around the skid trails (selection cutting silviculture), the logs from the bole of each felled tree formed a complete machine load (the logs of each felled tree were individually extracted in one skidding cycle). Due to the above conditions, the farther away from the landings the traffic intensity decreases (Figure 2A). The average length of skid trails in the study area was about 150 m, with the first 50 m (0–50 m) from the forest road considered as high traffic intensity (HST), the second 50 m (50–100 m) as medium traffic intensity (MST), and the third of 50 m (100–150 m) or subsidiary of the skid trail as low traffic intensity (LST) [35,36,50]. In the skid trails, three slope classes (0–10%, 10–20%, and 20–30%) were considered with regard to variations and maximum slope. Therefore, the research plots included three traffic-intensity classes and three slope-gradient classes, thus forming nine combinations of traffic frequency and trail gradients; each treatment combination was replicated three times at the forest level, totaling 27 treatment plots ($N = 27$ in each recovery period). On the skid trails, in each treatment (e.g., a combination of slope and traffic), sampling plots with the dimensions of 40 m^2 were designed. In each sampling plot, five sampling lines were designed at a distance of 2 m from each other and perpendicular to the skid trail, of which three lines were randomly selected for sampling. Soil samples were taken from a depth interval of 0–10 and 10–20 cm at three locations in each plot: (LW) the left wheel track, (BW) between the tracks, and (RW) the right wheel track (Figure 2B). To compare the soil properties between the skid trail and the undisturbed area, soil samples were taken inside the forest at least 20–30 m (the size of the average height of the dominant trees in the area) away from the skid trail, where the effects of the skidding operations on the soil completely disappeared ($n = 81$ in each recovery period). Ruts were measured on the skid trails at the site of the compaction measurement line, and ruts with a depth of 5 cm and a length of at least 2 m were considered to be caused by soil disturbance [57]. In each rut, at 25 mm horizontal intervals, depth was measured, and the mean was considered to be caused by soil disturbance. Obviously, the distance between the compaction and the rut sampling lines was not less than 2 m [58].

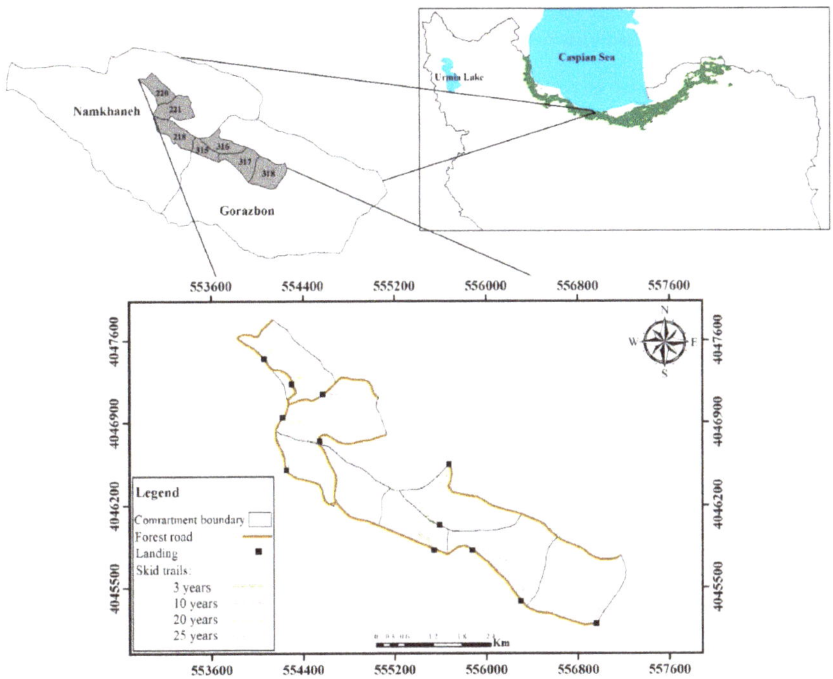

Figure 1. The study area in the Namkhaneh and Gorazbon districts in the Hyrcanian forests and the skid trails in different years after skidding operation (3, 10, 20, and 25 years).

2.3. Measurements and Laboratory Analysis

A steel ring (5 cm in diameter and 10 cm in height; 196.25 cm^3) was used to collect soil samples from depths of 0–10 and 10–20 cm. The soil samples were stored and coded in plastic bags. On the same sampling day, the wet weight of all samples was measured before transfer to the laboratory. In the laboratory, the soil samples were dried at 105 °C for 24 h to calculate the soil moisture content, dry bulk density, and porosity.

The bulk density and total porosity were calculated using Equations (1) and (2).

$$BD = \frac{WD}{VC} \quad (1)$$

where BD is the dry bulk density (g cm^{-3}), WD is the weight of the dry soil (g), and VC is the volume of the cylinder (cm^3).

$$\mathrm{TP} = 1 - \left(\frac{BD}{2.65}\right) \times 100 \quad (2)$$

where TP is the apparent total porosity (%), BD is the bulk density (g cm^{-3}), and 2.65 g cm^{-3} is the soil particle density [59].

Macroporosity and microporosity were calculated using Equations (3) and (4).

$$\mathrm{MIP} = \theta_m \times BD \quad (3)$$

where MIP is the microporosity (%), BD is the dry bulk density (g cm^{-3}), and θ_m is the water content on a mass basis (%).

$$MP = TP - MIP \quad (4)$$

where MP is the macroporosity (%), TP is the total porosity (%), and MIP is the microporosity (%).

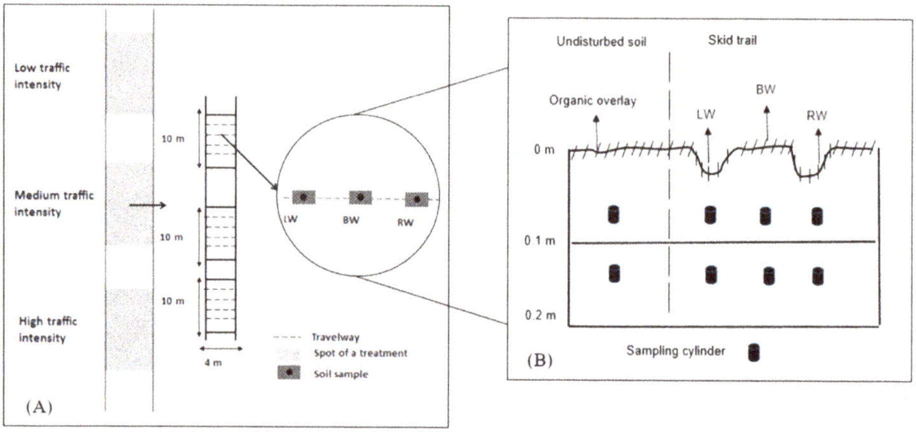

Figure 2. A sketch of the sampling design in the skid trails and the undisturbed area. (**A**): Each skid trail was divided into three traffic intensities (high, medium, and low traffic intensity), and in each traffic class three slope classes were considered (0–10%, 10–20%, and 20–30%). The sampling plots were designed with dimensions of 40 m^2 in each of the skid trail treatments, along with the sampling plots in the undisturbed area and the soil sampling point at depth intervals of 0–10 and 10–20 cm at the three locations in each plot (**B**): the left wheel track (LW), between the tracks (BW), and the right wheel track (RW).

The hydrometric method was used to determine the soil texture, whereas the determination of soil penetration resistance was made with a hand-held soil penetrometer (Eijkelkamp 06.01.SA penetrometer with a 60° cone and a 1 m maximum measuring depth) that was inserted vertically into the soil with equal pressure at the location of each sample. Since soil moisture influences the measurement of penetration resistance, all measurements were performed under similar conditions.

In order to measure the soil chemical properties, samples were taken from the mineral soil layer. Therefore, soil samples were air-dried and soil particles <2 mm were used for the experiments. The Orion Ionalyzer (Model 901, Cambridge, MA, USA) pH meter was used to measure soil pH in a soil/water ratio of 1:2.5. Electrical conductivity (EC) was determined using an Orion Ionalyzer EC meter in a 1:2.5 soil/water solution. The Walkley–Black procedure [60] was used to determine the organic carbon (OC) content in percentage, and the Kjeldahl method was used to measure total N [61]. The Olsen method was used by utilizing a spectrophotometer to determine the available phosphorus (P) of the soil, and an atomic absorption spectrophotometer (by ammonium acetate extraction at pH 9) was used to determine available potassium (K), calcium (Ca), and magnesium (Mg) of the soil [62]. To determine the earthworm density and biomass, sample plots of 25 × 25 cm were designed on the sampling lines and the number of earthworms were counted at soil depths of 0–10 and 10–20 cm. After collection, the earthworms were washed and weighed, and then to determine the dry weight, earthworms were dried for 24 h at 60 °C [62].

2.4. Statistical Analyses

Statistical analyses were performed using SPSS version 17 (Chicago, IL, USA) software. As a first step, data distribution was plotted and checked for normality (Kolmogorov–Smirnov) and homogeneity of variance (Levene test). One-way and two-way ANOVAs were used to assess the significance of the observed mean differences in the physical, chemical, and biological properties of the soil under different skidder traffic levels, trail gradients, wheel track locations, soil depths, and their interactions. The comparison between the physical, chemical, and biological properties of the soil was made by using one-way ANOVA (p-level $\alpha \leq 0.05$) and Duncan's multiple range tests. The relationships between soil properties were determined using Pearson correlation. Significant relationships between variables and principal components were determined by using principal component analysis (PCA) via the PC-ORD (Version 4, WILD BLUEBERRY MEDIA LLC, Corvallis, OR, USA) software.

3. Results

3.1. Physical Properties of the Soil

The age of a skid trail and the sampling depth treatments had a significant effect on all the physical properties of the soil. Different traffic intensities had a significant effect on BD, PR, and total porosity (TP), while having no significant effect on the other properties. Different slopes had no significant effect on the physical properties (except PR and TP), nor did wheel track location (except PR) or the interaction of treatments (Table 2).

Table 2. Analysis of variance (p-values) of the effect of abandoned skid trails, traffic intensity, slope, wheel track location, soil sampling depth, and their interactions on the physical properties of the soil.

Source of Variance	Age of Skid Trail		Traffic Intensity		Slope		Wheel Track		Depth		Interaction	
	d.f.	p-Value	d.f.	p-Value	d.f.	p-Value	d.f.	p-Value	d.f.	p-Value	d.f.	p-Value
BD (g cm^{-3})	3	0.000 **	2	0.000 **	2	0.418 ^{ns}	2	0.254 ^{ns}	1	0.000 **	24	0.63 ^{ns}
PR (MPa)	3	0.000 **	2	0.000 **	2	0.000 **	2	0.000 **	1	0.000 **	24	0.998 ^{ns}
TP (%)	3	0.000 **	2	0.000 **	2	0.041 *	2	0.224 ^{ns}	1	0.005 **	24	0.993 ^{ns}
MP (%)	3	0.000 **	2	0.546 ^{ns}	2	0.305 ^{ns}	2	0.350 ^{ns}	1	0.000 **	24	0.998 ^{ns}
MIP (%)	3	0.000 **	2	0.549 ^{ns}	2	0.481 ^{ns}	2	0.783 ^{ns}	1	0.000 **	24	0.999 ^{ns}
SM (%)	3	0.000 **	2	0.549 ^{ns}	2	0.481 ^{ns}	2	0.783 ^{ns}	1	0.000 **	24	0.999 ^{ns}

Note: * $p < 0.05$; ** $p < 0.01$; ^{ns} not significant; BD, bulk density, PR, penetration resistance; TP, total porosity; MP, macroporosity; MIP, microporosity; SM, soil moisture; df, degrees of freedom.

Changes in the physical properties of the soil showed that soil properties recover within different years of skidding activities (Table 3). From the three-year-old skid trail to the 25-year-old skid trail, the BD, PR, and MIP decreased. In contrast, the TP, MP, and SM increased and differed significantly from the undisturbed area. Three years after the skidding operation, the BD, PR, and MIP were 12.6%, 107%, and 69.51% higher than the undisturbed area, respectively. Meanwhile, 25 years after the skidding operations, they were 8.4%, 27.4%, and 50.44% higher than the undisturbed area, respectively. Furthermore, three years after the skidding operation, the TP, MP, and SM were 2.47%, 23.3%, and 10.23% lower than the undisturbed area, respectively. Meanwhile, 25 years after the logging operations, they were 1.96%, 17.1%, and 4.58% lower than the undisturbed area, respectively (Table 3).

The investigation of the rut depth under different traffic intensities showed an increasing trend with the increase of traffic intensity in the skid trails (Figure 3). The rut depth in all the skid trails (except the 25-year-old skid trail) was significantly different for all three traffic intensities. The rut depth improved over the years following the skidding operation, so that by 25 years after the skidding operation, low and medium traffic intensity was fully recovered. The highest rut depth was observed in the three-year-old skid trail under high traffic intensity, with the amount of 20.1 cm, while the lowest was observed in the 25-year-old skid trail under high traffic intensity, with the amount of 5.3 cm (Figure 3). The results of this study show that the rut depth recovered in less time than other physical properties of the soil.

Table 3. Changes in the physical properties of the soil (mean ± standard error) in different years after the skidding operation.

	Different Years After the Skidding Operation (Years)				
Soil Physical Properties	**3**	**10**	**20**	**25**	**Un**
BD (g cm^{-3})	1.07 ± 0.01 ^{a}	1.12 ± 0.01 ^{b}	1.05 ± 0.01 ^{a}	1.03 ± 0.01 ^{a}	0.95 ± 0.03 ^{c}
PR (MPa)	3.25 ± 0.06 ^{a}	3.04 ± 0.04 ^{b}	2.90 ± 0.04 ^{b}	2.0 ± 0.04 ^{c}	1.57 ± 0.12 ^{d}
TP (%)	84.03 ± 0.23 ^{b}	83.0 ± 0.15 ^{c}	84.04 ± 0.15 ^{b}	84.47 ± 0.15 ^{ab}	86.16 ± 0.48 ^{a}
MP (%)	51.23 ± 0.57 ^{c}	52.66 ± 0.38 ^{c}	53.14 ± 0.38 ^{b}	55.36 ± 0.38 ^{b}	66.78 ± 0.19 ^{a}
MIP (%)	32.8 ± 0.63 ^{a}	30.34 ± 0.42 ^{a}	30.9 ± 0.42 ^{b}	29.11 ± 0.42 ^{b}	19.38 ± 1.3 ^{c}
SM (%)	40.2 ± 1.24 ^{c}	41.2 ± 0.82 ^{b}	41.76 ± 0.82 ^{b}	42.73 ± 0.82 ^{b}	44.78 ± 2.58 ^{a}

Note: Different letters in a row indicate significant differences among the intensities of the physical properties of the soil ($p < 0.05$), based on the Duncan's multiple range tests. The values 3, 10, 20, and 25 denote the age of the skid trails; Un, undisturbed area; BD, bulk density; PR, penetration resistance; TP, total porosity; MP, macroporosity; MIP, microporosity; SM, soil moisture.

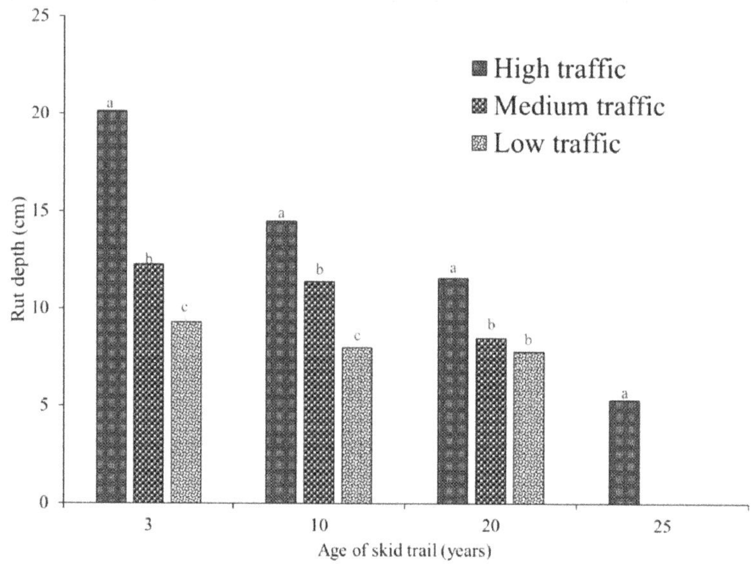

Figure 3. Average rut depth under different traffic intensities in different skid trail age classes, and results of the Duncan's multiple range tests.

3.2. The Chemical and Biological Properties of the Soil

The effect of the age of the skid trail and traffic intensity on all chemical properties was significant, while it was not significant on biological properties (Table 4). The slope of the skid trail had a significant effect on chemical properties (except C), while it had no significant effect on biological properties. The wheel track location had a significant effect on the pH and EC of the chemical properties. It also had a significant effect on biological properties at the 0.01% level. Furthermore, the soil depth has a significant effect on the chemical (except C) and biological properties. However, the interaction effects of treatments on the chemical (except EC and C/N ratio) and biological properties were not significant (Table 4).

Over the years following the skidding operation, the chemical and biological soil properties improved from the three-year-old skid trail to the 25-year-old skid trail (Table 5). Twenty-five years after the skidding operation, the values of pH, EC, C%, N%, available P, K, Ca, and Mg, earthworm density, and biomass soil were 6.4%, 20%, 22.77%, 38.8%, 11.2%, 14.2%, 10.3%, 15.1%, 28.3%, and 30.5% lower than the control area, respectively, and value of the C/N ratio was 26.2% higher than the control area (Table 5). The chemical and biological properties of the soil were significantly different from the control area 25 years after the skidding operation (Table 5).

Table 4. Analysis of variance (p-values) of the effect of abandoned skid trail, traffic intensity, slope, wheel track location, soil sampling depth, and their interaction on the chemical and biological properties of the soil.

Source of Variance	Age of Skid Trail		Traffic Intensity		Slope		Wheel Track		Depth		Interaction	
	d.f.	p-Value	d.f.	p-Value	d.f.	p-Value	d.f.	p-Value	d.f.	p-Value	d.f.	p-Value
pH (1:2.5 H_2O)	3	0.000 **	2	0.000 **	2	0.000 **	2	0.000 **	1	0.000 **	24	0.690 ns
EC (ds/m)	3	0.000 **	2	0.000 **	2	0.000 **	2	0.000 **	1	0.000 **	24	0.000 **
C (%)	3	0.000 **	2	0.000 **	2	0.107 ns	2	0.999 ns	1	0.722 ns	24	1.000 ns
N (%)	3	0.000 **	2	0.000 **	2	0.000 **	2	0.361 ns	1	0.000 **	24	1.000 ns
C/N ratio	3	0.000 **	2	0.000 **	2	0.000 **	2	0.302 ns	1	0.000 **	24	0.035 *
Available P (mg kg^{-1})	3	0.000 **	2	0.000 **	2	0.000 **	2	0.513 ns	1	0.006 **	24	1.000 ns
Available K (mg kg^{-1})	3	0.000 **	2	0.000 **	2	0.000 **	2	0.694 ns	1	0.020 *	24	1.000 ns
Available Ca (mg kg^{-1})	3	0.000 **	2	0.000 **	2	0.000 **	2	0.256 ns	1	0.000 **	24	1.000 ns
Available Mg (mg kg^{-1})	3	0.000 **	2	0.000 **	2	0.001 **	2	0.813 ns	1	0.050 *	24	1.000 ns
Earthworm density (n m^{-2})	3	0.199 ns	2	0.422 ns	2	0.090 ns	2	0.001 **	1	0.019 *	24	0.652 ns
Earthworm biomass (mg m^{-2})	3	0.215 ns	2	0.429 ns	2	0.115 ns	2	0.002 **	1	0.030 *	24	0.687 ns

Note: * $p < 0.05$; ** $p < 0.01$; ns not significant.

Table 5. Changes in the chemical and biological properties of the soil (mean ± standard error) in different years after the skidding operation.

Soil Chemical and Biological Properties	Different Years after the Skidding Operation (Years)				
	3	10	20	25	Un
pH (1:2.5 H_2O)	5.96 ± 0.04 c	6.07 ± 0.04 c	6.32 ± 0.04 b	6.59 ± 0.04 b	7.04 ± 0.07 a
EC (ds/m)	0.26 ± 0.19 c	0.26 ± 0.19 c	0.29 ± 0.19 b	0.32 ± 0.19 b	0.4 ± 0.35 a
C (%)	2.31 ± 0.05 d	2.78 ± 0.05 c	2.9 ± 0.05 c	3.12 ± 0.05 b	4.04 ± 0.09 a
N (%)	0.2 ± 0.05 d	0.27 ± 0.05 c	0.36 ± 0.05 b	0.41 ± 0.05 b	0.67 ± 0.01 a
C/N ratio	11.55 ± 0.07 a	10.29 ± 0.07 b	8.05 ± 0.07 c	7.61 ± 0.07 d	6.03 ± 0.13 d
Available P (mg kg^{-1})	9.59 ± 0.05 d	12.01 ± 0.05 c	14.49 ± 0.05 bc	17.11 ± 0.05 b	19.28 ± 0.09 a
Available K (mg kg^{-1})	138.1 ± 1.08 d	187.8 ± 1.08 cd	238.4 ± 1.08 c	275.7 ± 1.08 b	321.49 ± 2.03 a
Available Ca (mg kg^{-1})	105.7 ± 0.39 d	137.1 ± 0.39 c	176.6 ± 0.39 c	214.3 ± 0.39 b	239.05 ± 0.75 a
Available Mg (mg kg^{-1})	28.85 ± 0.23 d	37.23 ± 0.23 c	44.27 ± 0.23 bc	51.94 ± 0.23 b	61.18 ± 0.43 a
Earthworm density (n m^{-2})	0.22 ± 0.07 d	0.33 ± 0.07 c	0.41 ± 0.07 b	0.48 ± 0.07 b	0.67 ± 0.14 a
Earthworm biomass (mg m^{-2})	0.94 ± 0.31 d	1.41 ± 0.31 c	1.71 ± 0.31 bc	2.05 ± 0.31 b	2.95 ± 0.59 a

Note: Different letters in a row indicate significant differences among the intensities of the physical properties of the soil ($p < 0.05$), based on Duncan's multiple range tests. The values 3, 10, 20, and 25, denote the age of the skid trails; Un, undisturbed area.

Changes in the earthworm number and biomass under different traffic intensities and soil depths were lower at depths of 10–20 cm than depths of 0–10 cm, and lower under high rather than medium and low traffic intensities (Figure 4). The lowest earthworm number and biomass were obtained under high traffic intensity at soil depths of 10–20 cm. The earthworm number and biomass under different traffic intensities in both depths of soil were lower than the undisturbed area and were significantly different (Figure 4).

3.3. Principal Component Analysis (PCA)

The different traffic intensities (high, medium, and low) and different slopes (0–10%, 10–20%, and 20–30%) in the skid trails and the physical, chemical, and biological properties of the PCA analysis are presented in Figures 5 and 6. The PCA results show the relation between traffic intensity in the skid trails and the soil properties and reveal that the first and second axes explain 76.98% and 14.63% of the total variance, respectively (Figure 5). The PCA analysis results for the relationship between the different slopes in the skid trails and the soil properties show that the first and second axes explain 74.34% and 13.91% of the total variance, respectively (Figure 6). The analysis of the traffic intensity and the slope of the skid trails three and 25 years after the skidding operation show that as these two treatments increase, the effect on the soil increases. The results also show that high traffic intensity (H3) and slope classes of 20–30% (C3) in three-year-old skid trails had the greatest impact on soil properties (according to maximum distance from the coordinate center and proximity to axis 1) (Figures 5 and 6). The PCA analysis shows that 25-year-old skid trails under different traffic intensities and slopes had the least impact on soil properties (according to proximity to the coordinate center and away from the axis 1). The TP, MP, MIP, SM, and pH were close to the center axis, indicating that the effect of traffic and slope treatments was lower on these properties. However, the effects of these treatments on other physical, chemical, and biological properties were greater due to the distance from the coordinate center. The location of treatments affecting soil variables indicates that the undisturbed area is on the left side of the graph (Figures 5 and 6). The undisturbed area is slightly different, with older trails in low traffic and slope classes, whereas younger trails have high traffic and slope classes.

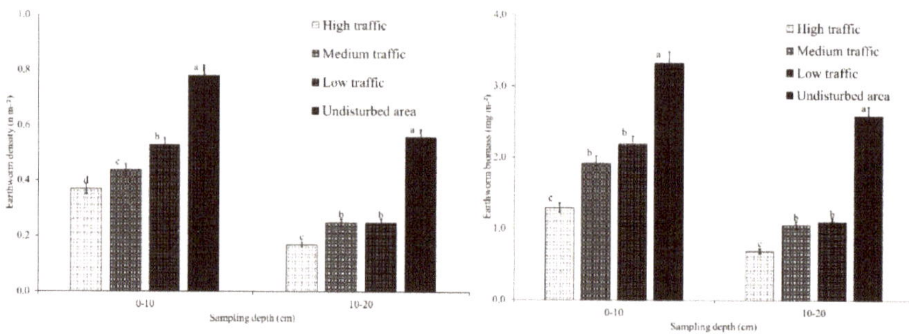

Figure 4. Changes in earthworm density (left) and biomass (right) under different traffic intensities in the two soil depths.

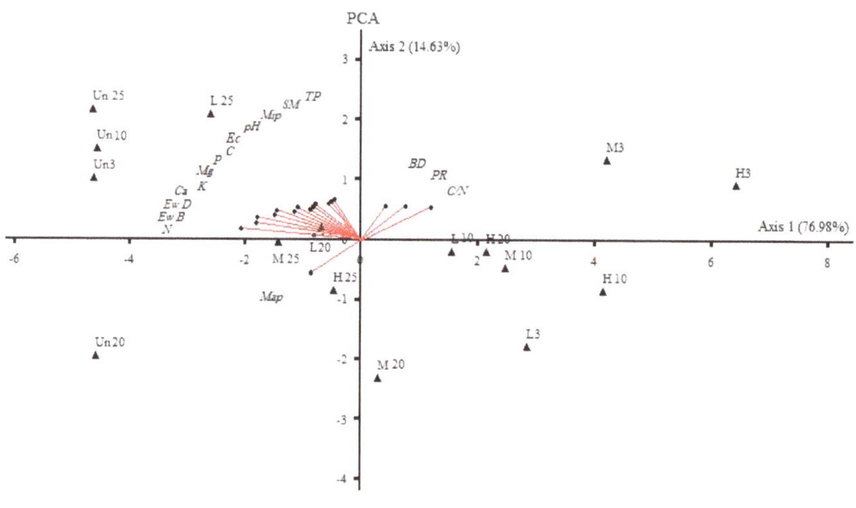

Figure 5. Principal component analysis (PCA) ordination of the traffic intensity in skid trails (3, 10, 20, and 25, age of the skid trails; H, M, L, and Un, high, medium, and low traffic intensity and undisturbed area, respectively), and soil physical (BD, bulk density; TP, total porosity; MP, macroporosity; MIP, microporosity; PR, penetration resistance; SM, soil moisture), chemical (pH; EC, electrical conductivity; C, organic C; N, nitrogen content; C/N, C/N ratio; P, available phosphorous; K, available potassium; Ca, available calcium; Mg, available magnesium), and biological (Ew D, earthworm density; Ew B, earthworm biomass) properties.

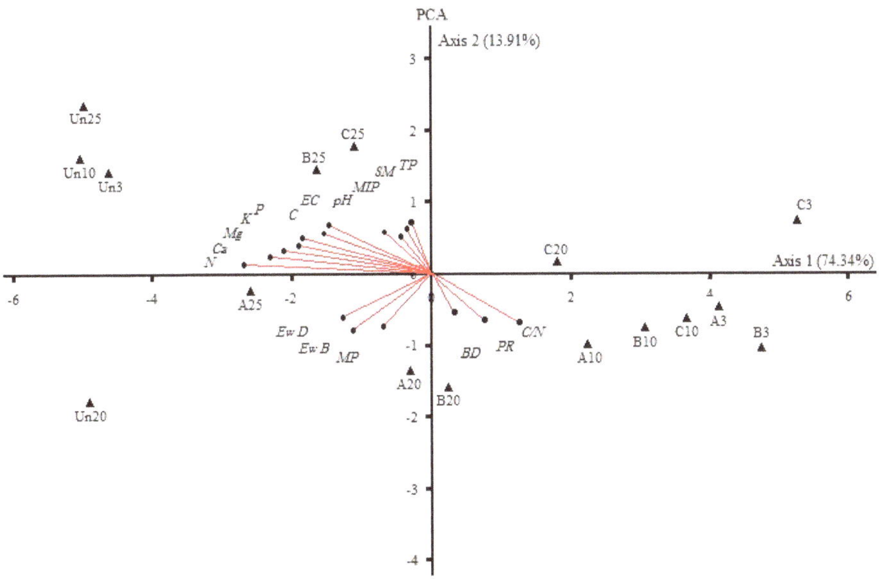

Figure 6. PCA ordination of the slope classes in the skid trails (3, 10, 20, and 25, age of the skid trails; A, B, C, and Un, 0–10%, 10–20%, and 20–30% slope classes and undisturbed area, respectively), and soil physical (BD, bulk density; TP, total porosity; MP, macroporosity; MIP, microporosity; PR, penetration resistance; SM, soil moisture), chemical (pH; EC, electrical conductivity; C, organic C; N, nitrogen content; C/N, C/N ratio; P, available phosphorous; K, available potassium; Ca, available calcium; Mg, available magnesium), and biological (Ew D, earthworm density; Ew B, earthworm biomass) properties.

The traffic intensity of three-year-old skid trails (H3) was positively correlated with BD ($r = 0.92$), PR ($r = 0.95$), and C/N ratio ($r = 0.84$), while it was negatively correlated with the other soil properties (Table 6). Also, the slope classes of three-year-old skid trails was positively correlated with BD ($r = 0.83$), PR ($r = 0.93$), and C/N ratio ($r = 0.67$), while it was negatively correlated with the other soil properties (Table 6).

Table 6. Pearson correlation coefficients between the soil properties and the effective treatments on soil (traffic intensity and slope classes of three-year-old skid trails).

Variable		Traffic Intensity		Slope Classes	
		Pearson Correlation (R)	*p*-Value	Pearson Correlation (R)	*p*-Value
Soil physical properties	BD	0.92 **	0.000	0.83 **	0.000
	PR	0.95 **	0.000	0.93 **	0.000
	TP	−0.76 **	0.000	−0.91 **	0.000
	MP	−0.52 *	0.021	−0.99 **	0.000
	MIP	−0.51 *	0.015	−0.58 *	0.020
	SM	−0.52 *	0.015	−0.58 *	0.020
Soil chemical properties	pH	−0.98 **	0.000	−0.98 **	0.000
	EC	−0.94 **	0.000	−0.95 **	0.000
	C	−0.97 **	0.000	−0.98 **	0.000
	N	−0.98 **	0.000	−0.98 **	0.000
	C/N	0.84 **	0.000	0.67 **	0.000

Table 6. *Cont.*

Variable		Traffic Intensity		Slope Classes	
		Pearson Correlation (R)	*p*-Value	Pearson Correlation (R)	*p*-Value
Soil chemical properties	Available P	−0.98 **	0.000	−0.98 **	0.000
	Available K	−0.98 **	0.000	−0.98 **	0.000
	Available Ca	−0.98 **	0.000	−0.97 **	0.000
	Available Mg	−0.99 **	0.000	−0.98 **	0.000
Soil biological properties	Ew D	−0.96 **	0.000	−0.88 **	0.000
	Ew B	−0.96 **	0.000	−0.90 **	0.000

** A significant difference at the 0.01% level. * A significant difference at the 0.05% level. BD, bulk density; TP, total porosity; MP, macroporosity; MIP, microporosity; PR, penetration resistance; SM, soil moisture; EC, electrical conductivity; C, organic C; N, nitrogen content; C/N, C/N ratio; P, available phosphorous; K, available potassium; Ca, available calcium; Mg, available magnesium; Ew D, earthworm density; Ew B, earthworm biomass.

4. Discussion

A comprehensive review of the natural recovery of the physical, chemical, and biological properties of the soil after skidding operations can complement the previous research chain and can be a guide to reducing the detrimental effects of skidding operations on forest soil and acquiring new knowledge on the sustainability of forest operations. In the past, there have been many studies conducted on the effects of skidding operations on soil of skid trails in controlled conditions. It is important to specify that the present research is retrospective, and the statistics and information obtained from the skidding operation were descriptive. Further experiments and sampling were performed on the desired trails and compared with the control area to evaluate the soil recovery process after skidding operations and to validate the descriptive data. Our study confirms that the effects of the age of the skid trail, traffic intensity, and sampling depth are more than those of other treatments. In this study, the slope of the skid trail changed the physical properties of the soil but had no significant effect due to the slope being divided into classes with less variation range (0–10%, 10–20%, and 20–30%). There appears to be a threshold to the slope gradient that has to be surpassed, however, before slope effects become visible [36]. Similarly, Najafi et al. [35] observed that the disturbance in soil physical properties was not significant in the slope classes of 0–10% and 10–20%, whereas for slopes of more than 20%, soil disturbance was significantly increased.

The change in the physical properties of the soil under the influence of skidding has been confirmed by other researchers, including Tavankar et al. [63], Ezzati et al. [50], Jourgholami at al. [32], Picchio et al. [9], and Sohrabi et al. [27]. The highest soil changes were obtained for high traffic intensity and slope, three years after the skidding operation. Further soil changes under high traffic intensity may be due to the high soil moisture during the skidding operation, the output timber volume of the area, and the number of passes. Skidding operations on steep terrain cause severe wheel slips of the machine, resulting in more puddling and dragging of the soil [44,64]. In addition, the slower speed of a machine on steeper terrain, especially upward skidding, causes more vibration of the surface soil and is disturbed more severely than compared to flat terrain [65]. With an increasing number of passes on steep terrain, the proportion of microporosity to macroporosity increases, and previous studies have confirmed this result [44,66]. During the skidding operation, the pores in the soil surface layer are compressed, which increases the soil strength, thereby reducing and changing the soil pores. The Soil moisture content decreases with the increasing traffic intensity and slope of the skid trail, so that the lowest moisture content is obtained under high traffic intensity and for a slope of 20–30%. The decrease in soil moisture is mainly due to a decrease in total porosity and macroporosity and an increase in the bulk density at the surface and in the soil profile [10,35,50].

Over the years after the skidding operation, the soil physical properties improve from the three-year-old trail to the 25-year-old trail, although there are significant differences with the control area. The results show that 25 years after the skidding operations, the BD, PR, and MIP were 8.4%,

27.4%, and 50.44% greater, respectively, and the TP, MP, and SM, which were 1.96%, 17.1%, and 4.58% lower than the undisturbed area, respectively. The results of this study are consistent with the results of the study by Ezzati et al. [50], which reported that, 20 years after skidding operations, under high traffic intensity and gentle slope class (<20%), the bulk density was 35–42% higher and the total porosity and macroporosity were 18–24% and 19–28% lower than the undisturbed area, respectively. After 25 years of skidding, the BD was lower than the threshold value of 1.40–1.55 g/cm^{-3} [38], indicating a recovery process for this feature. According to the findings of Ampoorter et al. [67], increasing the PR to more than 2 MPa limits the infiltration and growth of root trees in soil types. In this study, 25 years after skidding, the PR was less than 2 MPa, which indicates its recovery over time. One important effect of soil compaction is a decrease in total porosity by a decrease in macroporosity and an increase in microporosity [50]. The increase in the proportion of TP and MP over 25 years after the skidding operation is not unexpected. The MP in this study exceeded the threshold value of 10% [22], which is a prerequisite for airflow, microbial activity, and rooting. The soil moisture content increased in different years after skidding operation from the three-year-old trail to 25-year-old trail, which may be due to the reduced soil compaction and increased litter layer of the soil surface for water infiltration, as well as the reduced surface runoff through more pores in the soil [67].

The creation of soil ruts contributed to the increase in traffic intensity in this study, which was also reported by Botta [68], Eliasson [69], and Ezzati et al. [50]. When the soil moisture is saturated and soil cavities are filled with water, the skidding operation removes the surface layers of the soil and thus creates deep ruts in the soil. Soil ruts are more specified in the immediate years after skidding and high traffic intensity, leading to the disruption of natural drainage structures and reduced stability and soil aggregation. The tire size and average pressure on the soil are other effective factors in creating a rut in the soil during the skidding operation [50]. The rut depth improves in different years after the skidding operation, so that 25 years after the skidding operation, under low and medium traffic intensity, the soil is fully recovered. Consistent with the current study, Hatchel et al. [70] stated that rut recovery created after skidding operations requires a period of 18–19 years. The results of this study show that rut depth recovery requires less time than other physical properties of the soil.

Our study indicates that the effects of the treatments on the chemical properties were more intense than on the physical properties, which may be due to the mixing of the surface soil with the lower layers, the removal of soil, and soil sampling from the deep layers when compared to the undisturbed area [36]. On the other hand, the puddling and dragging of the soil of the skid trail after the skidding operation causes erosion and the loss of soil nutrients, which ultimately results in prolonged recovery for soil chemical properties. Changes in soil chemical properties under the influence of skidding operations have also been confirmed in previous studies [17,36,71].

In this study, the changes (decrease) in organic matter content and in different slopes and under different traffic intensities of the skidding trails compared to the control area are due to low levels of litter layer at the soil surface. Cutting down trees along the skidding trails for easier logging operation results in reduced tree density and, consequently, in reduced the litter layer of the forest floor [10,28,35,70]. C and N in soil have a close relationship with each other, in which, in this study, the trend changes were similar. A decrease in C and N levels after the skidding operations can be due to the displacement and mixing of organic and mineral soils, as well as the appearance of deep soil layers following the skidding operations [36]. The skidding operations reduce soil acidity on different slopes and under different traffic intensities of the skid trails compared to the undisturbed area. Consistent with the results of the current study, Hosseini et al. [71] reported that soil acidity on skid trails was lower than in undisturbed areas in the Hyrcanian forest. Soil compaction, disturbance, surface runoff, and soil erosion caused by skidder traffic are the main reasons for these changes [51]. Therefore, based on this hypothesis, the soil C, N, P, K, and pH concentrations in skid trails are significantly lower than in control areas [17,36]. The decrease in soil organic matter content and available P, K, Ca, and Mg, in soil after the skidding operations was reported in this study and in other studies [17,36,72,73].

The results show that soil chemical properties are not fully recovered after 25 years following skidding operations; thus, the difference compared to the undisturbed area is significant. The chemical soil properties improved over the years following the skidding operations. Accordingly, Hosseini et al. [69] reported that seven years after skidding operations in the Hyrcanian forest, the chemical properties of the soil were recovered and had no significant difference compared to the undisturbed area. The improvement of soil chemical properties can be due to climatic conditions, a decrease in bulk density over time, the litter layer thickness, the type and quality of the litter layer, and the activity of soil organisms. Since there is a direct relationship between the chemical properties of the soil and the vegetation type and density and the amount of litter, changes in the forest floor cover and the soil organic layer have direct effects on the chemical properties. In line with the current results, Jourgholami et al. [74] reported that, by adding a litter layer to the soil surface, the chemical properties can be partially recovered.

The results show that the earthworm number and biomass under different traffic intensities of the skid trails was lower than in the undisturbed area; thus, these changes were higher under a high traffic intensity. The soil habitat had a lower quality in terms of coarse porosity, temperature and moisture, organic matter content, and vegetation under different traffic intensities compared to the undisturbed area. In addition, the earthworm number and biomass at a soil depth of 10–20 cm was lower than at a depth of 0–10 cm. Earthworm immobilization plays an important role in the reduction of earthworm number and biomass due to the increase in penetration resistance and the reduction of coarse pores at a depth of 10–20 cm. The soil habitat quality and earthworm mobility are two effective factors in the recovery of earthworm communities after skidding operations [24]. Twenty-five years after a skidding operation, soil biological properties in skid trails will be improved, and will be significantly different from the undisturbed area. The earthworm density (0.48 n m^{-2}) and biomass (2.05 mg m^{-2}) were higher in the 25-year-old skid trail than the 20- > 10- > 3-year-old skid trails. Furthermore, Jourgholami et al. [74] reported that by adding a litter layer of different trees (beech + hornbeam + velvet maple) to the skid trail soil, 5 years after skidding operations, an increase in earthworm density (1.02 n m^{-2}) and biomass (13.26 mg m^{-2}) can be observed, but this differs significantly from the undisturbed area. The lack of full recovery of earthworm number and biomass after 25 years may be due to unfavorable soil properties and a low litter layer. The litter layer increases the organic matter content and the decomposition rate of organic matter at the soil surface, which increases soil respiration by positively impacting the microbial communities of the soil [51,74–76]. Previous studies have shown that increasing organic matter decomposition rate, decreasing C/N ratio in organic layers, and fine soil texture all have a significant relationship with earthworm biomass [17,77,78].

5. Conclusions

From a sustainable forest operation perspective, forest operational planning must consider all possible factors affecting environmental impacts, as well as their interactions. The evaluation of the time required for soil recovery is essential for managing forests on the basis of SFO, particularly in ground-based logging systems. This study was designed to determine the effects of different levels of traffic intensity, slope gradient, soil depth, and wheel track on the physical, chemical, and biological properties of the soil on the soil recovery assessment after timber harvesting. We found increased BD, PR and MIP and decreased TP, MP, SM, in the chemical and biological properties of the soil following ground-based skidding compared to the undisturbed areas, particularly on steep slopes with high traffic intensity and a sampling depth greater than 0.1 m. Twenty-five years after the skidding operations, the physical, chemical, and biological properties of the soil were somewhat recovered in the skid trails, but remained significantly different to the undisturbed area. The outcomes of this study indicate that the recovery time of the chemical properties of the soil is longer than that of the biological and physical properties, and that the complete recovery of these properties takes more than 25 years. Accordingly, forest managers need to consider the potential impacts of skidding operations on soil disturbance. Therefore, due to the long recovery process of compacted soils, reducing the area covered by skidding trails, avoiding or limiting skidding trails on steep slopes, using appropriate machinery, designing exact and specified skidding trails, and cutting down trees in the correct direction are all essential to avoiding unnecessary traffic. On the other hand, the use of improvement treatments (different foliage mulch and tree planting(, and comparing their performance in terms of the recovery of the soil properties of skid trails, are essential in order to accelerate biological activities and natural regeneration, prevent soil water erosion, and regulate the water cycle. A careful use of machines and tools, a mindful choice of operating season and meteorological conditions, well thought out logging planning and practices, and the training and skill development of forest operators can mitigate the negative environmental impacts of forest operations. The "reevaluation of logging practices" and "logging development programs" should be developed as the main aims of SFO to reduce soil compaction during timber skidding and extraction.

Author Contributions: Conceptualization, H.S., M.J. (Meghdad Jourgholami), M.J. (Mohammad Jafari), and N.S.; methodology, H.S., M.J. (Meghdad Jourgholami), M.J. (Mohammad Jafari), N.S., R.V., F.T., and R.P.; software, H.S., M.J. (Meghdad Jourgholami), M.J. (Mohammad Jafari), and N.S.; validation, H.S., M.J. (Meghdad Jourgholami), M.J. (Mohammad Jafari), N.S., R.V., F.T., and R.P.; formal analysis, H.S., M.J. (Meghdad Jourgholami), M.J. (Mohammad Jafari), and N.S.; investigation, H.S., M.J. (Meghdad Jourgholami), M.J. (Mohammad Jafari), and N.S.; data curation, H.S., M.J. (Meghdad Jourgholami), M.J. (Mohammad Jafari), N.S., R.V., F.T., and R.P.; writing—original draft preparation, H.S., M.J. (Meghdad Jourgholami), M.J. (Mohammad Jafari), N.S., and R.V.; writing—review and editing, H.S., M.J. (Meghdad Jourgholami), M.J. (Mohammad Jafari), N.S., R.V., F.T., and R.P.; supervision, M.J. (Meghdad Jourgholami), F.T., and R.P. All authors read and agreed to the published version of the manuscript.

Funding: This research received no external funding.

Acknowledgments: This work was supported by the Italian Ministry for education, University and Research (MIUR) for financial support (Law 232/2016, Italian University Departments of excellence)—UNITUS-DAFNE WP3.

Conflicts of Interest: The authors declare no conflicts of interest. The funders had no role in the design of the study; in the collection, analyses, or interpretation of data; in the writing of the manuscript; or in the decision to publish the results.

References

1. Heinimann, H.R. Forest operations engineering and management—The ways behind and ahead of a scientific discipline. *Croat. J. For. Eng.* **2007**, *28*, 107–121.
2. Marchi, E.; Picchio, R.; Spinelli, R.; Verani, S.; Venanzi, R.; Certini, G. Environmental impact assessment of different logging methods in pine forests thinning. *Ecol. Eng.* **2014**, *70*, 429–436. [CrossRef]
3. Marchi, E.; Chung, W.; Visser, R.; Abbas, D.; Nordfjell, T.; Mederski, P.S.; McEwan, A.; Brink, M.; Laschi, A. Sustainable Forest Operations (SFO): A new paradigm in a changing world and climate. *Sci. Total Environ.* **2018**, *634*, 1385–1397. [CrossRef] [PubMed]
4. Riala, M.; Athanassiadis, D.; la Hera, P.; Rodriguez, J. Development Potential of Inventions in Forest Biomass Harvesting. D 6.3. INFRES Project Report. 2015. Available online: http://www.forestenergy.org/observer:get_page/observer/action/details/itemid/706 (accessed on 28 March 2018).
5. Venanzi, R.; Picchio, R.; Piovesan, G. Silvicultural and logging impact on soil characteristics in Chestnut (*Castanea sativa* Mill.) Mediterranean coppice. *Ecol. Eng.* **2016**, *92*, 82–89. [CrossRef]
6. Dominati, E.; Patterson, M.; Mackay, A. A framework for classifying and quantifying the natural capital and ecosystem services of soils. *Ecol. Econ.* **2010**, *69*, 1858–1868. [CrossRef]
7. Picchio, R.; Tavankar, F.; Nikooy, M.; Pignatti, G.; Venanzi, R.; Lo Monaco, A. Morphology, growth and architecture response of beech (*Fagus orientalis* Lipsky) and maple tree (*Acer velutinum* Boiss.) seedlings to soil compaction stress caused by mechanized logging operations. *Forests* **2019**, *10*, 771. [CrossRef]
8. Picchio, R.; Mercurio, R.; Venanzi, R.; Gratani, L.; Giallonardo, T.; Lo Monaco, A.; Frattaroli, A.R. Strip clear-cutting application and logging typologies for renaturalization of pine afforestation—A case study. *Forests* **2018**, *9*, 366. [CrossRef]
9. Picchio, R.; Venanzi, R.; Tavankar, F.; Luchenti, I.; Iranparast Bodaghi, A.; Latterini, F.; Nikooy, M.; Di Marzio, N.; Naghdi, R. Changes in soil parameters of forests after windstorms and timber extraction. *Eur. J. For. Res.* **2019**, *138*, 875–888. [CrossRef]
10. Jourgholami, M.; Ghassemi, T.; Labelle, E.R. Soil physio-chemical and biological indicators to evaluate the restoration of compacted soil following reforestation. *Ecol. Indic.* **2019**, *101*, 102–110. [CrossRef]

11. Marchi, E.; Picchio, R.; Mederski, P.S.; Vusić, D.; Perugini, M.; Venanzi, R. Impact of silvicultural treatment and forest operation on soil and regeneration in Mediterranean Turkey oak (*Quercus cerris* L.) coppice with standards. *Ecol. Eng.* **2016**, *95*, 475–484. [CrossRef]
12. Klaes, B.; Struck, J.; Schneider, R.; Schüler, G. Middle-term effects after timber harvesting with heavy machinery on a fine-textured forest soil. *Eur. J. For. Res.* **2016**, *135*, 1083–1095. [CrossRef]
13. Tavankar, F.; Bonyad, A.E.; Nikooy, M.; Picchio, R.; Venanzi, R.; Calienno, L. Damages to soil and tree species by cable-skidding in Caspian forests of Iran. *For. Syst.* **2017**, *26*, 1–9. [CrossRef]
14. Matthews, D.M. *Cost Control in the Logging Industry*; McGraw-Hill: New York, NY, USA, 1942; p. 374.
15. Poltorak, B.J.; Labelle, E.R.; Jaeger, D. Soil displacement during ground-based mechanized forest operations using mixed-wood brush mats. *Soil Till. Res.* **2018**, *179*, 96–104. [CrossRef]
16. Flores Fernández, J.L.; Rubin, L.; Hartmann, P.; Puhlmann, H.; von Wilpert, K. Initial recovery of soil structure of a compacted forest soil can be enhanced by technical treatments and planting. *For. Ecol. Manag.* **2019**, *431*, 54–62. [CrossRef]
17. Jourgholami, M.; Labelle, E.R.; Feghhi, J. Efficacy of leaf litter mulch to mitigate runoff and sediment yield following mechanized operations in the Hyrcanian mixed forests. *J. Soils Sediments* **2019**, *19*, 2076–2088. [CrossRef]
18. Cambi, M.; Grigolato, S.; Neri, F.; Picchio, R.; Marchi, E. Effects of forwarder operation on soil physical characteristics: A case study in the Italian Alps. *Croat. J. For. Eng.* **2016**, *37*, 233–239.
19. Picchio, R.; Spina, R.; Calienno, L.; Venanzi, R.; Lo Monaco, A. Forest operations for implementing silvicultural treatments for multiple purposes. *Ital. J. Agron.* **2016**, *11*, 156–161.
20. Venanzi, R.; Picchio, R.; Grigolato, S.; Latterini, F. Soil and forest regeneration after different extraction methods in coppice forests. *For. Ecol. Manag.* **2019**, *454*, 117666. [CrossRef]
21. Flores Fernández, J.L.; Hartmann, P.; Schäffer, J.; Puhlmann, H.; von Wilpert, K. Initial recovery of compacted soil planting and technical treatments decrease CO_2 concentrations in soil and promote root growth. *Ann. For. Sci.* **2017**, *74*, 73. [CrossRef]
22. Ampoorter, E.; Goris, R.; Cornelis, W.M.; Verheyen, K. Impact of mechanized logging on compaction status of sandy forest soils. *For. Ecol. Manag.* **2007**, *241*, 162–174. [CrossRef]
23. Makineci, E.; Demir, M.; Comez, A.; Yilmaz, E. Chemical characteristics of the surface soil, herbaceous cover and organic layer of a compacted skid road in a fir (*Abies bornmulleriana* Mattf.) forest. *Transport. Res. Part D* **2007**, *12*, 453–459. [CrossRef]
24. Bottinelli, N.; Capowiez, Y.; Ranger, J. Slow recovery of earthworm populations after heavy traffic in two forest soils in northern France. *Appl. Soil Ecol.* **2014**, *73*, 130–133. [CrossRef]
25. Schweier, J.; Blagojević, B.; Venanzi, R.; Latterini, F.; Picchio, R. Sustainability assessment of alternative strip clear cutting operations for wood chip production in renaturalization management of pine stands. *Energies* **2019**, *12*, 3306. [CrossRef]
26. Brevik, E.; Fenton, T.; Moran, L. Effect of soil compaction on organic carbon amounts and distribution, South-Central Iowa. *Environ. Pollut.* **2002**, *116*, S137–S141. [CrossRef]
27. Sohrabi, H.; Jourgholami, M.; Tavankar, F.; Venanzi, R.; Picchio, R. Post-harvest evaluation of soil physical properties and natural regeneration growth in steep-slope terrains. *Forests* **2019**, *10*, 1034. [CrossRef]
28. Makineci, E.; Demir, M.; Comez, A.; Yilmaz, E. Effects of timber skidding on chemical characteristics of herbaceous cover, forest floor and topsoil on skidroad in an oak (*Quercus petrea* L.) forest. *J. Terramech.* **2007**, *44*, 423–428. [CrossRef]
29. Goutal, N.; Renault, P.; Ranger, J. Forwarder traffic impacted over at least four years soil air composition of two forest soils in northeast France. *Geoderma* **2013**, *193*, 29–40. [CrossRef]
30. Hartmann, M.; Niklaus, P.A.; Zimmermann, S.; Schmutz, S.; Kremer, J.; Abarenkov, K.; Lüscher, P.; Widmer, F.; Frey, B. Resistance and resilience of the forest soil microbiome to logging-associated compaction. *ISME J.* **2014**, *8*, 226. [CrossRef]
31. Epron, D.; Plain, C.; Ndiaye, F.K.; Bonnaud, P.; Pasquier, C.; Ranger, J. Effects of compaction by heavy machine traffic on soil fluxes of methane and carbon dioxide in a temperate broadleaved forest. *For. Ecol. Manag.* **2016**, *382*, 1–9. [CrossRef]
32. Jourgholami, M.; Soltanpour, S.; Etehadi Abari, M.; Zenner, E.K. Influence of slope on physical soil disturbance due to farm tractor forwarding in a Hyrcanian forest of northern Iran. *iForest* **2014**, *7*, 342–348. [CrossRef]

33. Cambi, M.; Hoshika, Y.; Mariotti, B.; Paoletti, E.; Picchio, R.; Venanzi, R.; Marchi, E. Compaction by a forest machine affects soil quality and *Quercus robur* L. seedling performance in an experimental field. *For. Ecol. Manag.* **2017**, *384*, 406–414. [CrossRef]
34. Picchio, R.; Latterini, F.; Mederski, P.S.; Venanzi, R.; Karaszewski, Z.; Bembenek, M.; Croce, M. Comparing accuracy of three methods based on the gis environment for determining winching areas. *Electronics* **2019**, *8*, 53. [CrossRef]
35. Najafi, A.; Solgi, A.; Sadeghi, S.H. Soil disturbance following four wheel rubber skidder logging on the steep trail in the north mountainous forest of Iran. *Soil Till. Res.* **2009**, *103*, 165–169. [CrossRef]
36. Jaafari, A.; Najafi, A.; Zenner, E.K. Ground-based skidder traffic changes chemical soil properties in a mountainous Oriental beech (*Fagus orientalis* Lipsky) forest in Iran. *J. Terramech.* **2014**, *55*, 39–46. [CrossRef]
37. Jourgholami, M.; Labelle, E.R.; Feghhi, J. Response of Runoff and Sediment on Skid Trails of Varying Gradient and Traffic Intensity over a Two-Year Period. *Forests* **2017**, *8*, 472. [CrossRef]
38. Kozlowski, T.T. Soil compaction and growth of woody plants. *Scand. J. For. Res.* **1999**, *14*, 596–619. [CrossRef]
39. DeArmond, D.; Emmert, F.; Lima, A.J.N.; Higuchi, N. Impacts of soil compaction persist 30 years after logging operations in the Amazon Basin. *Soil Till. Res.* **2019**, *189*, 207–216. [CrossRef]
40. Dexter, A.R. Amelioration of soil by natural processes. *Soil Till. Res.* **1991**, *20*, 87–100. [CrossRef]
41. Meyer, C.; Lüscher, P.; Schulin, R. Recovery of forest soil from compaction in skid tracks planted with black alder (*Alnus glutinosa* (L.) Gaertn.). *Soil Till. Res.* **2014**, *143*, 7–16. [CrossRef]
42. Lal, R. Effects of macrofauna on soil properties in tropical ecosystems. *Agric. Ecosyst. Environ.* **1988**, *24*, 101–116. [CrossRef]
43. Fründ, H.C.; Averdiek, A. Soil aeration and soil water tension in skidding trails during three years after trafficking. *For. Ecol. Manag.* **2016**, *380*, 224–231. [CrossRef]
44. Jourgholami, M.; Etehadi Abari, M. Effectiveness of sawdust and straw mulching on postharvest runoff and soil erosion of a skid trail in a mixed forest. *Ecol. Eng.* **2017**, *109*, 1–9. [CrossRef]
45. Mace, A.C., Jr. *Recovery of Forest Soils from Compaction by Rubber-Tired Skidders*; School of Forestry, University of Minnesota: St. Paul, MN, USA, 1971; Volume 226, p. 7688.
46. Reisinger, T.W.; Pope, P.E.; Hammond, S.C. Natural recovery of compacted soils in an upland hardwood forest in Indiana. *North. J. Appl. For.* **1992**, *9*, 138–141. [CrossRef]
47. Zenner, E.K.; Fauskee, J.T.; Berger, A.L.; Puettmann, K.J. Impacts of skidding traffic intensity on soil disturbance, soil recovery, and aspen regeneration in north central Minnesota. *North. J. Appl. For.* **2007**, *24*, 177–183. [CrossRef]
48. Mohieddinne, H.; Brasseur, B.; Spicher, F.; Gallet-Moron, E.; Buridant, J.; Kobaissi, A.; Horen, H. Physical recovery of forest soil after compaction by heavy machines, revealed by penetration resistance over multiple decades. *For. Ecol. Manag.* **2019**, *449*, 117472. [CrossRef]
49. Rab, M.A. Recovery of soil physical properties from compaction and soil profile disturbance caused by logging of native forest in Victorian Central Highlands, Australia. *For. Ecol. Manag.* **2004**, *191*, 329–340. [CrossRef]
50. Ezzati, S.; Najafi, A.; Rab, M.A.; Zenner, E.K. Recovery of soil bulk density, porosity and rutting from ground skidding over a 20-year period after timber harvesting in Iran. *Silva Fenni.* **2012**, *46*, 521–538. [CrossRef]
51. Jourgholami, M.; Khajavi, S.; Labelle, E.R. Mulching and water diversion structures on skid trails: Response of soil physical properties six years after harvesting. *Ecol. Eng.* **2018**, *123*, 1–9. [CrossRef]
52. Von Wilpert, K.; Schäffer, J. Ecological effects of soil compaction and initial recovery dynamics: A preliminary study. *Eur. J. For. Res.* **2006**, *125*, 129–138. [CrossRef]
53. Mariani, L.; Chang, S.X.; Kabzems, R. Effects of tree harvesting, forest floor removal, and compaction on soil microbial biomass, microbial respiration, and N availability in a boreal aspen forest in British Columbia. *Soil Biol. Biochem.* **2006**, *38*, 1734–1744. [CrossRef]
54. Macedo, M.O.; Resende, A.S.; Garcia, P.C.; Boddey, R.M.; Jantalia, C.P.; Urquiaga, S.; Campello, E.F.C.; Franco, A.A. Changes in soil C and N stocks and nutrient dynamics 13 years after recovery of degraded land using leguminous nitrogen-fixing trees. *For. Ecol. Manag.* **2008**, *255*, 1516–1524. [CrossRef]
55. Hope, G.D. Changes in soil properties, tree growth, and nutrition over a period of 10 years after stump removal and scarification on moderately coarse soils in interior British Columbia. *For. Ecol. Manag.* **2007**, *242*, 625–635. [CrossRef]

56. Li, Q.; Allen, H.L.; Wollum II, A.G. Microbial biomass and bacterial functional diversity in forest soils: Effects of organic matter removal, compaction, and vegetation control. *Soil Biol. Biochem.* **2004**, *36*, 571–579. [CrossRef]
57. Curran, M.; Dubé, S.; Bulmer, C.; Berch, S.; Chapman, B.; Hope, G.; Currie, S.; Courtin, P.; Kranabetter, M. *Forest and Range Evaluation Program, Ministry of Forests and Range and Ministry of the Environment*; Protocol for Soil Resource Stewardship Monitoring: Victoria, BC, Canada, 2009.
58. Nugent, C.; Kanali, C.; Owende, P.M.; Nieuwenhuis, M.; Ward, S. Characteristic site disturbance due to harvesting and extraction machinery traffic on sensitive forest sites with peat soils. *For. Ecol. Manag.* **2003**, *180*, 85–98. [CrossRef]
59. Picchio, R.; Neri, F.; Petrini, E.; Verani, S.; Marchi, E.; Certini, G. Machinery-induced soil compaction in thinning two pine stands in central Italy. *For. Ecol. Manag.* **2012**, *285*, 38–43. [CrossRef]
60. Walkley, A.; Black, I.A. An examination of the Degtjareff method for determining soil organic matter, and a proposed modification of the chromic acid titration method. *Soil Sci.* **1934**, *37*, 29–38. [CrossRef]
61. Salehi, A.; Ghorbanzadeh, N.; Kahneh, E. Earthworm biomass and abundance, soil chemical and physical properties under different poplar plantations in the north of Iran. *J. For. Sci.* **2013**, *59*, 223–229. [CrossRef]
62. Kooch, Y.; Zaccone, C.; Lamersdorf, N.P.; Tonon, G. Pit and mound influence on soil features in an Oriental Beech (*Fagus orientalis* Lipsky) forest. *Eur. J. For. Res.* **2014**, *133*, 347–354. [CrossRef]
63. Tavankar, F.; Majnounian, B.; Bonyad, A.E. Logging damages on forest regeneration and soil compaction using ground-based system (case study: Asalem forest area, Guilan). *Water Soil Sci.* **2009**, *13*, 449–456.
64. Gayoso, J.; Iroume, A. Compaction and soil disturbances from logging in Southern Chile. *Ann. Forest Sci.* **1991**, *48*, 63–71. [CrossRef]
65. Majnounian, B.; Jourgholami, M. Effect of rubber-tired cable skidder on soil compaction in Hyrcanian forest. *Croat. J. For. Eng.* **2013**, *34*, 123–135.
66. Greacen, E.L.; Sands, R. Compaction of forest soils. A review. *Soil Res.* **1980**, *18*, 163–189. [CrossRef]
67. Ampoorter, E.; De Schrijver, A.; Van Nevel, L.; Hermy, M.; Verheyen, K. Impact of mechanized harvesting on compaction of sandy and clayey forest soils: Results of a meta-analysis. *Ann. For. Sci.* **2012**, *69*, 533–542. [CrossRef]
68. Botta, G.F.; Jorajuria, D.; Rosatto, H.; Ferrero, C. Light tractor traffic frequency on soil compaction in the Rolling Pampa region of Argentina. *Soil Till. Res.* **2006**, *86*, 9–14. [CrossRef]
69. Eliasson, L. Effects of forwarder tyre pressure on rut formation and soil compaction. *Silva Fenn.* **2005**, *39*, 549. [CrossRef]
70. Hatchell, G.E.; Ralston, C.W.; Foil, R.R. Soil disturbance in logging. *J. For.* **1970**, *68*, 772–775.
71. Hosseini, S.A.O.; Nasiri, M.; Akbarimehr, M. Skidders traffic assessment on forest soil properties. *Int. J. Civ. Eng.* **2015**, *13*, 372–377.
72. Demir, M.; Makineci, E.; Yilmaz, E. Investigation of timber harvesting impacts on herbaceous cover, forest floor and surface soil properties on skid road in an oak (*Quercus petrea* L.) stand. *Build. Environ.* **2007**, *42*, 1194–1199. [CrossRef]
73. Demir, M.; Makineci, E.; Comez, A.; Yilmaz, E. Impacts of repeated timber skidding on the chemical properties of topsoil, herbaceous cover and forest floor in an eastern beech (*Fagus orientalis* Lipsky) stand. *J. Environ. Biol.* **2010**, *31*, 477–482.
74. Jourgholami, M.; Nasirian, A.; Labelle, E. Ecological restoration of compacted soil following the application of different leaf litter mulches on the skid trail over a five-year period. *Sustainability* **2018**, *10*, 2148. [CrossRef]
75. Prescott, C.E.; Grayston, S.J. Tree species influence on microbial communities in litter and soil: Current knowledge and research needs. *For. Ecol. Manag.* **2013**, *309*, 19–27. [CrossRef]
76. Jourgholami, M.; Labelle, E.R. Effects of plot length and soil texture on runoff and sediment yield occurring on machine-trafficked soils in a mixed deciduous forest. *Ann. For. Sci.* **2020**, *77*, 19. [CrossRef]
77. Jourgholami, M.; Fathi, K.; Labelle, E.R. Effects of litter and straw mulch amendments on compacted soil properties and Caucasian alder (*Alnus subcordata*) growth. *New For.* **2020**, *51*, 349–365. [CrossRef]
78. Ampoorter, E.; De Schrijver, A.; De Frenne, P.; Hermy, M.; Verheyen, K. Experimental assessment of ecological restoration options for compacted forest soils. *Ecol. Eng.* **2011**, *37*, 1734–1746. [CrossRef]

sustainability

MDPI

Article

Postural Risk in Manual Planting Operations of Poplar: Two Options Compared

Tiberiu Marogel-Popa, Marina Viorela Marcu and Stelian Alexandru Borz *

Department of Forest Engineering, Forest Management Planning and Terrestrial Measurements, Faculty of Silviculture and Forest Engineering, Transilvania University of Braşov, Şirul Beethoven No.1, 500123 Braşov, Romania; marogel@unitbv.ro (T.M.-P.); viorela.marcu@unitbv.ro (M.V.M.)

* Correspondence: stelian.borz@unitbv.ro; Tel.: +40-742-042-455

Received: 5 June 2020; Accepted: 6 July 2020; Published: 9 July 2020

Abstract: Poplar forests are cultivated worldwide on extended areas, contributing to the provision of wood for industries. Their management is intensive, especially in planting operations which are done, in many parts of the world, by the use of manual labor. This situation raises the question on their sustainability from an ergonomics point of view. Particularly, the postural risk is in question, as uncomfortable work postures may cause musculo-skeletal disorders. Two types of planting operations (large cutting—CP and bare-root seedling—SP) were selected as representatives for the evaluation of postural risks which was carried out for 14 subjects. Based on the analysis of approximately 14,500 images (approximately 67 h of field study), the postural risk indexes were estimated at 259 and 250 for the CP and SP, respectively. No significant differences were found between the operations, but the high share of effective planting tasks and their associated postural risk indexes generated these concerning results. The main conclusion is that these kinds of planting operations need postural improvement and ways for doing so should be researched in the future.

Keywords: agro-forestry operations; poplar planting; manual labor; postural risk; OWAS method; sustainability; reengineering; improvement

---

1. Introduction

The cultivated poplar forests stand for a valuable source of wood for industrial and individual house-hold use, and they hold an important potential to contribute to the satisfaction of the increasing demand of lignocellulosic raw materials. They are cultivated on extended areas [1,2] using fast-growing trees by operations which could be regarded as being more labor intensive compared to those specific to the traditional close-to-nature management of forests; that's because the typical sequence includes soil preparation, planting, fertilization, irrigation, and weed control operations [3], which are followed by regular harvesting operations such as thinning and clear-cutting [4]. Among these, planting is crucial in the operational management because on its quality and timely implementation depends the success of new forests [5].

Depending on the local forest management, terrain morphology, and the available technology, to name just few factors, planting of forests may be implemented either by mechanized [6,7], partly mechanized [8,9], or manual operations solely [5]. In Romania, as well as in other Eastern European countries, partly mechanized systems are still dominating the scene of forest operations [10]. This applies also to the planting operations of fast-growing poplar forests where the use of manual labor is still intensive. As such, at least problems in the sphere of ergonomics may arise, especially in the context in which many operations are still carried out manually around the world, and they have been proven to be physically demanding [11]. Therefore, scientific information is still required to engineer the operations to an extent which will ensure their sustainability. This approach is congruent with recent research that describes the main pillars of sustainability in forest operations [12,13], with the ergonomics and risk awareness being among them [12]. In particular, exposure to risks of developing musculo-skeletal disorders may be one of the most critical currently-faced problems of manual work, especially in forest operations, which are developed outdoors and are generally recognized to be among the most difficult ones [14].

A particular area of research in ergonomics is that referring to postural analysis which has provided important means and ways for research and improvement, enabling attempts to ensure the sustainability of various industrial sectors. As such, several methods were developed, validated, and used in time to evaluate the postural status-quo of different operations and to take corrective measures [15]. Such approaches are of invaluable importance since the musculoskeletal disorders (MSDs) are seen nowadays to be the most common cause of severe pain and physical disability, which in turn have important negative economic effects [16]; therefore, targeted studies have been conducted in many sectors, showing a strong relation between work factors and tasks and the development of MSDs. From these points of view, it seems that manual planting operations have received less attention by scientific research in the area of postural assessment. However, it is likely for manual planters to face awkward working postures given the tasks they are required to undertake. Such postures may refer to frequent situations in which the workers may hold their back bent and twisted, as well as to tasks requiring working with the knees bent. For instance, the Romanian standards describe working tasks that have an increased potential to generate deviant working postures in seedling planting operations [17]. In addition, the Romanian technique of poplar planting which uses large poplar cuttings, may generate unsustainable situations from a postural point of view given the size of planting material and the ways used to handle and plant it into the pits. Both techniques have been proven to be very intensive from the point of view of cardiovascular workload [9], and this outcome has provided some hints on the potential postural difficulties, given the fact that body posture changes are causing an increased cardiovascular activity [18–20]. However, the postures themselves adopted during the work may not be the sole drivers of an increased cardiovascular workload while they could inflict other problems related to the health of workers. In our knowledge, postural assessments have not been carried out for the typical case of manual poplar planting operations.

Based on the above, the main goal of this study was to characterize, from a postural assessment point of view, the manual planting operations of poplar forests to be able to set the stage for their improvement. Two types of planting operations were selected for this study (large cutting planting and bare-root seedling planting) and the objectives were set to: (i) describing and characterizing the distribution of tasks per subjects and type of operations, (ii) describing and characterizing postural risks on subjects, tasks, and type of operations, and (iii) comparing at task and operation type level the postural risks as specific to the two planting options taken into study.

2. Materials and Methods

2.1. Study Location and Planting Material

The Romanian cultivated poplar forests are dominantly established in flatlands, and they are harvested at the age of 25 years. It is typical that planting operations for them are undertaken by a two-pass system [17]. As such, in a first pass, regular farming tractors equipped with drilling devices are used to configure the planting pits at the dimensions required to plant either large cuttings or bare-root seedlings. From this point of view, the option used depends on the poplar species to be planted [5]. Specific to the studied conditions was that the first option required pits having approximately 2 m in depth and 0.2 m in diameter; the second option required pits of 0.6 m in depth and 0.6 m in diameter. Then, in a second pass, the effective planting operations were carried out exclusively by manual labor. As the scope of this study was the manual planting operations, field observations were carried out during November and December of 2018 in eight planting sites (hereafter L1 to L8) which were chosen based on representativity criteria of the operations taken into study (Table 1). By doing so, the aim was to cover the variability given by the planting schemes and the planting material used.

Two poplar species have been used in operations, namely *Populus × canadensis*, which is a hybrid poplar, (hereafter PLEA), and *Populus alba* (hereafter PLA), respectively. These two species are commonly used in the study area (Dolj county) and in Romania to establish fast-growing poplar forests [21].

Table 1. Description of planting sites, species used, planting scheme, and planting material. PLEA: *Populus × canadensis*. PLA: *Populus alba*.

Study Location	Forest District	Geographical Coordinates	Forest Compartment and Its Area (ha)	Species Used	Planting Scheme (m)	Planting Material
L1	Sadova	43°45′05.47″ N 23°54′56.23″ E	106A% (0.47)	PLEA	6 × 4	cuttings
L2	Sadova	43°45′33.59″ N 23°52′39.86″ E	88B% (0.89)	PLEA	6 × 4	cuttings
L3	Calafat	43°57′58.03″ N 22°52′55.42″ E	91B (0.96)	PLEA	4 × 4	cuttings
L4	Poiana Mare	43°50′51.35″ N 23°36′06.75″ E	22C (2.91)	PLA	3 × 2	seedlings
L5	Poiana Mare	43°50′12.70″ N 23°11′10.63″ E	43A (1.45)	PLA	3 × 2	seedlings
L6	Dăbuleni	43°45′01.60″ N 23°57′33.55″ E	7A,2D (1.81)	PLEA	5 × 4	seedlings
L7	Poiana Mare	43°50′49.75″ N 23°10′19.30″ E	31F,31N (1.61)	PLA	3 × 2	seedlings
L8	Dăbuleni	43°44′49.21″ N 23°59′33.56″ E	9D,10A (4.16)	PLEA	5 × 4	seedlings

In the planting sites L1–L3, the used planting material consisted of large poplar cuttings (PLEA, average length, diameter at the thick end, and weight estimated at approximately 6 m, 70 mm, and 5 kg, respectively); in the rest of planting sites, bare-root seedlings of both PLEA (2.5 m in height and 15 mm in diameter) and PLA (1.5 m in height and 8 mm in diameter) species were used for planting. Cuttings and seedlings used as planting material were procured from the plant reproduction facilities located in Zăval (Dolj county) and were transported to the planting sites on distances ranging from 10 to 110 km.

At the field study dates, all the planting sites were prepared in the sense that pits were already configured. In most of the planting sites, the existing stumps were previously removed and the soil was mechanically mobilized before the configuration of the pits. The soils of the planting sites were specific to the area, which is located near the Danube river, being either of alluvial origin or having a high content of sand.

2.2. Selection of Subjects and Organization of Work

Selection of the subjects for field monitoring was based on several criteria that covered, mainly, the availability and technical limitations of the monitoring devices, the informed consent of the subjects to participate in the study, and a specific distribution and inclusion of the work elements that were designed to emulate all the tasks specific to manual planting. As the existing Romanian planting rates for poplar seedlings indicate that the teams may be composed of 2 to 5 workers [17], and given the technical limitations of the devices used in this study, a number of two subjects were randomly selected for field monitoring in each study location. Following the selection, one subject was observed in three study locations; the final sample of subjects covered 14 male individuals (Table 2). Before observation, each subject was interviewed to collect data on his basic anthropometric features (Table 2) and to explain to him in detail the scope of the study and the intended use of the data. Based on this data, the body mass index (BMI) of each subject was computed in the office phase of the study. The sample of subjects selected for observation was characterized by an age of 43.3 ± 11.0 years, a body weight of 78.1 ± 9.5 kg, a body height of 1.73 ± 0.06 m, and a body mass index of 26.8 ± 2.7. The selected subjects were experienced in planting operations and they were instructed to carry on their jobs as usual. All the subjects participating in the study expressed their wish to remain anonymous.

Table 2. Basic anthropometric characteristics of the study group and locations of observation.

Subject	Abbreviation in This Study	Age (years)	Body Weight (kg)	Body Height (m)	Body Mass Index	Location of Field Observation
Subject 1	S1	53	81	1.75	26.45	L1
Subject 2	S2	38	78	1.65	28.65	L1
Subject 3	S3	27	74	1.73	24.73 [1]	L2
Subject 4	S4	50	95	1.82	28.68	L2, L4, L5
Subject 5	S5	45	87	1.85	25.42	L3
Subject 6	S6	20	69	1.74	22.79 [1]	L3
Subject 7	S7	49	78	1.70	26.99	L4
Subject 8	S8	48	80	1.70	27.68	L5
Subject 9	S9	51	83	1.75	27.10	L6
Subject 10	S10	40	85	1.69	29.76	L6
Subject 11	S11	54	75	1.70	25.95	L7
Subject 12	S12	52	55	1.65	31.93	L7
Subject 13	S13	28	70	1.80	21.60 [1]	L8
Subject 14	S14	51	83	1.75	27.10	L8

[1] Denotes a normal weight according to Body Mass Index.

Typical organization of the work was deduced in the office phase of the study based on the field collected data. In this regard, there were some differences as specific to the two types of planting material used. Cutting planting operations (hereafter CP) consisted of several work tasks such as the cutting sharpening, which is typically done to enhance the survival rate and for a better placement into the pit by pushing (Figure 1d), cutting handling and distribution within the planting site, cutting placement into the pit, which involved some pushing of the planting material (Figure 1c), pit filling with soil, gradual soil compaction by a handled tool and the subjects' movement between the pits; these were complemented by technical, personal, meal, and study-caused delays. In the case of seedling planting operations (hereafter SP), the differences consisted of the way used to place the seedlings into the pits as well as the way used to compact the soil. In both cases, the soil was gradually placed and compacted into the pits, but in the case of SP, soil compaction was done using the feet (Figure 1a).

(a) (b) (c) (d)

Figure 1. Examples of typical planting operations and planting material used: (**a**) Manual planting of PLA seedlings; (**b**) manual planting of PLEA seedlings; (**c**) manual planting of PLEA cuttings; (**d**) sharpened PLEA cuttings.

Other events consisted of moving by foot between nearby planting sites and unloading the planting material from the transportation means. These accounted for low shares and were not typical for all the subjects taken into the study. All the tasks described above and which supposed planting were done manually and, procedurally, the study was designed to cover at least 100 min of operation per day of observation, subject, and study site. This was considered to be sufficient to cover different time-windows of the day, several repetitions of all the tasks, and the observation of all the body postures assumed by subjects during their work. In addition, the sites to be planted by the described operations are typically small in area. Once the workers have finished the work on one site, they move to the next one. Nevertheless, to produce more reliable data, and since the operations allowed it, in some sites field data collection was extended to more than 200 min. Given the limited availability of planting sites for the studied operations, it was not possible to implement a balanced approach to get the same number of sites for each condition (planting material and planting scheme used). To accommodate this situation, the approach used in data processing and analysis enabled comparisons for unbalanced designs.

2.3. Data Collection

The primary field data was collected by video monitoring of operations. A small-sized (10 × 6 × 2 cm) Schwartz B1080 video camera was mounted on a tripod and set to continuously collect video files of 20 min in length each; this media file length is the maximum one enabled by the camera and, when reaching it, the camera saves it and automatically starts the next recording session of 20 min. This automatic procedure of recording and saving media files was applied until the work of the subjects had been finished in each planting site taken into the study. Due to a small setting error, in one site the length of consecutively collected video files was set accidentally to 3 min, but this did not affect the approach of the study, since in some sites the last of the collected video files taken into analysis were shorter than 20 min, anyway, because the planting work was finished before reaching the full length of the media file. The camera collected full HD video data (1920 × 1080 pixels) using a field of view of 90° oriented towards the two subjects monitored on each site. It was equipped with an internal accumulator of 4000 mAh that enables a video recording duration of up to 8 h, and the data was stored on a memory card of 32 GB. The camera was successively moved within each planting site as the operations progressed and it was placed each time at up to ca. 15 m from the observed subjects, enabling a clear vision on the subjects' postures. However, in most of the cases, the distance between the camera and the subjects was much less, close to ca. 5 m. As such, the approach provided the extraction of high-quality pictures, which were similar to those shown in Figure 1. On each site, after the completion of daily data collection, video files were transferred on a computer and saved in folders named by the date and site of data collection.

2.4. Data Processing and Statistical Analysis

A first step of data processing consisted of a sampling procedure that aimed to extract representative images for postural analysis. Given the previous experience with sampling and handling such data [22], and under the assumption that some of the data extracted from the video files will not be suitable for analysis, the targeted sample was designed to cover a number of exactly 100 randomly extracted images per 20 min of video recording (approximately 8% of the video sampled data). For this reason, pseudo-random numbers [23] were generated in Microsoft Excel (Microsoft Excel 2013, Microsoft, Redmond, WA, USA) in an interval that covered the length of each video file measured in seconds. Irrespective of the planting site, for those video files that had 20 min in length, 100 random numbers were generated, while for shorter files the numbers were generated proportionally with their length. This procedure was applied to all the video files collected in each plot as shown in Table 3. The extracted numbers were saved as lists for each video file taken into analysis, then each video file was broken into frames extracted at a rate of 1 Hz. In a following step, based on the list of random numbers, the corresponding frames were selected and organized in folders by considering the video file, subject, and study site to which they belonged. Postural analysis was implemented by using the Ovako Working posture Analysis System (OWAS) postural analysis method as described in [24], and it aimed also to document the tasks to which the analyzed frames belonged. The concept used in the implementation of the postural analysis method was based on the typical characteristics of a frequency study carried at random intervals as defined in [25]. To do so, and where the case allowed, each frame taken into study was analyzed in conjunction with the video file to which it belonged, to be able to accurately describe the work task it depicted. This supposed a detailed analysis of the video file some seconds before and after the occurrence of a given frame in it, by playing the video file and seeing the events depicted by them. For instance, if a subject was identified in the video file in a still state following a previous work task that supposed body movement, it was assumed that the frame in question belonged to a rest pause. Similarly, if a worker has been seen in a media file to handle either a cutting or a seedling by putting it into a pit, then it was assumed that the frame occurring in that part of the movie belonged to the effective planting. This approach was used due to the specific limitations of the OWAS method, as described, for instance, in [26].

Table 3. Description of the field sampled and office processed data.

Subject and Location	Type of Planting Operation	Number of Video Files	Length of Video Files (s)	Number of Analyzed Frames	Number of Valid and Reorganized Frames	Share of Valid and Reorganized Frames in the Collected Data (%)
S1 × L1	CP	6	7200	600	419 (417 [3])	5.82 (5.79 [4])
S2 × L1	CP	6	7200	600	410 (406 [3])	5.69 (5.64 [4])
S3 × L2	CP	12	14,400	1200	787 (781 [3])	5.47 (5.42 [4])
S4 × L2	CP	12	14,400	1200	833 (832 [3])	5.78 (5.78 [4])
S5 × L3	CP	9	10,800	900	659 (653 [3])	6.10 (6.05 [4])
S6 × L3	CP	9	10,800	900	707 (707 [3])	6.55 (6.55 [4])
S7 × L4	SP	20	23,400 [1]	1772	1522 (1510 [3])	6.50 (6.45 [4])
S4 × L4	SP	20	23,400 [1]	1772	1358 (1349 [3])	5.80 (5.76 [4])
S8 × L5	SP	81	14,580 [2]	1215	864 (818 [3])	5.93 (5.61 [4])
S4 × L5	SP	81	14,580 [2]	1215	820 (679 [3])	5.62 (4.66 [4])
S9 × L6	SP	12	13,700 [1]	1140	732 (722 [3])	5.34 (5.27 [4])
S10 × L6	SP	12	13,700 [1]	1140	738 (729 [3])	5.39 (5.32 [4])
S11 × L7	SP	14	16,800	1400	967 (967 [3])	5.76 (5.76 [4])
S12 × L7	SP	14	16,800	1400	898 (880 [3])	5.35 (5.24 [4])
S13 × L8	SP	17	19,860 [1]	1650	1373 (1197 [3])	6.91 (6.03 [4])
S14 × L8	SP	17	19,860 [1]	1650	1377 (1211 [3])	6.93 (6.10 [4])

[1] One video file was shorter than 20 min. [2] Video files were collected at 3 min in length each. [3] Number of frames used in the final analysis after the reorganization of work tasks. [4] Share of frames used in the final analysis in the total number of frames extracted at 1 Hz rate from the video files.

The data was analyzed by four researchers who had an extensive experience in handling the method and in the procedures used in data processing and analysis. Frames that were not suitable for analysis were disregarded. The exclusion process considered those frames that, for some reason, failed to clearly show all the body components taken into analysis due to the position of the worker in the field of view, those in which other persons obstructed the view on the worker under study by crossing the camera's field of view, as well as those frames which failed to capture useful data due to the need to move forward the tripod and the camera. The rest were kept as valid and were used in the postural analysis. The analyzed data, consisting of the code attributed to each body part according to the OWAS method, as well as a descriptive code to indicate the tasks observed, was included into Microsoft Excel sheets for each location and subject taken into study. Table 3 shows the basic statistics of the data used in this study. A total number of 5400 (CP) and 14,354 (SP) frames were extracted, covering and describing an observation time of approximately 18 and 49 h for CP and SP, respectively. Of these, 3815 (71%) were kept as valid for CP, and 10,649 (74%) were kept as valid for SP.

While all the work tasks and delays were monitored and separated based on the video data, some conceptual readjustments were necessary to target only the tasks specific and common to the two types of planting operations (CP and SP, respectively), as well as to enable the comparability between them in terms of postures assumed during the work. For that reason, the concept used in data analysis covered only those tasks that referred strictly to the studied operations. As such, frames that described meal, study-caused, and technical delays were excluded, and only those describing personal delays (hereafter Rest) were retained for analysis because these corresponded to resting pauses taken by the subjects in the planting sites. On the other hand, delays caused by meal taking were specific only to a part of subjects and study sites, therefore keeping them in the analysis would have affected the comparability of data. An additional step, which was undertaken to enhance the comparability, consisted of task regrouping and recoding to describe only four categories of tasks that were observed to be common to both types of operations: Moving (hereafter Move), which grouped all the reorganized frames (Table 3) in which the subjects were identified as moving between planting pits, moving within the planting site to bring planting material and so on, distributing (hereafter Distribute), which grouped the reorganized frames depicting actions of movement and placement of planting material near the pits, planting (hereafter Plant), that supposed the grouping of all the reorganized frames describing actions such as cutting sharpening, placement of cuttings or seedlings into the pits, filling the pits with soil, and compacting the soil, etc., and Rest, as described before.

Data analysis consisted of implementing the common procedures to see if the studied operations are sustainable from the point of view of postural condition. To do so, data was aggregated and analyzed at three levels: Tasks, subjects, and operation type. The distributions of the reorganized frames per tasks was analyzed by considering the last two levels, by the means of a relative frequency study which was implemented first at subject (site) level, then at the operation type (CP, SP) level. This approach allowed for a raw comparison in terms of task shares between the subjects (sites) and types of operations, respectively. Postural analysis aimed to estimate the postural risk index (hereafter PRI), as described by Zanuttini et al. [27] and used in this kind of research in forest operations [22,28–31], at task, subject, and operation type level. To do so, the input data was aggregated into datasets describing these three levels. From this point of view, the codes attributed when using the OWAS method consist of four digits [16,26] that can be further documented by a description of the task [16] to which a given instance (i.e., a frame extracted from a video file, a field observation by visual means) belongs to. The codes describing given instances by the posture of the back (4 postures), arms (3 postures), legs (7 postures), and force exertion (3 categories) are leading to a number of 252 possible combinations [26,32]. Of these, 72 are considered to be non-deviant from the regular posture of the body and are included in action category 1 (hereafter AC1) that does not require intervention for improvement, 53 are included in action category 2 (hereafter AC2), which assumes some redesign in the near future, 55 are included in action category 3 (hereafter AC3), which assumes the necessity of intervention as soon as possible, and the remaining 72 are included in action category 4 (hereafter AC4), which supposes immediate intervention for improvement. The postural risk index is calculated based on the relative frequency of observations falling into the four action categories, and it may take values in the range of 100–400 [27]. Based on the above, the postural risk index (PRI) stands for a good metric to check if the data sampled from a population (i.e., a subject working in a given site or an operation type) describes either a sustainable state from an postural point of view or it indicates the need for improvement. As such, values close to 100 characterize a good state from postural point of view, while values close to 400 reflect an urgent need for improvement [27]. While PRIs were calculated at task, subject, and operation level, and they gave the opportunity to visually compare the data, a third step consisted of a statistical comparison between the two operation types (CP and SP) at task and global level, respectively. The comparison was made by the means of Pearson's χ^2 statistic test due to the type of the data used and its distribution, which provided an unbalanced design. The input data used in comparison was the relative frequency of the reorganized frames per action categories, and the assumed confidence threshold was set at $\alpha = 0.05$. Based on the obtained results, the last step consisted of comparing the data generated by this study with the data coming from similar studies. Acknowledging the fact that the method could have been implemented in many industries, for this last step, however, only studies coming from forest and wood processing operations were selected for comparison. As such, the comparison was made based on the postural risk index as a common metric used to describe the need for intervention, and the comparisons were included in the discussion section.

3. Results

3.1. Task Share per Subjects and Types of Operations

Following data reorganization, the share of the remaining frames used to characterize the planting tasks (Move, Distribute, Plant, and Rest) was close to 100% in most of the cases compared to the set of valid frames.

The exception was that of S13 × L8 and S14 × L8, where the shares of the remaining frames were of approximately 87%. This situation was the effect of meal pauses as well as of a more pronounced presence of delays caused by the study in the last two cases. Even so, the share of the remaining frames, compared to the original set extracted at a rate of 1 Hz, was the greatest, accounting for more than 6% (Table 3). Figure 2 shows the distribution of frames per tasks at subject and operation type level. As it can be observed, there was a variation among the subjects and operation types, but planting (Plant) accounted for the greatest share in the analyzed samples. It represented approximately 60% in the case of CP (ranging from 53 to 70%, Figure 2a) and approximately 67% in the case of SP (ranging from 55 to 89%, Figure 2b). Differences among the subjects were obvious in both cases (CP and SP) and they were the effect of different degrees of involvement in specific tasks and in particular to those events characterizing the resting pauses (Rest). Distribution tasks (Distribute) accounted for a greater share in the case of CP, and this could be related to the size of the planting material, which required more movements to deliver it to the planting pits. Apparently, there were no correlations between the planting scheme and the share of the movement and distribution tasks. Nevertheless, Move and Distribute accounted for relatively similar shares for most of the subjects working in the same SP site (Figure 2).

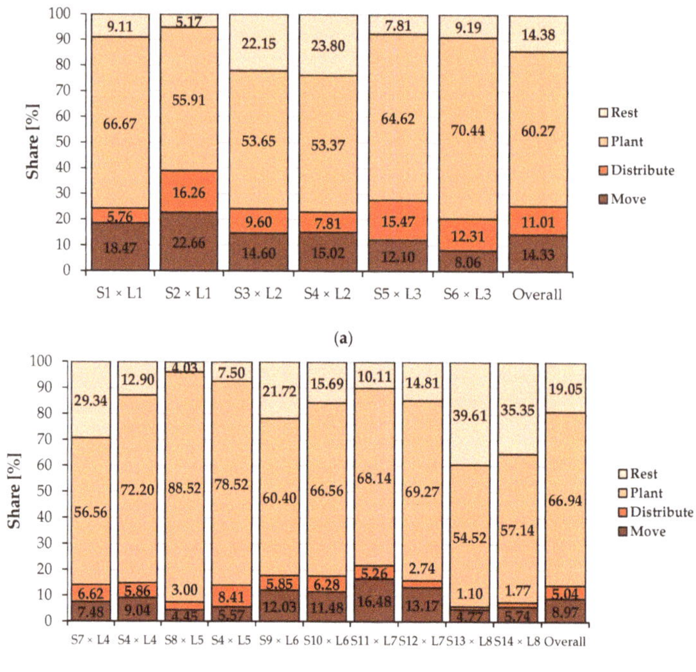

Figure 2. Share of the reorganized frames on tasks at the subject and operation type level: (**a**) Cutting planting operations (CP), (**b**) Seedling planting operations (SP).

3.2. Postural Assessment

3.2.1. Cutting Planting

In the case of CP, postural analysis by the means of OWAS method revealed a high inter-task and inter-subject variability of data categorized on action categories, as shown in Figure 3.

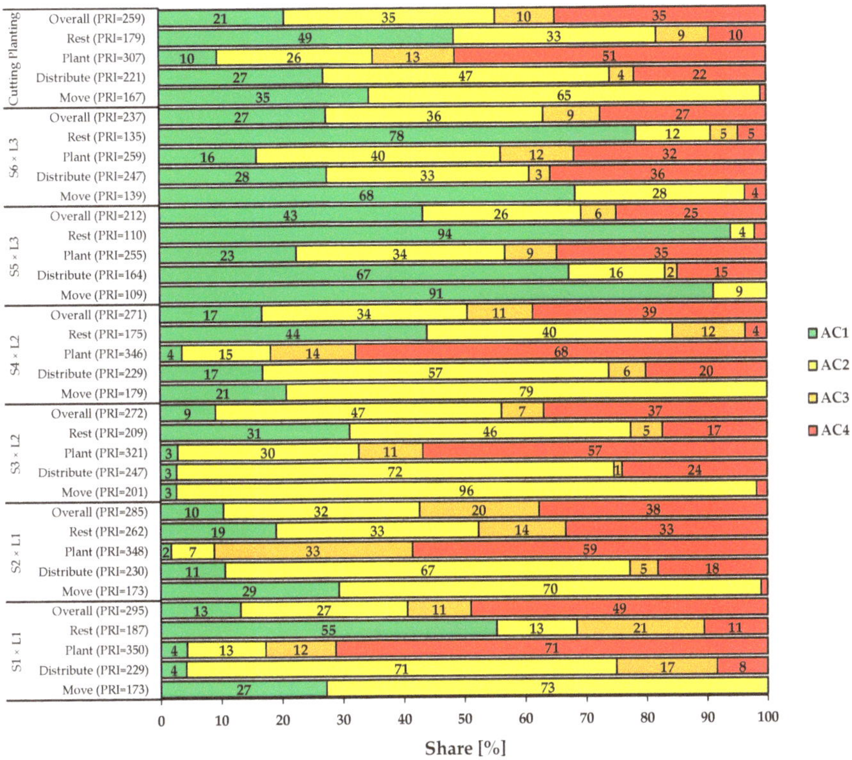

Figure 3. Data distribution on action categories (AC) and the estimated postural risk indexes (PRIs) at task, subject, and operation level for CP operations.

The frames characterizing the Move task were categorized, predominantly in AC1 and AC2. As such, the PRI of this task was evaluated to range, at the subject level, between 109 and 201, averaging 167 at the operation type level. Next in line were the Rest events, whose PRIs ranged from 110 to 262, averaging 179 at the operation level. Distribute task was characterized by values of PRI in between 164 and 247, averaging 221 at the operation level, while the most problematic task was that of planting (Plant), which accounted for PRI values in between 255 and 350, averaging 307 at the operation level. At the subject level, postural risk indexes varied between 212 and 295, while the overall PRI for CP was estimated at 259.

3.2.2. Seedling Planting

The situation on postural analysis as specific to SP operations is given in Figure 4. In this case, Rest events were less problematic as the PRIs were estimated at values in range of 107–189, averaging a value of 148 at the SP operation level.

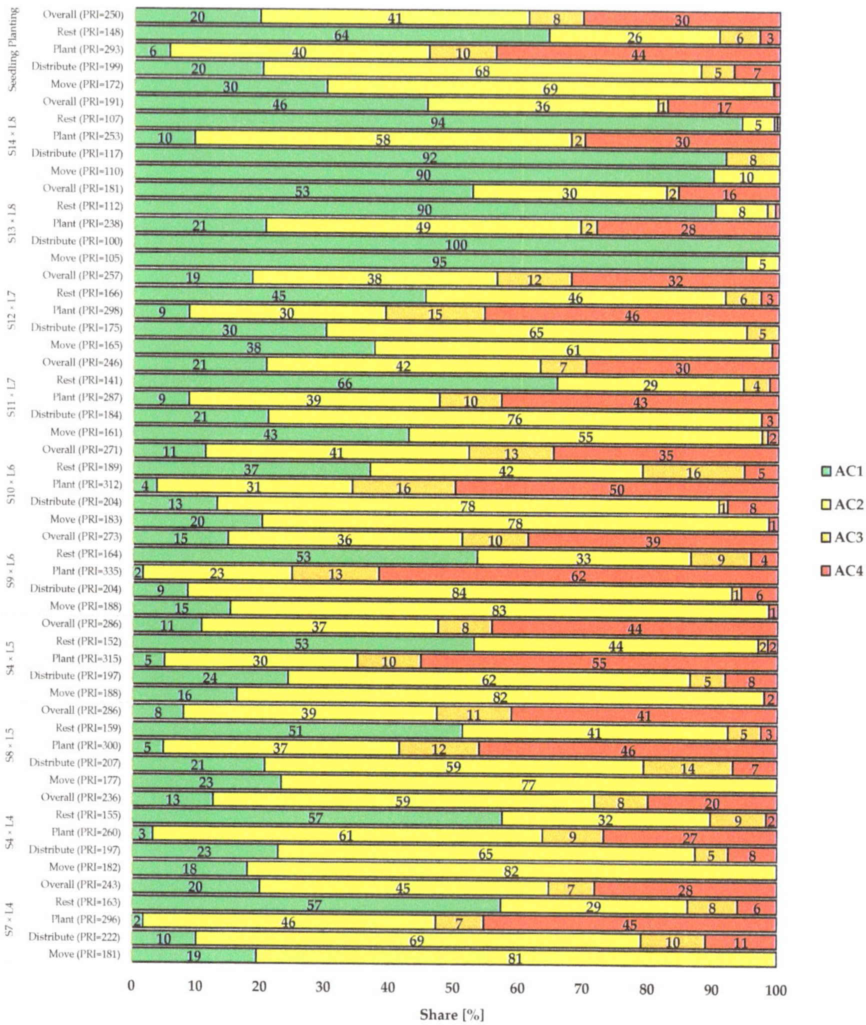

Figure 4. Data distribution on action categories (AC) and the estimated postural risk indexes (PRIs) at task, subject, and operation level for SP operations.

Next in line were the Move, Distribute, and Plant tasks, which were characterized by PRIs in range of 105–188, 100–222, and 238–355, respectively. At the subject level, PRIs varied in between 181 and 286. The overall PRI, calculated for the SP operation, was of 250, but it also varied at task level between 148 (Rest) and 293 (Plant). Therefore, similar to the CP operation, the most problematic task in the case of SP operations was the Plant task.

The results shown in Figures 3 and 4 were related, to some extent, with the distribution of the analyzed frames per tasks (Figure 2). For instance, S13 and S14, which were observed in L8 in the case of SP, were found to have the greatest share of Rest events (40 and 35%, respectively) in the task breakdown. For these two subjects, the PRIs were found to be the lowest (181 and 191, respectively). At the opposite side were the subjects S4 and S8, which were observed in L5 and for which were identified the lowest shares of Rest events in the task breakdown (Figure 2). Accordingly, for these two subjects, the PRIs were the greatest in the case of SP (286). As such, the way in which the work tasks are approached may affect the outcomes in terms of postural risks.

3.3. Comparison between Operations

Given the inter-subject variability, a comparison at this level on tasks was not feasible due to the high amount of data to be processed, analyzed, and reported. In turn, an analysis at the global level as well as an analysis at the task level, by considering all the data specific to an operation type, seemed to be more approachable and applicable since the aim of the study was to see if there are differences between the two types of operations and to compare the data with that coming from other studies. As such, the approach was to integrate the variability coming from inter-subject variation. Table 4 shows the results of the comparison tests carried out at the task and operation type level.

Table 4. Results of comparison tests between CP and SP at the task and operation type level.

Task Name and Tests Results	Operation Type	Action Category [%]				Postural Risk Index
		AC1	AC2	AC3	AC4	
Move	CP	34.56	64.52	0.00	0.92	167.28
$\chi^2 = 0.582, p = 0.901$	SP	30.01	68.99	0.11	0.89	171.87
Distribute	CP	27.03	47.13	4.07	21.77	220.57
$\chi^2 = 12.260, p = 0.007$ [1]	SP	20.12	67.65	5.13	7.10	199.21
Plant	CP	9.57	25.66	13.46	51.31	306.51
$\chi^2 = 5.226, p = 0.156$	SP	5.60	40.16	10.33	43.90	292.55
Rest	CP	48.53	33.33	8.61	9.52	179.12
$\chi^2 = 6.641, p = 0.084$	SP	64.32	26.29	6.26	3.13	148.20
Overall	CP	20.68	34.69	9.80	34.83	258.77
$\chi^2 = 1.039, p = 0.792$	SP	19.71	41.49	8.38	30.42	249.51

[1] Denotes significant differences between the relative frequencies per action categories among the two tested types of operations assuming a confidence threshold of $\alpha = 0.05$.

Before comparing the data, it is worth mentioning that the PRIs shown in Table 4 were calculated based on the frequencies (number) of the analyzed frames falling in a given action category per task and operation types. By the way of calculation, this metric does not support mean or median values, since it is based on a formula applied to the frequencies of frames. For instance, the PRI of 167.28 as specific to the Move task in CP was calculated based on the number of frames per action categories coming from all the subjects surveyed in CP operations.

As shown in Table 4, excepting the Distribute task, for which the test identified significant statistical differences, all the tasks were similar from the point of view of relative frame frequencies per action categories, as specific to the two types of operations. For the Rest events, the specific distribution on action categories failed closely ($p = 0.08$, $\alpha = 0.05$, $p > \alpha$) to indicate significant differences between the two operation types. Compared to the Distribute task, where the PRIs were close as values, in the case of Rest events, the PRIs were rather different, indicating that the outcome for SP has no serious problems, while the outcome for CP may have problems related to work postures assumed by the workers. While the postures assumed by the subjects in these events (Rest) are not constrained by the specific job tasks, the results indicate that further research is needed to be able to compare and differentiate more clearly between the inputs and outcomes of specific data sets analyzed by using the OWAS method.

In general, the PRIs calculated for the Move events were close to the value of 200, a value which can be interpreted in the way that some improvement may be required in the near future. The same may be applied to the Rest events, but here the improvements may rest in the free will of the subject to implement them, and they may be less related to the organization and technology of work. Similar figures (PRI close to 200) were found for the Distribute tasks, but the greatest problem lies in the results found for Plant tasks which were close to 300, and which indicate that immediate actions need to be undertaken for improvement. In addition, the Plant tasks were dominant in the analyzed data sets, being also, in general, dominant in planting operations. The PRIs found at the operation type level indicate that measures for improvement are needed rather urgently. Even if the differences between the two types of operations were not statistically significant, it is worth mentioning that CP tasks received, in most of the cases, higher PRIs compared to SP. This was more evident in the case of Plant tasks (difference of approximately 6 points), indicating also a similar problem related to data comparability, which should be researched in the future.

4. Discussion

The postural risk associated with the studied manual planting operations was found to be high, irrespective of the analyzed option (CP and SP, respectively). At the task level, the effective planting (Plant) was found to generate the worst postural situation because the PRIs were estimated to be close to 300. Unfortunately, by keeping the current situation, there is little room for improvement, as the effective planting stands for the main task specific to the studied planting operations and it accounts for the greatest share in the analyzed data. Even if not presented in the results section, the dominant body postures specific to AC4, that generated the current situation, were 4141 and 4151, respectively (where the first digit—4 stands for the back bent and twisted or bent forward and sideways, 1—both arms below the shoulder level, 4 and 5—standing or squatting with both or one knee bent, respectively, while the last digit stands for force exertion less than 10 kg). These were specific to the Plant task and they accounted for 1147 and 2949 observations for CP and SP, respectively, representing 30% and 28% of the analyzed data sets. Therefore, the main problems were related rather to the back (code 4) and leg (codes 4 and 5) postures, indicating that the back was either bent and twisted or bent forward and sideways. Accordingly, the legs were found to be with one (code 4) or both (code 5) knees bent in a standing or squatting posture. These postures of the back and legs are among the most deviant according to the OWAS method, and they were imposed by the actions and motions needed to place the cuttings and seedlings into the pits as well as those needed to fill the pits and to compact the soil in them. Elucidating for these tasks are also the examples given in Figure 1.

Worth mentioning that in the task breakdown, the Rest events were found to present shares of 5–24 and 4–40% for CP and SP, respectively. While these events were kept into analysis because they were related to the on-site work, and were intercalated in the typical work sequence, it is highly questionable if they could be eliminated or reduced in the general operational conditions. If so, then based on the results and task distribution presented by this study, the postural risk situation would become even worse, while the productivity will not be significantly enhanced. For instance, the total operated area (results not presented herein) in CP sites was of 0.55 ha, while in the case of SP it was of 1.46 ha. Nevertheless, these areas were operated by cohorts containing many more workers than those taken into study. Taking into consideration that the observation time was of 18 and 49 h, respectively, the global productivity could be estimated at approximately 0.03 ha per hour for both CP and SP, which was very low for the observed conditions, and it would become even lower if higher rest pauses will be taken. This situation indicates the limitations of manual work as being one of the current problems in forest operations.

Another approach which is seen as a potential ergonomic improvement is that of job rotation in the sense that work stations or tasks could be more wisely distributed between workers with the aim to reduce the postural risk. Spinelli et al. [31] have proposed such an approach for wood debarking jobs as a measure to balance the effect of difficult work postures. Nevertheless, in the case of manual planting this approach is less useful due to the task sequence and share in the typical work, in which the effective planting will dominate irrespective of how the tasks will be redistributed among workers. In addition, the job itself is different compared to assembly lines or jobs in similar industries in which, anyway, recent work has revealed that job rotation could be less effective [33].

Probably, one of the important factors that could have affected the results from an anthropometric point of view was the subjects' body height, which varied between 165 and 185 cm. For comparison, these two body heights were found in the case of CP for S2 × L1 and S5 × L3, respectively, cases in which the Rest events also accounted for low shares (approximately 5 and 8%, respectively). As such, the PRIs for the Plant tasks were found to be of 348 (165 cm) and 255 (185 cm), respectively, indicating that lower body heights could be associated with higher postural risks. However, one may just speculate that part of these outcomes may be the results of the places at which some of the subjects have chosen to grip the cuttings with their arms when they worked to introduce them into the pits, as the arms postures coded by 2 (one arm above the shoulder level) and 3 (both arms above the shoulder level) accounted for less than 200 events in the case of CP.

A comparison with other studies helps in understanding and categorizing these kinds of operations by considering the postural risk. Studies by Marogel et al. [22] and Cheţa et al. [29] have addressed the problem in cultivated forests located in the same region for manual cultivation and motor-manual tree felling and processing operations, respectively. The study by Marogel et al. [22] estimated a global PRI at 178 based on a cohort containing 14 subjects but, at the subject level, the PRI varied between 151 and 212. The operations surveyed by them were done manually by hoes, and the share of observations falling in AC4 was much lower (approximately 5%) compared to that from this study (30–35%). The study of Cheţa et al. [29], on the other hand, estimated the PRI of motor-manual tree felling and processing at 275, which was close to other similar forest operations as described by Calvo [32]. Therefore, it seems that operations that involve more leg movement such as those described by Marogel et al. [22], Borz et al. [28], as well as those that show a wider and more diverse succession of tasks [27] present lower postural risks, which are characterized by PRIs of up to 200. In comparison, wood processing operations were characterized by different classes of PRIs which range from non-threatening situations such as those specific to wood debarking [31], for which PRIs were found to be of up to 150, to those requiring postural improvement, such as in the case of firewood processing [30] and sawmilling [34], for which the PRIs were identified to be up to or even higher than 200. From this analysis, it seems that manual planting operations surveyed in this study hold an intermediary position, as they indicated PRIs of 250–259. Nevertheless, this situation requires immediate intervention for improvement, which is also supported by the statistics developed for the effective planting tasks (Plant) and by the assumption that, in other cases, the Rest events could have a lower share in the operations. Additionally, it is worth mentioning that, even if it seems to be unreasonable to find PRIs higher than 100 for Rest events, as shown by this study, and also found by other studies [27], this situation is real in operations. In addition, comparison of jobs by considering the estimates of PRI has its own limitations because is quite difficult to infer the typical postures of body parts from which the PRIs are obtained. A better task or job comparison would have been enhanced under the assumption that all the reported studies indicate the shares per action categories, which would enable the comparison by nonparametric tests. Nevertheless, this approach doesn't say much about the type and frequency of body part postures, which makes the analysis basis, and could be important in the work redesign effort.

Given the situation identified by this study, a good approach for improvement will be that of completely mechanizing the operations. This approach could be feasible, since some studies have shown that performance of mechanized planting reaches acceptable limits, and seedling planting machines could handle plants of different sizes [6]. Nevertheless, for the time being, the complete mechanization of poplar planting operations could be difficult since, at least for the cutting planting, special machines need to be developed. However, machines able to handle both soil preparation and planting using the described planting material and its size need to be developed in a short time given the limited availability of work force, low productivity, poor postural conditions of manual planting operations, and the spatial characteristics of such operations (i.e., sites having areas of up to 3 hectares, widely dispersed in the territory).

In regards to the method used in this study, one could appreciate the fact that even though it is resource-intensive in data processing and analysis, a fact that has been found by the authors of this study by their experience with other studies [22,29], it still provides the option of carrying on the studies by using affordable technology. While there are many other observational methods [15], the use of OWAS has gained attention, in particular in forest operations [see the reference list], due to its ability to evaluate the whole body and the possibility to provide comparable results. However, the method produces categorical data as outputs, which is difficult to address by the estimation of some advanced postural variability and diversity metrics that could work well for data measured on continuous scales [35]. Such approaches may be used to characterize given jobs and to better connect their postural condition to the development of MSD. They worth exploring further to see the extent to which they could be used to account for categorical data variability. Last, but not least, there is an increasing body of studies that bring evidence on less association between biomechanical postures and the development of MSDs [36]. While the topic is still debated, further exploration is required to see if the postures adopted during the work are the sole factor causing MSDs. This is important especially in forest operations where the workers are exposed, in addition to job-related risks, to other harmful environmental and technology-related factors [14,37].

5. Conclusions

The main conclusion of this study is that the manual poplar planting operations need interventions to correct the postures assumed during the work. As the results of this study have shown that the effective planting accounts for the greatest share in the observed operations, the intervention should be oriented towards finding feasible ways to improve the working postures in this kind of task because it also returned the most unfavorable situation from a postural point of view. While this study does not explicitly address the mechanization or training issues, it is likely that by following these two paths, improvements will be brought in the ergonomics of poplar planting operations.

Author Contributions: Conceptualization, S.A.B.; data curation, T.M.-P., M.V.M., and S.A.B.; formal analysis, T.M.-P., M.V.M., and S.A.B.; investigation, T.M.-P.; methodology, S.A.B.; project administration, S.A.B.; resources, T.M.-P., M.V.M., and S.A.B.; software, M.V.M.; supervision, S.A.B.; validation, T.M.-P., M.V.M., and S.A.B.; visualization, M.V.M. and S.A.B.; writing—original draft, S.A.B.; writing—review & editing, S.A.B. All authors have read and agreed to the published version of the manuscript.

Funding: This research received no external funding.

Acknowledgments: The authors acknowledge the technical and logistic support of Department of Forest Engineering, Forest Management Planning and Terrestrial Measurements, Faculty of Silviculture and Forest Engineering, Transilvania University of Brasov. The authors would like to thank the local managers of National Forest Administration—RNP Romsilva—for agreeing to carry on this study in the forests managed by them, to Eng. Marius Cheţa and Eng. Diana Chiriloiu for their help in data handling and analysis, and to the subjects who agreed to participate in the study, for their support. This work is part of a PhD thesis developed under the supervision of Doctoral School of the Transilvania University of Braşov. The authors acknowledge the support of the aforementioned organization.

Conflicts of Interest: The authors declare no conflict of interest.

References

1. Dickmann, D.I.; Kuzovkina, J. Poplars and willows of the world, with emphasis on silviculturally important species. In *Trees for Society and the Environment*; Isebrands, J.G., Richardson, J., Eds.; The Food and Agriculture Organization of the United Nations and CABI: Rome, Italy, 2014; Volume 2, pp. 8–12.
2. FAO. Improving Lives with Poplar and Willows. Synthesis of Country Progress Reports—Activities Related to Poplar and Willow Cultivation and Utilization—2008 Through 2011. In *24th Session of the International Poplar Commission, Dehradun, India, 29 October–2 November 2012*; Working Paper Ipc/12; 93p; Forest Assessment, Management and Conservation Division, FAO: Rome, Italy, 2012; 104p. Available online: http://www.fao.org/forestry/ipc2012/en/ (accessed on 24 February 2019).
3. Drew, A.P.; Zsuffa, L.; Mitchell, C.P. Terminology relating to woody plant biomass and its production. *Biomass* **1987**, *12*, 79–82. [CrossRef]
4. Florescu, I.; Nicolescu, N.V. *Silviculture*; Lux Libris Publishing House: Brasov, Romania, 1996.
5. Abrudan, I.V. *Împăduriri*; Editura Universității Transilvania din Brașov: Brasov, Romania, 2006.
6. Niuwenhuis, M.; Egan, D. An evaluation and comparison of mechanised and manual tree planting on afforestation and reforestation sites in Ireland. *Int. J. For. Eng.* **2002**, *3*, 11–23. [CrossRef]

7. Laine, T.; Saksa, T. Mechanized tree planting in Sweden and Finland: Current state and key factors for future growth. *Forests* **2018**, *9*, 370.
8. Boja, N.; Boja, F.; Teușdea, A.; Vidrean, D.; Marcu, M.V.; Iordache, E.; Duţă, C.I.; Borz, S.A. Resource allocation, pit quality, and early survival of seedlings following two motor-manual pit-drilling options. *Forests* **2018**, *9*, 665. [CrossRef]
9. Marogel-Popa, T.; Marcu, M.V.; Cheța, M.; Nuţă, I.S.; Borz, S.A. Dificultatea muncii în operații de plantare a puieților și a sadelor de plop: Analiza ritmului cardiac. *Rev. Pădur.* **2020**, *135*, 43–62. (In Romanian)
10. Moskalik, T.; Borz, S.A.; Dvořák, J.; Ferencik, M.; Glushkov, S.; Muiste, P.; Lazdiņš, A.; Styranivsky, O. Timber harvesting methods in Eastern European countries: A review. *Croat. J. For. Eng.* **2017**, *38*, 231–241.
11. Åstrand, P.O.; Rodahl, K. *Textbook of Work Physiology—Physiological Bases of Exercise*, 3rd ed.; Human Kinetics: Champaign, IL, USA, 1986.
12. Marchi, E.; Chung, W.; Visser, R.; Abbas, D.; Nordfjell, T.; Mederski, P.S.; McEwan, A.; Brink, M.; Laschi, A. Sustainable forest operations (SFO): A new paradigm in a changing world and climate. *Sci. Total Environ.* **2018**, *634*, 1385–1397. [CrossRef]
13. Heinimann, H.R. Forest operations engineering and management—The ways behind and ahead of a scientific discipline. *Croat. J. For. Eng.* **2007**, *28*, 107–121.
14. Potočnik, I.; Poje, A. Forestry ergonomics and occupational safety in high ranking scientific journals from 2005–2016. *Croat. J. For. Eng.* **2017**, *38*, 291–310.
15. David, G.C. Ergonomic methods for assessing exposure to risk of factors for work-related musculoskeletal disorders. *Occup. Med. C.* **2005**, *55*, 190–199. [CrossRef]
16. Helander, M. *A Guide to Human Factors and Ergonomics*, 2nd ed.; CRC Press: Boca Raton, FL, USA, 2006; pp. 224–255.
17. Lupuşoru, V.; Diaconu, M.; Diaconu, S.; Ionele, A. *Norme de Timp şi Producţie Unificate Pentru Lucrări din Silvicultură*; S.C. ROF S.A. Suceava: Suceava, Romania, 1997.
18. Hanson, J.A.; Jones, P.P. Heart rate and small postural changes in man. *Ergonomics* **1970**, *13*, 483–487. [CrossRef] [PubMed]
19. Jones, A.Y.M.; Kam, C.; Lai, K.W.; Lee, H.Y.; Chow, H.T.; Lau, S.F.; Wong, L.M.; He, J. Changes in heart rate and R-wave amplitude with posture. *Chin. J. Physiol.* **2003**, *46*, 63–69. [PubMed]
20. Šipinková, I.; Hahn, G.; Meyer, M.; Tadlánek, M.; Hájek, J. Effect of respiration and posture on heart rate variability. *Physiol. Res.* **1997**, *46*, 173–179.
21. Şofletea, N.; Curtu, L. *Dendrologie*, 2nd ed.; Pentru Viaţă Publishing House: Braşov, Romania, 2008; pp. 335–336.
22. Marogel-Popa, T.; Cheța, M.; Marcu, M.V.; Dută, C.I.; Ioraș, F.; Borz, S.A. Manual cultivation operations in poplar stands: A characterization of job difficulty and risks of health impairment. *Int. J. Environ. Res. Public Health* **2019**, *16*, 1911. [CrossRef]
23. Zar, J.H. *Biostatistical Analysis*, 5th ed.; Pearson Prentice-Hall: Upper Saddle River, NJ, USA, 2010.
24. Karhu, O.; Kansi, P.; Kuorinka, I. Correcting working postures in industry: A practical method for analysis. *Appl. Ergon.* **1977**, *8*, 199–201. [CrossRef]
25. Björheden, R.; Apel, K.; Shiba, M.; Thompson, M. *IUFRO Forest Work Study Nomenclature*; The Swedish University of Agricultural Science: Garpenberg, Sweden, 1995.
26. Corella-Justavino, F.; Jimenez Ramirez, R.; Meza Perez, N.; Borz, S.A. The use of OWAS in forest operations postural assessment: Advantages and limitations. *Bull. Transilv. Univ. Braşov Ser. II For. Wood Ind. Agric. Food Eng.* **2015**, *8*, 7–16.
27. Zanuttini, R.; Cielo, P.; Poncino, D. The OWAS method. Preliminary results for the evaluation of the risk of work-related musculo-skeletal disorders (WMSD) in the forestry sector in Italy. *For. Riv. Selvic. Ecolog. For.* **2005**, *2*, 242–255.
28. Borz, S.A.; Talagai, N.; Cheța, M.; Chiriloiu, D.; Gavilanes Montoya, A.V.; Castillo Vizuete, D.D.; Marcu, M.V. Physical strain, exposure to noise and postural assessment in motor-manual felling of willow short rotation coppice: Results of a preliminary study. *Croat. J. For. Eng.* **2019**, *40*, 377–388. [CrossRef]
29. Cheţa, M.; Marcu, M.V.; Borz, S.A. Workload, exposure to noise, and risk of musculoskeletal disorders: A case study of motor-manual tree felling and processing in poplar clear cuts. *Forests* **2018**, *9*, 300. [CrossRef]
30. Spinelli, R.; Aminti, G.; de Francesco, F. Postural risk assessment of firewood processing. *Ergonomics* **2016**, *60*, 1–9. [CrossRef]

31. Spinelli, R.; Aminti, G.; Magagnotti, N.; De Francesco, F. Postural risk assessment of small-scale debarkers for wooden post production. *Forests* **2018**, *9*, 111. [CrossRef]
32. Calvo, A. Musculoskeletal disorders (MSD) risks in forestry: A case study to propose an analysis method. *Agric. Eng. Int.* **2009**, *11*, 1–9.
33. Comper, M.L.C.; Dennerlein, J.T.; dos Santos Evangelista, G.; da Silva, P.R.; Padula, R.S. Effectiveness of job rotation for preventing work-related musculoskeletal diseases: A cluster randomized controlled trial. *Occup. Environ. Med.* **2017**, *74*, 543–544. [CrossRef] [PubMed]
34. Aragón-Vásquez, A.Y.; Silva-Lugo, E.D.; Nájera-Luna, J.A.; Méndez-González, J.; Hernández, F.J.; de la Cruz-Carrera, R. Postural analysis of the forestry worker in sawmills of El Salto, Durango, Mexico. *Madera Bosques* **2019**, *25*, e2531904.
35. Mathiassen, S.E. Diversity and variation in biomechanical exposure: What is it, and what would we like to know? *Appl. Ergon.* **2006**, *37*, 419–427. [CrossRef] [PubMed]
36. Da Costa, B.R.; Vieira, E.R. Risk factors for work-related musculoskeletal disorders: A systematic review of recent longitudinal studies. *Am. J. Ind. Med.* **2010**, *53*, 285–353. [CrossRef] [PubMed]
37. Borz, S.A.; Ignea, G.; Vasilescu, M.M. Small gains in wood recovery rate when disobeying the recommended motor-manual tree felling procedures: Another reason to use proper technical prescriptions. *Bioresources* **2014**, *9*, 6938–6949. [CrossRef]

sustainability

MDPI

Communication

Sustainability of Forest-Based Bioenergy—A Case Study of Students Surveyed at a University in Finland

Tapio Ranta *, **Antti Karhunen** and **Mika Laihanen**

School of Energy Systems, Lappeenranta-Lahti University of Technology LUT, Yliopistonkatu 34, 53850 Lappeenranta, Finland; antti.karhunen@lut.fi (A.K.); mika.laihanen@lut.fi (M.L.)
* Correspondence: Tapio.Ranta@lut.fi; Tel.: +358-408644994

Received: 30 April 2020; Accepted: 13 July 2020; Published: 15 July 2020

Abstract: With the increasing use of forest biomass, concerns about negative impacts have been raised in the debate. The aim of this study was to find out the attitude of university students towards the energy use of forest-based biomass and how different areas of sustainable forest operations were addressed. The survey was conducted over two years (2018–2019) with both full-time students at university and distance learning students who study alongside their work. Background information such as gender, nationality and field of study was collected from students. Most of the students currently considered the energy use of forest biomass to be sustainable. Many replies stressed that the situation could change if the use of forests is increased from the present circumstances. The main factors mentioned that led to forest-based bioenergy being sustainable were positive felling balance, compliance with forest certification, use of waste fractions and implementation of the Renewable Energy Directive (RED II) directive, while the loss of biodiversity, over-exploitation of forests, C debt and the cascading principle were factors that led to forest-based bioenergy being unsustainable. Student background variables had no effect on responses except for the field of study.

Keywords: sustainability; forestry; forest residues; climate change; education

1. Introduction

1.1. Sustainability in Forest Management

This study seeks to find out the opinions of university students on the sustainability of forest-based energy biomass. They will be future decision makers in the industry, so their views will have an impact on future choices. With the increasing use of forest biomass, concerns about negative impacts have been raised in the debate. In order to assess possible negative impacts, we need to understand the concept of sustainability. Sustainability is defined by the Brundtland report as development that meets the needs of the present without compromising the ability of future generations to meet their own needs [1]. The continuous growth of forests needs to be ensured through sufficient investments in forests. Sustainable forest management addresses a great challenge in matching the increasing demands of a growing human population while maintaining ecological functions of healthy forest ecosystems [2]. The climate change perspective calls for a change in perspective for forest operations ecology from the local and regional scales to the global scale [3]. The effective implementation of sustainable forest management practice depends on carrying out forest operations in a sustainable manner. In fact, the silvicultural practices of forest operations may also have a strong effect on environmental, economic and social performances, and hence sustainability. Therefore, the focus of sustainable forest operations (SFO) should be widened from forest and environmental ecology. A recently renewed concept of SFO is based on a broader focus and different scales of sustainable development, consisting of economy, environment, society, ergonomics and quality optimisation. Its purpose is to balance the performance of forest operations across economic, environmental and social sustainability objectives [4]. It includes five performance areas: environment, ergonomics, economics, quality optimisation, people and society, all of which contain more specific key indicators.

Sustainability indicators include forest area, growth/loss ratio or maximum logging volume, which can be maintained without compromising future felling opportunities. During the last ten years, about 0.1% of the forest area, or almost 19,000 hectares per year, has been taken up for other land use, but on the other hand, growth is clearly higher than drain [5]. Other indicators to be monitored include forest biodiversity, health and vitality [6]. Today, common storm damage and diseases affecting forests also alter the condition of forests. The forthcoming EU Biodiversity Strategy of the European Commission proposes a number of measures to enhance biodiversity. One of the most important means of biodiversity management of commercial forests is to preserve valuable habitats and to exclude some trees from felling. The Forest Act defines particularly important habitats, the characteristics of which must be preserved in forest management [7].

In addition to the sustainability of wood supply, ecological, cultural and social sustainability has been required for forest use. Natural Resource Institute of Finland (LUKE) has developed an online tool called Metsämittari, which allows users to view the impact of forest usage scenarios uploaded to the service in a multidimensional way [8]. In its calculations, LUKE also takes dimensions of sustainability into account, which give an estimate of the maximum maintained harvest, which gives a somewhat lower felling estimate than the previous mentioned sustainable harvest level.

1.2. Use of Forest Chips in Finland

In its climate and energy policy, the European Union is committed to the objectives of reducing greenhouse gas emissions. Bioenergy will play a key role in achieving the targets and its use will increase significantly in the next few years [9]. Therefore, ensuring the sustainability of forest-based bioenergy is essential for the acceptability of its use. In 2017, bioenergy was the largest source of renewable energy in the EU, accounting for 57% of renewable energy [10]. In Finland, bioenergy accounted for 78% of renewable energy and consisted mainly of wood-based energy [10]. In recent years, wood fuels have produced over a quarter of total energy consumption in Finland. Indeed, wood fuels are now the single most important source of energy in Finland since they have been more important than oil, coal or natural gas since 2012 [11]. The energy wood produced in Finland is largely based on the by-products of the wood processing industry and forestry. Typically, wood energy resources are used in highly efficient district heating (DH) systems and combined heat and power (CHP) plants [12].

Over the last seven years, the use of forest chips in thermal and power plants has varied between 7.2 and 8.0 million cubic metres and has averaged 7.5 million cubic metres per year (approximately 15 TWh) (Figure 1). In recent years, more than half of the forest chips have been chipped from small sized-trees, i.e., small-diameter energy wood, and are harvested especially in connection with silvicultural activities on young forests. The next most used are logging residues. The rest consist of stumps and rotten log wood, whose uses have remained low. The use of forest chips in Finland is thus based in particular on wood parts that are not suitable for pulpwood or dimensional lumber in mechanical forest industry. Forest silvicultural activities ensure the supply of marketable wood. In addition, sensitive and valuable natural sites are typically protected and are not included in felling.

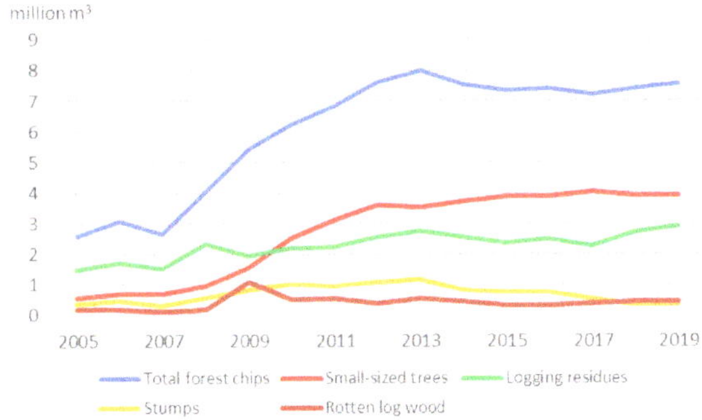

Figure 1. Forest chips use by raw material (2005–2019) [11].

1.3. The Aim of the Study

The aim of this study is to find out the attitudes of university students towards the energy use of forest-based biomass and how different areas of sustainable forest operations were addressed. Students are asked to justify their attitudes. The questionnaire is used to assess the impact of students' background variables on their opinions. The research is topical because there are many different views on the sustainability of energy use of forest biomass. The EU has defined a set of sustainability criteria to ensure the sustainability. Rigorous requirements are set in the Renewable Energy Directive (RED II) on the minimum level of GHG savings, appropriate land use, as well as monitoring requirements for any potentially adverse effects. There criteria will cover all biomass use in transport and in large-scale heat and power to ensure the sustainability. However, especially the carbon neutrality of biomass use has received a lot of attention in the debate. Depending on their characteristics and local circumstances, forests can play different roles in the carbon cycle, from net emitters to net sinks of carbon. Wood energy can be considered carbon-neutral if the source forest is sustainably managed and sequesters at least the same amount of CO_2 released during the production process of energy from the wood. This sink effect is determined in the LULUCF (Land Use, Land Use Change and Forest) sector [13]. Therefore, changes in carbon sinks must be considered in national and international climate policies more extensively than before, since the goals of the Paris Agreement cannot be achieved without strengthening global carbon sinks [14]. Moreover, it is necessary to find out the attitudes of students from time to time in order to see how attitudes change in relation to previous research.

Similar studies on the sustainability of forest energy have been conducted for Chinese university students [15] and school students in Finland [16] and some other countries [17]. Chinese students showed a positive attitude towards renewable energy in general, but slightly less positive towards forest-based bioenergy. They expected to receive more information and knowledge about renewable energy and forest-based bioenergy [15]. The school students in Finland appeared to be very critical toward bioenergy production from forest biomass, particularly with the issues related to its sustainability, environmental friendliness, and the future role of wood energy in overall bioenergy production. This study showed statistically significant gender and residential (urban–rural) differences among the students related to their perceptions of forest-based bioenergy production [16]. The international study among young students toward bioenergy in Finland, Slovakia, Taiwan, and Turkey found statistically significant differences in students' bioenergy knowledge with respect to the countries. Most knowledgeable students appeared to be very critical of bioenergy and especially of the issues related to bioenergy production from forests [17]. Various surveys on attitudes towards bioenergy [18], forest bioenergy [19] or certain fractions [20] have also been conducted in the past or, more specifically, to implement sustainability criteria for bioenergy [21–23]. An interesting observation in the study, "Young people's acceptance of bioenergy and the influence of attitude strength on information provision" was that the school lecture weakly contributed to building attitude strength, rendering opinion changes less likely in the future [17].

2. Materials and Methods

2.1. Implementing the Survey on Moodle

The survey was conducted over two years (2018–2019) for both full-time students at Lappeenranta-Lahti University of Technology (LUT) University and distance learning students who study alongside their work. LUT University (Lappeenranta-Lahti University of Technology (LUT)) is a science university in Finland, bringing together the fields of science and business. The questionnaire was targeted at both full-time and distance learning students in the Bioenergy course, who were given the opportunity to answer questions electronically at the end of the course. Responses were not given anonymously, and respondents were aware of this. Full-time students have regular face-to-face lectures, while distance learners self-study through the digital platform Moodle. Both groups had the same lecture material available; only the teaching method was different. Distance learning students also had the opportunity to ask questions of the teacher through a digital platform. The survey was conducted in a digital learning environment. Moodle™, "Modular Object-Oriented Dynamic Learning Environment", is a free, open-source object-based learning platform, i.e., virtual learning environment (VLE). Moodle can be used to build courses that allow you to publish material (e.g., in a timed fashion) and conduct experimental tests. Moodle provides tools for interaction, content production and material sharing, among other things. The Bioenergy course was conducted in this environment for both student groups. One of the topics of the course was the sustainability of bioenergy supply and use. All dimensions of sustainability, such as environmental, social and economical, were included in the definition of sustainability. In a digital learning environment, each group of students was able to answer the question option in their own time and justify their answer. Answers were not graded and there was no length limit to the justification of the answers. The choice question offered three options, "yes", "no" and "don't know", from which the respondent chose one.

The questions were:

- How do you see the current woody biomass use for energy in Finland? Is it sustainable?
- Can you justify your previous choice of sustainability?

There was no attempt to guide the students in their answers, because there could be reasons to be either positive, negative or not know opinion. In practice, the course material and the way it was delivered may affect responses compared to a hypothetical reference group that would not have seen the teaching material. Therefore, the group of students taking the course are not representative of a broader population of students that didn't just learn about sustainability and bioenergy. However, students were encouraged to think critically and source criticism. The course does not tell whether the use of forest-based bioenergy is sustainable or not but highlights various methodological choices which may influence to the outcome. Those are, e.g., definition of reference (no bioenergy) scenario, time frame of evaluation period, spatial scale, scope (like one product life cycle or system level assessment) and metric choice. The survey gave an idea of how these different groups view the use of forest-based bioenergy and how different options were justified.

The number of distance learning students was larger than the number of full-time students each year, and the number of students increased in 2019 (Table 1). The response rate for full-time students was 70% and 100% for distance learners who completed the query as part of the course. A total of 273 students participated in the study. For full-time students, answering was voluntary, resulting in a lower response rate.

Table 1. Student population.

Group	2018	2019	Total
Full-time students	43	59	102
Distance learners	77	94	171
Total	120	153	273

Background information such as gender, nationality and field of study was collected from students (Figures 2 and 3). Nationality was classified into groups of Finnish and foreigners because the group of foreigners was divided into so many nationalities. The background information was obtained directly from the system and did not need to be requested separately. Full-time students could also be graded according to the progress of their studies, either bachelor's (BSc) or master's (MSc). All distance learning students were master's students. The group of distance learning students clearly differed from the group of full-time students. It was more male dominated, the majority were Finns, and the study fields were more concentrated. The group of distance learning students were working people who had a bachelor's degree and were generally older and more experienced than full-time students. However, age was not collected separately, but there was a difference in age between groups.

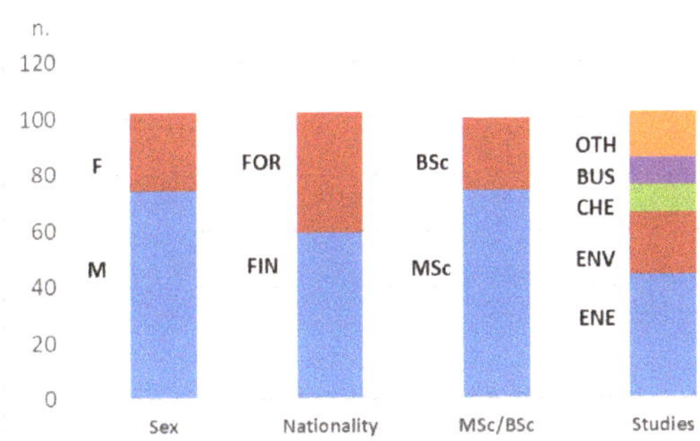

Figure 2. Background variables for full-time students (2018–2019). Sex: male (M), female (F); nationality: Finnish (FIN), foreigners (FOR), master's students (MSc), bachelor's students (BSc); studies: energy (ENE), environmental (ENV), chemical (CHE), business (BUS), other (OTH).

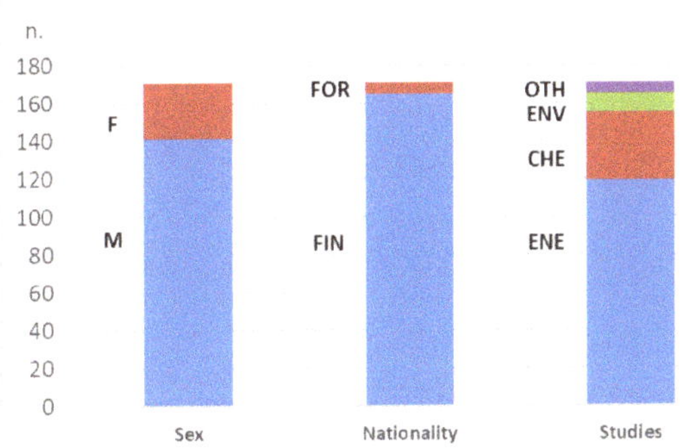

Figure 3. Background variables for distance learning students (2018–2019). Sex: male (M), female (F); nationality: Finnish (FIN), foreigners (FOR); studies: energy (ENE), environmental (ENV), chemical (CHE), business (BUS), other (OTH).

2.2. Testing the Background Variables

Responses were cross tabulated for background variables. This was to determine whether the background variable had an effect on the response. In cross tabulation, observations were presented in number distributions and a hypothetical tabulation with no dependence on background variables was constructed. The numbers in the hypothetical table are called expected frequencies. In order to make the frequencies high enough (>5), the answers "no" and "don't know" were combined, as were the individual observations in the field of studies, under the group "other". Testing the difference between the observed table and the hypothetical table is not reliable if the numbers of the hypothetical table, i.e., the expected frequencies, are too small as in the groups "don't know" and "no". "Don't know" is not a same as "no" answer, but they were expected to be more sceptical than those who answered "yes". However, the justification for all response options was further analysed. The justification was categorised under certain keywords which describe the answers in the best possible way, but students were not encouraged to use any keywords.

The difference between the observed and the hypothetical table was measured by the Chi-square test variable (X^2). The Chi-square test variable is known to follow a roughly Chi-square probability distribution, whose exact shape depends on the degree of freedom (df). The degrees of freedom are calculated from the table (number of rows−1) × (number of columns−1). The value of the Chi-square test variable is higher the more the observed frequencies deviate from the expected frequencies.

3. Results

3.1. Student Response to the Query

Most full-time students considered the use of forest biomass in energy production sustainable (82%) (Figure 4). There was no significant difference between the years. There was also no significant difference in the other background variables except for the field of studies (Table 2). Environmental engineering (ENV) students were more sceptical about sustainability.

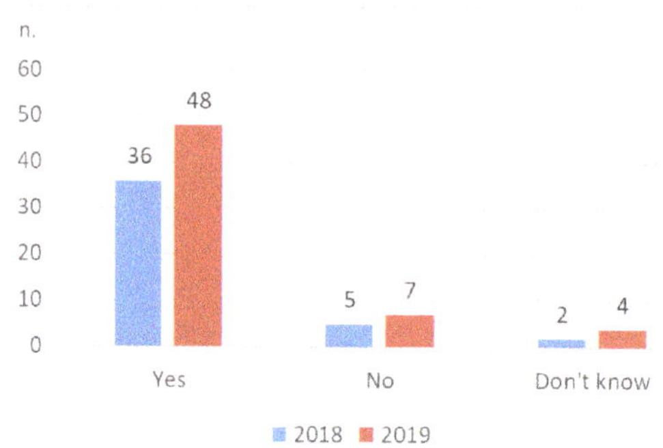

Figure 4. Full-time students' views on the sustainability of forest biomass use in energy production in 2018 and 2019.

Table 2. The significance of background variables for full-time students.

	df	X^2-Value	p-Value
sex	1	0.001	0.973
nationality	1	0.047	0.829
BSc/MSc	1	0.614	0.433
studies	4	14.268	0.006

The use of forest biomass in energy production was also considered sustainable by most distance learning students (87%) (Figure 5). Equally, for them, of all the background variables, only the study fields influenced attitudes towards sustainability (Table 3). Chemical engineering (CHE) students were more sceptical about sustainability. Within each group, the majority were energy technology students, and they found forest biomass use sustainable. Moreover, there was no significant difference in the attitudes towards sustainability between full-time and distance learning student groups.

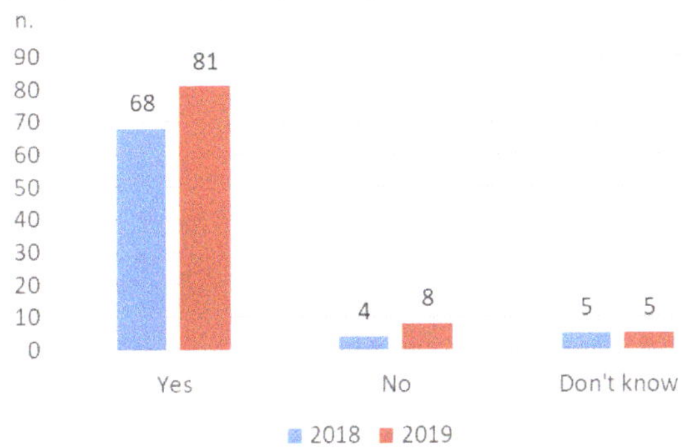

Figure 5. Distance learning students' views on the sustainability of forest biomass use in energy production in 2018 and 2019.

Table 3. The significance of background variables for distance learning students.

	df	X^2-Value	p-Value
sex	1	6.182	0.013
nationality	1	0.080	0.777
studies	3	19.94	0.0002

3.2. Students' Justification

The open answers were categorised into the four most mentioned topic themes, which were given keywords that best describe the theme. Synonyms can be found for keywords that mean the same thing. However, no other themes with more mentions were found. In practice, several themes could appear in a single response. The lengths of the responses also varied considerably. Students were not given keywords as answer options as they could influence student choices. The four most mentioned themes described by keywords in the "yes" answers were: felling balance, certification, waste fraction, directive (REDII). Whereas the four most mentioned themes described by keywords in the "no" answers were: biodiversity, overcapacity, C debt, cascading. The content of each theme is opened in more detail in the following chapters. Table 4 summarises the named keywords, which describes both positive and negative responses to sustainability among students. Keywords are ranked according to how many answers could be placed under them. However, no detailed analysis of their mutual significance was made.

Table 4. Summary table of justifications.

Sustainable	Not Sustainable
Felling balance	Biodiversity
Certification	Overcapacity
Waste fraction	C debt
Directive (RED II)	Cascading

3.3. Students' Justification of Choices: Yes, Sustainable

Initially, the fact that forests grow more than they are felled, or forest felling fell below the sustainable felling level, were the most frequently cited reasons for a sustainable alternative. It was also mentioned that forests continue to function as a significant carbon sink. Several justifications stressed that this is the case at present, but the increasing use of wood may make the situation unsustainable.

Second, different policy instruments, such as forest certificates (PEFC, FSC) and laws mean that woody biomass use in Finland is largely sustainable. Forest laws require, among other things, that forests must be replanted after felling. Finland also has a long history of wood use and forestry, whereby knowledge and know-how of the long-term balance of forest management is known. For example, the Finnish Forest Centre produces forestry guides that have the aim of maintaining biodiversity and reducing the impact of forestry on the environment.

Third, it was mentioned that the sustainability of forest bioenergy production is always context- and feedstock-dependent. In Finland, bioenergy is mainly produced from process by-flows, residues and small diameter stems from forest operations as a result of integrated systems that deliver bioenergy and other forest products. Forests are not grown for bioenergy but are a by-product of cultivation and processing.

Fourth, the new Renewable Energy Directive (RED II) has sustainable criteria for all biomass used for energy use or biofuels [24]. Respondents saw this as a guarantee of sustainability. The Renewable Energy Directive (RED II) was published in December 2018. The directive contains binding EU-wide sustainability criteria for biomass used for energy production. The sustainability criteria aim to ensure that the increasing use of bioenergy in the EU produces significant reductions in greenhouse gas emissions compared to fossil fuels. In addition, sustainability criteria include requirements for growing biomass in forests, fields and grasslands. The new RED II Directive will extend the criteria to include solid energy biomass for electricity and heat production from July 2021. In addition to environmental and ecological sustainability, social and economic sustainability were also mentioned as measures of sustainability. Many studies emphasise social sustainability and the employment opportunities offered by forest biomass for energy use [25–27].

3.4. Students Justification of Choices: No, Not Sustainable

First, loss of biodiversity, flora and fauna was mentioned in several justifications as a factor negatively affecting the sustainability of forest biomass. Forest management practices favour monoculture, and the proportion of unmanaged forests is decreasing. The preservation of native species of wood, the amount of decayed wood, valuable habitats and the structure of forests are linked to the maintenance of biodiversity.

Second, the increasing use of forests was seen as a threat to sustainability. There are many projects in Finland that aim at increasing the use of wood in the forest and energy industries. The use of wood can only be increased to a limited extent and there is a great deal of regional variation in supply and demand in Finland. Increasing the use of energy in biomass also means extending harvesting to sites with a lower biomass balance than before, which may imply a deterioration of the nutrient balance. Increasing the use of wood can also lead to an increase in wood imports, whose sustainability might be a challenge.

Third, biogenic carbon dioxide emissions were also mentioned, which raises atmospheric CO_2 and accelerates the rise in temperatures. The CO_2-balance of bioenergy is neutral from a long-term perspective, but the warming impacts of forest bioenergy needs consideration. C debt is a measure for the global warming potential (GWP) of biogenic CO_2 as it has a longer atmospheric residence time. This is affected by the type of biome (boreal, temperate, tropic) and type of forest product (residues, thinning, low- or high-quality stem wood). Logging residues are better in this regard compared to stumps or stem wood.

Lastly, other responses that questioned sustainability included the cascade principle and poor biomass efficiency in energy use. The cascade principle means prioritising the use of raw materials for resource efficiency. For example, wood is first made into higher-value products that are reused or recycled and lastly converted into energy. In particular, electricity can be generated from other renewable sources (solar, wind) and biomass can be used to replace fossil-based chemicals and products. Renewable electricity can be used in conjunction with Power to X-technologies where atmospheric CO_2 and hydrogen from water are converted into synthetic fuels (methane, methanol, dimethyl ether). The poor efficiency of biomass use was mentioned in the context of electricity generation in condensing plants and long biomass supply chains. Potential GHG emission reductions will decrease as transport distances increase, but this is dependent on the logistics system.

3.5. *Students Justification of Choices: Don't Know*

The reason for not being able to respond to either option was the conflicting views of the different parties. The lack of consensus within the science community regarding the essentiality of carbon sinks raised questions about forestry and sustainability. Different methodological approaches exist, and the perspective has shifted from narrow to system-level thinking, which has led to different conclusions. Biomass supply is characterised by numerous alternative value chains from forest to end product, which means that the impact of forest biomass on net GHG emission savings is context- and feedstock-specific, due to the fact that many factors vary across regions and time.

4. Discussion

4.1. *Student Responses and Justification*

Most of the students considered the current energy use of forest biomass to be sustainable. Many replies stressed that the situation could change if the use of forests is increased from the present. Student background variables had no effect on the response to the first question, except for the field of studies, where environmental technology and chemical engineering students were more sceptical about the energy use of forest biomass. The effect of background variables on the second question, the justification, was not assessed. Attitudes about the sustainability of forest-based bioenergy were more positive than in a study of Chinese students who were more sceptical [15]. Moreover, the schoolchildren appeared to be very critical of bioenergy and especially of the issues related to bioenergy production from forests [16,17]. In those studies of schoolchildren, gender and residential (urban–rural) distribution played a role in attitudes, whereas in this study, gender had no effect. For university students, gender seems to have less of an impact than for schoolchildren. A university of technology may have more like-minded students than the general population. These earlier studies emphasised also the need to increase knowledge among students and it would seem that increasing knowledge is changing attitudes in a more positive direction based on this survey.

The balance of forest use was highlighted in the responses of positive attitude. LUKE's calculations on forest growth and use were considered to be reliable sources [28–30]. In many responses, students used external sources of information to justify their answer. The justification ranged from single sentences to multi-paragraph responses. Forest laws, certifications and various recommendations and guidelines were considered to ensure the sustainability of forest use. The preservation of forests as a sink is a prerequisite for their sustainability [30]. Part of the forest must be outside economic use.

The current use of forests was considered to reduce biodiversity, when most of the forests are commercially exploited. This is linked to the current level of forest use, which some respondents felt was too high and therefore unsustainable. Even though the biomass going into energy would come from side fractions, it does increase the carbon dioxide in the atmosphere after combustion, causing global warming potential (GWP). The phenomenon called C-debt was mentioned, which is a measure of the amount of C released from the soil and plant C stocks following the conversion of an area for bioenergy production. The cascading principle was pointed as a relevant argument for side fractions for which alternative end uses can be found instead of energy use, such as in the chemical industry.

4.2. Emphasis of Student Responses to the SFO Concept

Of the five performance areas of SFO-concept, the environment area was most emphasised in the student's responses, being either a positive or negative attitude towards sustainability. The keywords (cf. Table 4) felling balance, overcapacity, C-debt and biodiversity collected and named from the students' responses fall into this environment area. These terms can be seen key responses relative to the SFO concept indicators. The second most emphasised areas were economics and quality optimisation, where the keywords certification, waste fraction cascading and directive (RED II) can be included in this area. The people and society area, which includes a wide range of ecological, political, economic, social and cultural systems and processes that are necessary for people and society, was very poorly addressed. It may not have been perceived as important in forestry in industrialised countries compared to developing countries, where the well-being of local communities should be prioritised. Moreover, the ergonomics area, which focuses on individual aspects of the work conditions of forest workers, was not addressed by respondents. On the other hand, this topic was not covered in the bioenergy course, as it relates in detail to forest harvesting operations. This may also be more relevant to forestry in developing countries, where the mechanisation of logging is at a lower level and a shortage of training is more obvious. It seems that the more traditional environmental, economic and social designations of sustainability, will be better suited to the definition of sustainability discussed in the course with an emphasising on the first two areas.

4.3. Validity and Reliability of the Study

Validity expresses how well the measurement method used in the study measures the characteristic of the phenomenon being investigated, which is what it is intended to measure. This study was conducted in the form of multiple-choice questions and justification for the answers was required. The concept of sustainability in relation to the research question is not necessarily unambiguous and involves different interpretations that emerged in the justification of the answers. Measuring sustainability requires previous research data and related studies and calculations. The need to rely on previous models and calculations is typical when measuring sustainability concepts.

The answers were not graded, and the query was obligatory for distance learners and voluntary for full time students. Thus, the grading had no effect on the results, which is important for validity. As a result, the study provided information on the factors on which the sustainability of forest biomass was chosen. This was in fact a more important result from the study than the proportions of the alternatives. This study was valid for this university student group at this time and the results cannot be generalised more widely. This is because the results of these type of studies are time and place dependent. Extending the survey to a wider population, for example among university students, would improve validity.

Reliability refers to how reliably and reproducibly a metric is used to measure a desired phenomenon. If the reliability of measurement is poor, it also results in poor validity. The opposite may not be true. Query surveys can have many random errors that need to be eliminated beforehand. In this survey, the respondents were not anonymous, but the answers could be associated with the respondent and this was known to the participants. In this study, anonymity would not have added value to the responses, since it would not have changed the answers. Thus, it has no effect on reliability. It was stressed to the students that all the options are "equally correct" if the answer is justified. Now the study was conducted twice in the same way in 2018 and 2019, and there was no significant difference in the results between the different years. This repetition improves reliability.

One possible factor that influences a respondent's answers is the time it takes to respond, such as a quick interview or a written response within a certain time. The students had several weeks to answer the question and justify their answer, so the time had no effect on the answer, and the students could study the matter in their own time. Questions must relate to the respondent's field of experience or motivation in order for the respondent to be able to answer. In this case, "no opinion" responses will decrease. The students were also well placed to answer the questions because they had studied the subject in the course. This makes the group more homogeneous with respect to the lecture material on the topic. Some students had more experience-based knowledge because they were working and engaged in distance learning compared to full-time students at university. No significant difference was found between these groups in the responses. Possibly applying for the same type of technical education means that the group of students is more homogeneous regardless of the way the study takes place. Distance students are, on average, older, so age does not change attitudes. The research question was related to the situation and circumstances in Finland, so one would assume that students with a foreign background may have different attitudes towards sustainability, but no difference was found in the answers to this background variable either. The study of similar course material may have been the reason for this. The respondent must have the motivation to answer the questions, which was not a problem in this study. The questionnaire was part of the course, so all the distance learners and most of the full-time students answered the questionnaire.

4.4. Possible Recommendations and Future Research

If it is to be assumed that some of the respondents are not in a position to form an opinion of the argument, the response options on the Likert scale could be: 'totally agree, somewhat agree, somewhat disagree, disagree, no opinion'. This kind of questioning could have added value for a study where there is justifiable disagreement. It is assumed that responses to the different categories would have been more evenly distributed than in the current survey, because there are also answers between yes-no opinions. The field of study influenced the opinion, but the reason remained open. There are more female students in environmental and chemical engineering, but based on the crosstabulation, gender had no effect. Further clarifications could be made as to what factors influenced opinion in these fields of study. Since this group of students already enrolled in this class it shows a level of interest and possibly bias on the topic relative to the general population.

Finally, examining sustainability in connection with the use of bioenergy will become more important in the future as the use of bioenergy increases. The targets set for Finland's use of bioenergy are a good example of this, as the additional use of renewable energy in traditional energy production (heat/electricity) and transport will rely heavily on bioenergy over the next decade.

Author Contributions: Conceptualization, T.R.; methodology, T.R.; validation, T.R., A.K. and M.L.; formal analysis, T.R.; data curation, T.R.; writing—original draft preparation, T.R.; writing—review and editing, T.R, A.K., M.L.; visualization, T.R. All authors have read and agreed to the published version of the manuscript.

Funding: This research received no external funding.

Conflicts of Interest: The authors declare no conflict of interest.

References

1. World Commission on Environment and Development (WCED). *Our Common Future*; Oxford University Press: Oxford, UK, 1987; Available online: http://www.un-documents.net/our-common-future.pdf (accessed on 23 June 2020).
2. MacDicken, K.; Sola, P.; Hall, J.E.; Sabogal, C.; Tadoum, M.; de Wasseige, C. Global progress toward sustainable forest management. *For. Ecol. Manag.* **2015**, *352*, 47–56. [CrossRef]
3. Heinimann, H.R. Forest operations engineering and management–the ways behind and ahead of a scientific discipline. *Croat. J. For. Eng.* **2007**, *28*, 107–121.
4. Marchi, E.; Chung, W.; Visser, R.; Abbas, D.; Nordfjell, T.; Mederski, P.S.; McEwan, A.; Brink, M.; Laschi, A. Sustainable Forest Operations (SFO): A new paradigm in a changing world and climate. *Sci. Total Environ.* **2018**, *634*, 1385–1397. [CrossRef] [PubMed]
5. Total Roundwood Removals and Drain, 2019 (Provisional). Natural Resources Institute Finland. Available online: https://stat.luke.fi/en/total-roundwood-removals-and-drain-2019-provisional_en (accessed on 23 June 2020).

6. Lier, M.; Korhonen, K.T.; Packalen, T.; Savula-Seppälä, T.; Tuomainen, T.; Viitanen, J.; Mutanen, A.; Vaahtera, E.; Hyvärinen, J. Suomen metsät 2019: Kestävän metsänhoidon kriteereihin ja indikaattoreihin perustuen. Available online: http://urn.fi/URN:NBN:fi-fe2019091628401 (accessed on 23 June 2020).
7. Forest Act. 1093/1996. Ministry of Agriculture and Forestry. Available online: https://www.finlex.fi/fi/laki/smur/1996/19961093 (accessed on 23 June 2020).
8. Miina, J.; Hallikainen, V.; Härkönen, K.; Merilä, P.; Packalen, T.; Rautio, P.; Salemaa, M.; Tonteri, T.; Tolvanen, A. Incorporating a model for ground lichens into multi-functional forest planning for boreal forests in Finland. *For. Ecol. Manag.* **2020**, *460*. [CrossRef]
9. Calderón, C. (Ed.) *Statistical Report*; Bioenergy Europe: Brussels, Belgium, 2019; Available online: https://bioenergyeurope.org/statistical-report.html (accessed on 23 June 2020).
10. Ranta, T.; Laihanen, M.; Karhunen, A. Development of the bioenergy as a part of renewable energy in the Nordic Countries: A comparative analysis. In Proceedings of the 6th Central European Biomass Conference, Graz, Austria, 22–24 January 2020.
11. Wood in Energy Generation 2019. Natural Resources Institute Finland. Available online: https://stat.luke.fi/en/wood-energy-generation-2019-provisional_en (accessed on 23 June 2020).
12. IRENA. *Bioenergy from Finnish Forests: Sustainable, Efficient, Modern Use of Wood*; International Renewable Energy Agency: Abu Dhabi, UAE, 2018; ISBN 978-92-9260-012-9; Available online: https://www.irena.org/media/Files/IRENA/Agency/Publication/2018/Mar/-IRENA_Bioenergy_from_Finnish_forests_2018.pdf (accessed on 23 June 2020).
13. Glossary of Climate Change Acronyms and Terms. UNFCCC. Available online: https://unfccc.int/process-and-meetings/the-convention/glossary-of-climate-change-acronyms-and-terms#l (accessed on 23 June 2020).
14. Hildén, M.; Soimakallio, S.; Seppälä, J.; Liski, J. *Forest Carbon Sinks Must Be Included in Bioeconomy Sustainability Assessments*; SYKE Policy Brief; Suomen ympäristökeskus: Helsinki, Finland, 2016; ISBN 978-952-11-4580-3; Available online: http://hdl.handle.net/10138/164797 (accessed on 23 June 2020).
15. Qu, M.; Ahonen, P.; Tahvanainen, L.; Gritten, D.; Mola-Yudego, B.; Pelkonen, P. Chinese university students' knowledge and attitudes regarding forest bio-energy. *Renew. Sustain. Energy Rev.* **2011**, *15*, 3649–3657. [CrossRef]
16. Halder, P. Perceptions of energy production from forest biomass among school students in Finland: Directions for the future bioenergy policies. *Renew. Energy* **2014**, *68*, 372–377. [CrossRef]
17. Halder, P.; Prokop, P.; Chang, C.-Y.; Usak, M.; Pietarinen, J.; Havu-Nuutinen, S.; Pelkonen, P.; Cakir, M. International Survey on Bioenergy Knowledge, Perceptions, and Attitudes Among Young Citizens. *Bioenergy Res.* **2012**, *5*, 247–261. [CrossRef]
18. van Dael, M.; Lizin, S.; Swinnen, G.; van Passel, S. Young people's acceptance of bioenergy and the influence of attitude strength on information provision. *Renew. Energy* **2017**, *107*, 417–430. [CrossRef]
19. Mayfield, C.A.; Foster, C.D.; Smith, C.T.; Gan, J.; Fox, S. Opportunities, barriers, and strategies for forest bioenergy and bio-based product development in the Southern United States. *Biomass Bioenergy* **2007**, *31*, 631–637. [CrossRef]
20. Rahman, A.; Khanam, T.; Pelkonen, P. People's knowledge, perceptions, and attitudes towards stump harvesting for bioenergy production in Finland. *Renew. Sustain. Energy Rev.* **2017**, *70*, 107–116. [CrossRef]
21. Stupak, I.; Joudrey, J.; Smith, T.; Pelkmans, L.; Chum, H.; Cowie, A.; Englund, O.; Goh, C.S.; Junginger, M. A global survey of stakeholder views and experiences for systems needed to effectively and efficiently govern sustainability of bioenergy. *WIREs Energy Environ.* **2016**, *5*, 89–118. [CrossRef]
22. Buchholz, T.; Luzadis, V.A.; Volk, T.A. Sustainability criteria for bioenergy systems: Results from an expert survey. *J. Clean. Prod.* **2009**, *17*, S86–S98. [CrossRef]
23. van Dam, J.; Junginger, M. Striving to further harmonization of sustainability criteria for bioenergy in Europe: Recommendations from a stakeholder questionnaire. *Energy Policy* **2011**, *39*, 4051–4066. [CrossRef]
24. Report from the Commission to the Council and the European Parliament on Sustainability Requirements for the Use of Solid and Gaseous Biomass Sources in Electricity, Heating and Cooling. COM(2010)11 final. Available online: https://eur-lex.europa.eu/LexUriServ/LexUriServ.do?uri=COM:2010:0011:FIN:EN:PDF (accessed on 23 June 2020).
25. Eker, M.; Spinelli, R.; Gürlevik, N. Recovering energy biomass from sustainable forestry using local labor resources. *J. Clean. Prod.* **2017**, *157*, 57–64. [CrossRef]

26. Valente, C.; Spinelli, R.; Hillring, B.G. LCA of environmental and socio-economic impacts related to wood energy production in alpine conditions: Valle di Fiemme (Italy). *J. Clean. Prod.* **2011**, *19*, 1931–1938. [CrossRef]
27. McKay, H. Environmental, economic, social and political drivers for increasing use of woodfuel as a renewable resource in Britain. *Biomass Bioenergy* **2006**, *30*, 308–315. [CrossRef]
28. Forest Resources. Natural Resources Institute Finland. Available online: https://stat.luke.fi/en/forest-resources (accessed on 23 June 2020).
29. NFI Computing Service. Natural Resources Institute Finland. Available online: https://vmilapa.luke.fi/#/?locale=fi-FI (accessed on 23 June 2020).
30. Official Statistics of Finland (OSF): Greenhouse Gases. Available online: http://www.stat.fi/til/khki/index_en.html (accessed on 23 June 2020).

sustainability

MDPI

Article

Soil Disturbance and Recovery after Coppicing a Mediterranean Oak Stand: The Effects of Silviculture and Technology

Rachele Venanzi [1,2,*], Rodolfo Picchio [1], Raffaele Spinelli [3] and Stefano Grigolato [2]

1 Department of Agricultural and Forest Sciences, University of Tuscia, 01100 Viterbo, Italy; r.picchio@unitus.it
2 Department of Land, Environment, Agriculture and Forestry, Università degli Studi di Padova, 35020 Legnaro, Padova, Italy; stefano.grigolato@unipd.it
3 CNR Institute of Bioeconomy, Via Madonna del Piano 10, 50019 Sesto Fiorentino FI, Italy; spinelli@ivalsa.cnr.it
* Correspondence: venanzi@unitus.it; Tel.: +39-0761357400; Fax: +39-0761357250

Received: 24 April 2020; Accepted: 12 May 2020; Published: 15 May 2020

Abstract: Traditional coppice management system is one of the most debated topics in the Mediterranean area, as it is a forest management system that accounts for over 23 million hectares. Coppicing is considered the oldest form of sustainable forest management. Its past and current widespread popularity is mainly due to its capacity to positively contribute to the rural economy and ecosystem services. This research aimed at assessing the effect of coppicing on soil characteristics, understanding a possible treatment return time, and evaluating the implementation of proper sustainable forest operations (SFOs) in order to have a better understanding of the disturbance caused by silvicultural treatment and forest operations with two different harvesting techniques. The results demonstrated that physical, chemical, and biological soil features were partially disturbed by the coppicing. Both silvicultural treatment and forest operations influenced soil disturbance. The least impactful technique was extraction by winch, while forwarding resulted in heavier alterations of soil characteristics. It took about five years for the soil to recover its original pre-harvest conditions when the disturbance was caused by the silvicultural treatment alone (non-trafficked areas) and about eight to nine years when the disturbance was the cumulated effect of silvicultural treatment and logging activity (trafficked areas).

Keywords: forest operation; skidding-winch; forwarding; soil resilience; Mediterranean area

---

1. Introduction

The coppice management system is one of the most debated topics in the Mediterranean area, as it is a forest management system that accounts for over 23 million hectares [1]. Coppicing is considered the oldest form of sustainable forest management and for this reason it is considered as a natural forest management system [2]. Its past and current popularity is mainly due to its capacity to positively contribute to the rural economy and ecosystem services [3]. Even if this management system presents environmental shortcomings, coppicing represents a valid and flexible management system that requires low inputs and guarantees maintenance of many aesthetic, environmental, social, and economic functions [1,4,5].

Recent findings in ecological and forestry research have highlighted that coppice forests contribute to soil protection and biodiversity conservation [6,7], showing good resilience and significant adaptability to climate change [1,8].

In the past, Mediterranean oak coppice stands were an important source of timber, firewood, and charcoal [9], as well as litter and pasture [10]. Today, they are mostly invested in the production of wood biomass for energy use due to their capacity to yield sustained amount of raw material at short intervals (on average one cutting cycle is every 12–18 years) [2].

These short cutting cycles could negatively affect soil quality and regeneration vigor [11,12]. Special concern is aroused by the risk for soil degradation connected with frequent machine traffic, which may cause compaction, topsoil removal, and general disturbance [13–16]. Nevertheless, not all logging techniques have the same impact potential and the specific characteristics of any given operation depend on site characteristics, silvicultural management, technological level, and product strategy [17,18]. Furthermore, technological innovations in forest logging and mechanization [19,20] could positively contribute to the improvement of work conditions, compared with traditional logging systems [21,22].

Assessing ground disturbance and minimizing possible damage due to silvicultural treatments and forest operations remain the main focus of sustainable forest management (SFM) [23–25]. In order to reach this aim, numerous suggestions have come from recent research, namely: to minimize the area of soil disturbance and compaction by appropriate operation planning [18,26,27]; to make careful execution of logging operations [28,29]; and to use suitable mechanization [12,30–34]. All this is in consideration of the fact that adequately managed forest ecosystems are highly resilient in the long-term [25,35].

Focusing the attention on coppice systems, there is a need to acquire more information about the impacts due to silvicultural treatment, i.e., actual logging and their interactions. These topics are very often the subject of heated arguments and detailed scientific results are needed to better understand the issue and provide best practice suggestions [2]. For these reasons, the concept of sustainability is frequently overlooked and is not considered as a clear instrument to assess the impact of global change and development.

SFM is based on continuous improvements of silvicultural practice and logging methods. In particular, better knowledge is needed on the recovery time of managed forest ecosystems after the inevitable disturbance cause by forest operations, however well they are managed. This is one of the key factors for sustainable use and an important issue both for high forests and coppices [1].

Recent studies [4,5,36] on coppicing in the Mediterranean area have highlighted that within a short time after harvesting (0–3 years), soil and regeneration characteristics show clear signs of recovery. These findings demonstrated that physical, chemical, and biological soil features were only marginally affected by the silvicultural treatment applied, but strongly impacted by harvesting operations.

Coppicing maintains a cyclical pattern of extreme changes in ground-level light penetration [1,37,38], producing heterogeneous mosaics of forest in various stages of succession that harbor a rich variety of animals and vascular plants [39–43].

Only efficient planning and management of forest operations and accurate knowledge of the environmental dynamics of the forest will offer high social and environmental benefits and provide various ecosystem services in the long term [44]. These aspects can be guaranteed only through SFM in synergy with sustainable forest operations (SFO) [45]. These tools are essential for proper environmental protection and they are mandatory in order to maintain forests and their multiple functions [45]. In particular, forest operations and coppice management are interesting but delicate issues to be analyzed and evaluated in order to achieve real sustainability.

Starting from this background to increase scientific knowledge on the effects of coppicing, the present experiment was designed with four specific goals:

- to investigate the impact of the silvicultural treatment on soil condition;
- to find out how both silvicultural treatment and forest operations influence soil characteristics;
- to compare the impact of two different harvesting techniques on soil condition;
- to assess the recovery capacity of soil after harvesting in order to project a possible treatment return time; and evaluating the existence of a proper SFO.

To this end, soil conditions in a Turkey oak coppice located in central Italy were monitored every year for five years after harvesting.

2. Materials and Methods

2.1. Study Sites

The study stand was Turkey oak (*Quercus cerris* L.) forest managed as coppice with standards. The stand was located in Central Italy, Lazio Region, Tarquinia municipality (42°34′37.07″ N, 11°76′40.00″ E). The whole forest covered about 100 ha, with homogeneous elevation, slope gradient, and roughness: 100 m a.s.l., 25% slope gradient and ca 5% of the surface presenting obstacles to machine traffic, respectively. The accessibility of the forest was therefore fairly good (12 m/ha of main forest roads) but the road network in the area included few permanent skid trails.

The soil was an Abruptic Luvisol (EpiArenic Cutanic) (WRB, 2014) alluvial, with good depth (ranging from 0.6 to 0.9 m) non-hydromorph and with a neutral reaction. Soil texture was defined as Silty-Loam (SL), due to the high silt content (52%), moderate sand content (38%) and low clay content (10%). Soil field capacity (CC) was 24%, determined using the soil water method [46].

2.2. Silviculture and Harvesting Technique

The study coppice was clear-cut at the age of 35 years, releasing 140 standards per hectare. Standards belonged to three age classes: 35-years-old (60%), 45-years-old (30%), and 60-years-old or older (10%). The harvesting operation was completed within approximately 150 days for the about 100 ha studied.

Only one harvesting system was applied, the tree length system (TLS) [47]. One control study area was selected of about 20 ha of coppice unharvested and not impacted for more than 20 years. Felling was performed motor-manually by three teams of two operators equipped with Husqvarna 550 XP chainsaws.

About one half of the wood (about 50 ha) was skidded to roadside landing using a forestry fitted farm tractor equipped with a winch. This was a four-wheel drive LANDINI tractor with a rated power of 80 kW and a total weight of 4200 kg, including the forestry winch. The tractor stationed on the main forest road reached the trees with its winch and pulled them to the road. Once it had assembled a full load, the tractor drove on the forest road all the way to the landing. The average tractor load was 0.9 t.

The remaining half of the wood (about 50 ha) was carried to the roadside landing using a John Deere 1410D eco III eight-wheeled purpose-built forwarder. This machine had a rated power of 136 kW, an empty weight of 16,600 kg and payload capacity of 14,000 kg. The forwarder left the forest road and entered the stand, travelling on the forest floor and picking the delimbed stems with its 7.5 m hydraulic loader. Once a full load had been assembled, the forwarder drove back to the forest road and then to the landing. The average single load size was 11.7 t.

Pre-harvest stand characteristics (Table 1) were obtained using standard forest mensuration techniques on thirty randomly selected circular plots with a 20 m radius, each therefore covering an area of 1256 m^2. These measurements confirmed the substantial uniformity of the three study areas: control, winch, and forwarder.

The stand data collected on the three-study area before harvesting showed the same average values with similar growth trends (Table 1). The post-harvest measurements of soil characteristics were taken every year for five years.

Table 1. Pre-harvest stand characteristics for winching area (W), forwarding area (F), and control (C).

Area	Age (Years)		DBH * (cm)		Height * (m)		Density * (Trees/ha)	Basal Area (m^2/ha)	Above-Ground Biomass Stock (m^3/ha)	Above-Ground Biomass Harvested (m^3/ha)
	Shoots	Standards	Shoots	Standards	Shoots	Standards				
W	36	52	14.6 ± 2.1	26.0 ± 1.4	10.5 ± 3.2	13.8 ± 1.2	1168 ± 75	24.1	132.9	109.5
F	34	54	14.1 ± 5.2	28.1 ± 4.1	10.9 ± 1.2	14.2 ± 2.3	1178 ± 47	24.6	136.4	112.8
C	20	45	12.8 ± 1.5	22.3 ± 7.2	10.1 ± 2.7	13.1 ± 5.2	1188 ± 66	23.7	121.9	-

* (average ± SD); W = Winch; F = Forwarder; C = Control.

2.3. Analytical Methods

Soil analyses were conducted on 30 randomly selected sample plots (SP) on each study area (W, F, and C). Each SP consisted of a circular area of 113 m^2. The analyses were done in order to assess the presence/absence of soil impacts due to the silvicultural treatment and to the forest operations and followed the research protocols proposed by Picchio et al. [21,31,48] and Venanzi et al. [2]. In particular, for the SPs in the harvested areas (W and F), two different strata were selected based on a visual assessment of impact (e.g., the presence or absence of bent understory, crushed litter, ruts, or soil mixing) to represent trafficked and untrafficked soil conditions, respectively. In the case of the winch treatment, these signs derived from the sliding of the tree bunches on the forest floor, as they were pulled to the road by the winch. In the case of the forwarders, these signs were derived from the passage of the wheels. On each stratum, the soil was analyzed for texture, bulk density (BD), penetration resistance (PR), shear resistance (or strength) (SR), organic matter content (OM), pH, and QBS-ar (soil biological quality index referred to micro arthropod community). For particle size distribution, three soil samples in each SP were randomly collected from the top 30 cm of soil [4].

Rock fragments (particles with >2 mm diameter) were removed from the air-dried samples by sieving. Silt, clay, and sand were determined using the Andreasen pipette method [48]. These fractions were used to find the soil classes using the textural USDA triangle [49].

BD, PR, and SR were determined through the methods proposed by Marchi et al. [4] and expressed in Mg m^{-3}, MPa, and t m^{-2}, respectively. The pH value was measured using potentiometric analysis in soil/saline solution suspensions (soil-KCl 1 mol) in a 1:2.5 proportion. OM measurement was performed by incineration in a mitten at 400 °C for 4 h following the thorough elimination of water and pre-treatment at 160 °C for 6 h.

As described in Marchi et al. [4] and Spinelli et al. [50], linear transect were laid down according to a systematic pattern. Each transect was rectangular in shape (1 m x 50 m) and was established using a compass and tape measure. These linear transect were used to assess the relative proportions of the two strata (i.e., trafficked and untrafficked).

The QBS-ar index was used as a biological indicator for the soil analysis, being extremely sensitive to environmental variations caused by disturbance. This index is mainly qualitative and evaluates the presence and complexity of the soil microarthropod community. The methodology applied was reported in Venanzi et al. [5] and Marchi et al. [4].

2.4. Statistical Analyses

Statistical analyses were carried out with StatisticaTM version 7.1 (TIBCO Software Inc.). Data distribution was plotted and checked for normality and homogeneity of variance using the Lilliefors and Levene tests, respectively. To check differences between treatments *t*-test and ANOVA, were applied, followed by Tukey's post-hoc test when necessary. The elected significance level was $\alpha < 0.05$. Data that violated the normality and/or the homoscedasticity assumptions were analyzed using the nonparametric Kruskal–Wallis test. Non-linear regression analysis was done for all the soil variables in relation to the time (in years) post-harvesting. This analysis was applied to assess the possible existence of a recovery trend for the main soil characteristics. The polynomial approach was applied because these natural dynamics can be better described by a non-linear function [51]. Non-metric multidimensional scaling (NMDS) was used to show the differences in the average soil parameters for the different treatments over the five observed periods.

3. Results

3.1. *Analysis of the Impacted Surface*

The proportion of the total surface affected by machine traffic was significantly different for the two harvesting systems (Table 2).

Table 2. Soil area impacted by bunching and extraction activities (*t*-test results; average ± SD). W: winching; F: forwarding.

Area	*p*-Value	Trafficked Soil	Untrafficked Soil
W	<0.05	21.2 ± 4.1% [a]	78.8%
F		31.7 ± 2.9% [b]	68.3%

Note: Different letters after means within each treatment indicate significant differences by Tukey test ($p < 0.05$).

3.2. *Physical and Chemical Analyses of Soil*

Soil data was collected every year for five years, always in the same month, in particular during the last fortnight of September, in order to operate always under the same weather conditions. These analyses showed no statistically significant differences regarding soil moisture in the harvested forests between sampling periods (average moisture ± standard deviation: 34 ± 2%; 32 ± 5%; 35 ± 7%; 32 ± 6%; 36 ± 4% for 1, 2, 3, 4, and 5 years after coppicing, respectively). In the control areas, soil moisture showed the same trend, but with slightly higher values (difference range of about +6% to +9%).

3.2.1. Soil Bulk Density

The soil bulk density (BD) data showed statistically significant differences between the two strata (trafficked and untrafficked), the two harvesting techniques and the five periods (in Table 3 were reported only the values of two periods, i.e., 1 and 5 years after coppicing). In particular, BD was higher in the trafficked areas than the untrafficked ones, in the first four periods, and within these areas BD was higher for the F than for the W treatment. Data collected in the fifth year showed a recovery of soil BD for all treatments.

A clear BD recovery was shown for the trafficked soil typologies (Figure 1) that started three years after harvesting, while for the untrafficked soil typology it started two years after coppicing. BD was affected by the uncovering effect due to coppicing and during the first two years after coppicing it was higher in the harvested but untrafficked areas than in the control ones (Table 3 and Figure 1). The regression models built (Table 4 and Figure 1) are all statistically significant but with a relatively weak explanatory value ($r^2_{adj} \sim 0.4$) for the control and the untrafficked treatments, and somewhat stronger ($r^2_{adj} \sim 0.6$) for the two trafficked treatments. During the fourth and fifth year post-harvesting the untrafficked treatment and the trafficked winch treatment showed a complete recovery of soil BD (Figure 1). Conversely, soil BD trafficked by the forwarder did not show a complete recovery five years after harvesting (Figure 1), but the improving trend indicated that full recovery could be achieved in the sixth or seventh year.

Table 3. Results of the ANOVA and Tukey test for Bulk Density (average ± SD), difference tested between trafficked, untrafficked, and control soil for the two mechanization levels (data showed in columns). Results of the *t*-test for BD, difference tested between two time periods (data showed in rows). W: winching; F: forwarding; C: control.

Area	Soil Condition	Bulk Density (g/cm^3)		*p*-Value
		1 Year	5 Years	
W	Untrafficked	1.07 ± 0.07^{a}	0.97 ± 0.14^{a}	<0.05
	Trafficked	1.15 ± 0.12^{b}	0.95 ± 0.11^{a}	<0.05
F	Untrafficked	0.89 ± 0.15^{c}	0.79 ± 0.09^{b}	<0.05
	Trafficked	1.19 ± 0.29^{d}	0.98 ± 0.03^{a}	<0.05
C	Control	$1.00 \pm 0.17^{a,c}$	1.00 ± 0.17^{a}	>0.05
p-value		<0.05	<0.05	

Note: Different letters after means within each treatment indicate significant differences by Tukey test ($p < 0.05$).

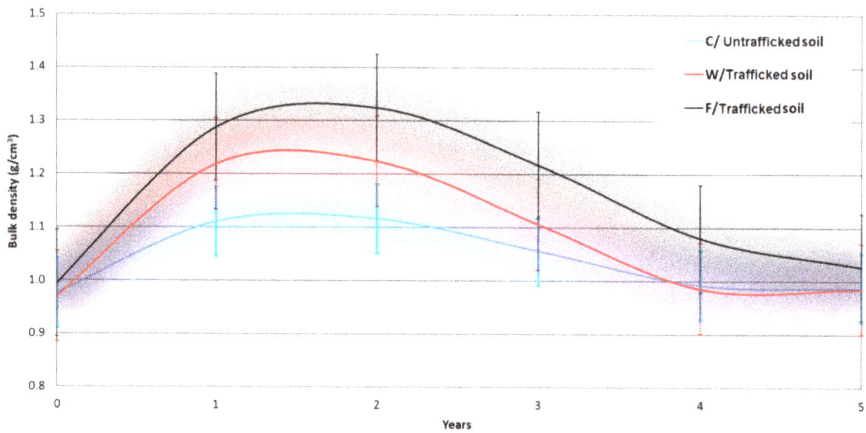

Figure 1. Graphical interpretation of the non-linear regression analysis for BD in relation to the time (in years) post-harvesting. The polynomial curves are showed with an area describing the ± regression standard estimate error. C/Untrafficked $r^2_{adj} = 0.438$, $F(3,176) = 39.685$, $p < 0.001$. W/Trafficked $r^2_{adj} = 0.615$, $F(3,176) = 40.408$, $p < 0.001$. F/Trafficked $r^2_{adj} = 0.623$, $F(3,176) = 41.686$, $p < 0.001$. W: winching; F: forwarding; C: control.

3.2.2. Soil Penetration Resistance

The results for soil penetration resistance (PR) matched those for BD and showed statistically significant differences between the two mechanization levels, traffic strata (trafficked and untrafficked), and five years (in Table 5 were reported only the values of years 1 and 5). In particular, PR was clearly higher in the trafficked areas than in the untrafficked ones in the first four periods and within those groups higher PR was recorded in F than in W. Analysis of the last period (five years after logging) showed a recovery of soil PR for the W area, while for the F area there were again higher values in the trafficked areas.

Table 4. Results of the non-linear regression analysis for bulk density (BD) (dependent variable, g cm^3) in relation to the time (in years) post-harvesting. W: winching; F: forwarding; C: control.

Area/Soil Condition	Regression	Beta	Beta Std. Err.	B	B Std. Err.	t	p-Level
C/Untrafficked	Intercept	–	–	0.976	0.012	83.162	<0.001
	Years	4.395	0.488	0.219	0.024	9.007	<0.001
	Years^2	−9.756	1.265	−0.095	0.012	−7.714	<0.001
	Years^3	5.320	0.838	0.010	0.002	6.347	<0.001
W/Trafficked	Intercept	–	–	0.970	0.022	44.779	<0.001
	Years	5.251	0.573	0.411	0.045	9.164	<0.001
	Years^2	−11.895	1.485	−0.182	0.023	−8.010	<0.001
	Years^3	6.586	0.984	0.020	0.003	6.691	<0.001
F/Trafficked	Intercept	–	–	0.994	0.025	39.096	<0.001
	Years	4.955	0.567	0.460	0.053	8.735	<0.001
	Years^2	−10.193	1.471	−0.185	0.027	−6.932	<0.001
	Years^3	5.213	0.975	0.019	0.004	5.349	<0.001

Table 5. Results of the ANOVA (average ± SD) and Tukey test for penetration resistance (PR) data. Difference tested between trafficked, untrafficked, and control soil for the two mechanization levels (data showed in columns). Results of the *t*-test for PR, difference tested among two time periods (data showed in rows). W: winching; F: forwarding; C: control.

Area	Soil Condition	Penetration Resistance (MPa)		*p*-Value
		1 Year	5 Years	
W	Untrafficked	0.59 ± 0.07 ^{a}	0.45 ± 0.05 ^{a}	<0.05
	Trafficked	1.01 ± 0.12 ^{b}	0.44 ± 0.10 ^{a}	<0.05
F	Untrafficked	0.64 ± 0.03 ^{a,c}	0.46 ± 0.08 ^{a}	<0.05
	Trafficked	1.18 ± 0.07 ^{d}	0.59 ± 0.02 ^{b}	<0.05
C	Control	0.49 ± 0.09 ^{a}	0.49 ± 0.09 ^{a}	>0.05
p-value		<0.05	<0.05	

Note: Different letters after means within each treatment indicate significant differences by Tukey test ($p < 0.05$).

A clear PR recovery trend was visible for the untrafficked and trafficked soil typologies (Figure 2), starting two years after logging. PR was affected by the uncovering effect due to coppicing during the first two years after harvesting and was higher in the harvested but untrafficked area than in the control ones (Table 5 and Figure 2). The regression models built (Table 6 and Figure 2) are all statistically significant with a reasonably good explanatory value ($r^2_{adj} \sim 0.5$) for the C and untrafficked treatments, a better explanatory value ($r^2_{adj} \sim 0.6$) for the W/trafficked typology, and a high explanatory value ($r^2_{adj} \sim 0.7$) for F/trafficked typology. During the third year post-harvest, untrafficked soil showed a complete recovery of soil PR (Figure 2). The same occurred for the soil trafficked in the winch treatment, but only one year later. Finally, the PR of soil trafficked by the forwarder showed a complete recovery five years after harvesting (Figure 2).

3.2.3. Soil Shear Resistance

The soil shear resistance (SR) data showed statistically significant differences between the two mechanization levels, soil strata (trafficked and untrafficked), and five periods (in Table 7 were reported only the values of two periods, 1 and 5 years after coppicing). In particular, SR was clearly higher in the trafficked treatments than in the untrafficked ones in all the five periods observed, with higher impacts in the F treatment than in the W treatment. Data from the last period (5 years after logging) showed full recovery of soil SR for the W treatment, but not for the F treatment in the trafficked areas.

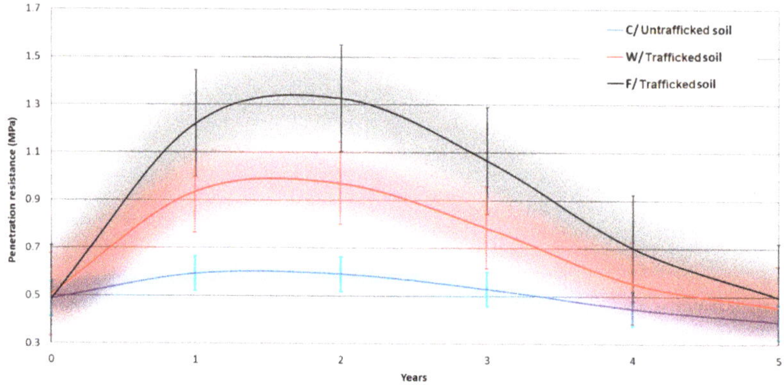

Figure 2. Graphical interpretation of the non-linear regression analysis for PR in relation to the time (in years) post-harvesting. The polynomial curves showed with a halo composed by ± regression standard estimate error. C/Untrafficked $r^2_{adj} = 0.502$, $F(3,176) = 30.917$, $p < 0.001$. W/Trafficked $r^2_{adj} = 0.589$, $F(3,176) = 43.425$, $p < 0.001$. F/Trafficked $r^2_{adj} = 0.695$, $F(3,176) = 68.571$, $p < 0.001$. W: winching; F: forwarding; C: control.

Table 6. Results of the non-linear regression analysis for penetration resistance (PR) (dependent variable, MPa) in relation to the time (in years) post-harvesting. W: winching; F: forwarding; C: control.

Area/Soil Condition	Regression	Beta	Beta Std. Err.	B	B Std. Err.	T	p-Level
C/Untrafficked	Intercept	–	–	0.486	0.018	26.880	<0.001
	Years	2.978	0.596	0.176	0.035	5.001	<0.001
	Years^2	−6.731	1.542	−0.076	0.018	−4.366	<0.001
	Years^3	3.335	1.021	0.008	0.002	3.266	<0.01
W/Trafficked	Intercept	–	–	0.504	0.043	11.610	<0.001
	Years	4.433	0.542	0.690	0.084	8.187	<0.001
	Years^2	−9.599	1.402	−0.287	0.042	−6.849	<0.001
	Years^3	4.961	0.928	0.029	0.006	5.344	<0.001
F/Trafficked	Intercept	–	–	0.487	0.057	8.628	<0.001
	Years	4.827	0.466	1.136	0.110	10.354	<0.001
	Years^2	−9.864	1.207	−0.446	0.055	−8.173	<0.001
	Years^3	4.886	0.799	0.044	0.007	6.112	<0.001

Table 7. Results of the ANOVA (average ± SD) and Tukey test for shear resistance (SR) data. Difference tested between trafficked, untrafficked, and control soil for the two mechanization levels (data showed in columns). Results of the *t*-test for SR, difference tested among two time periods (data showed in rows). W: winching; F: forwarding; C: control.

Area	Soil Condition	Shear Resistance (t/m^2)		*p*-Value
		1 Year	5 Years	
W	Untrafficked	4.53 ± 0.38 ^{a}	5.16 ± 1.36 ^{a}	<0.05
	Trafficked	11.42 ± 1.20 ^{b}	6.91 ± 2.57 ^{b}	<0.05
F	Untrafficked	4.29 ± 0.91 ^{c}	4.51 ± 2.94 ^{a,c}	<0.05
	Trafficked	12.67 ± 2.57 ^{d}	11.00 ± 3.76 ^{d}	<0.05
C	Control	7.04 ± 2.68 ^{e}	7.04 ± 2.68 ^{b}	>0.05
p-value		<0.05	<0.05	

Note: Different letters after means within each treatment indicate significant differences by Tukey test ($p < 0.05$).

A clear SR recovery trend was visible for the trafficked strata (Figure 3), starting two years after logging. SR was not affected by the uncovering effect due to coppicing and was higher in the control area than in the harvested ones (Table 7 and Figure 3). The estimated regression models (Table 8 and Figure 3) are statistically significant only for trafficked soils, with a medium explanatory value ($r^2_{adj} \sim 0.5$) for the W/Trafficked treatment and a low one ($r^2_{adj} \sim 0.4$) for the F/Trafficked treatment. During the fourth year post-harvesting the soil trafficked by winching showed a complete recovery of SR (Figure 3). On the contrary, the SR of the soil trafficked by forwarding did not show a complete recovery five years after harvesting (Figure 3). However, the trends indicated that full recovery should be achieved between the eighth and ninth year.

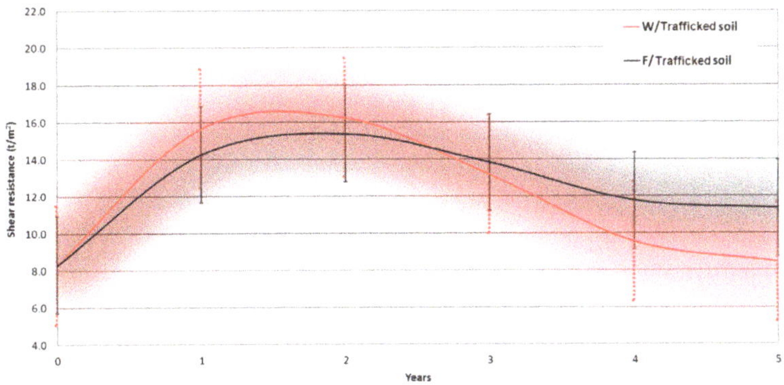

Figure 3. Graphical interpretation of the non-linear regression analysis for SR in relation to the time (in years) post-harvesting. The polynomial curves showed with a halo composed by ± regression standard estimate error. W/Trafficked $r^2_{adj} = 0.508$, $F(3,176) = 31.638$, $p < 0.001$. F/Trafficked $r^2_{adj} = 0.437$, $F(3,176) = 23.995$, $p < 0.001$. W: winching; F: forwarding.

Table 8. Results of the non-linear regression analysis for shear resistance (SR) (dependent variable, t/m^2) in relation to the time (in years) post-harvesting. W: winching; F: forwarding; C: control.

Area/Soil Condition	Regression	Beta	Beta Std. Err.	B	B Std. Err.	t	p-Level
C/Untrafficked	Intercept	–	–	8.305	1.094	6.180	<0.001
	Years	1.195	0.845	3.005	2.126	1.414	>0.05
	Years^2	−2.445	2.189	−1.180	1.056	−1.117	>0.05
	Years^3	1.296	1.450	0.124	0.139	0.894	>0.05
W/Trafficked	Intercept	–	–	8.305	0.811	9.239	<0.001
	Years	4.425	0.592	11.779	1.576	7.474	<0.001
	Years^2	−9.655	1.533	−4.934	0.783	−6.300	<0.001
	Years^3	5.102	1.015	0.517	0.103	5.027	<0.001
F/Trafficked	Intercept	–	–	8.305	0.660	12.588	<0.001
	Years	4.497	0.634	9.096	1.282	7.098	<0.001
	Years^2	−9.016	1.640	−3.501	0.637	−5.498	<0.001
	Years^3	4.684	1.086	0.361	0.084	4.313	<0.001

3.2.4. Soil Organic Matter Content

The soil organic matter content (OM) showed statistically significant differences between the two mechanization levels, soil strata (trafficked and untrafficked) and five periods (data partially shown in Table 9 for one and five years after coppicing). OM content was clearly lower in the trafficked stratum than in the untrafficked one in the five observed periods, with higher impact in the F treatment than in the W treatment. The data indicates that even five years after logging, OM content was not back to the original values in the trafficked stratum of both the W and the F treatments, while it was achieved in the untrafficked strata.

Table 9. Results of the ANOVA and Tukey test for organic matter content (OM) (average ± SD). Difference tested between trafficked, untrafficked, and control soil for the two mechanization levels (data showed in columns). Results of the *t*-test for OM, difference tested among two time periods (data showed in rows). W: winching; F: forwarding; C: control.

Area	Soil Condition	Organic Matter (%)		*p*-Value
		1 Year	5 Years	
W	Untrafficked	15.6 ± 0.019 [a]	18.4 ± 0.001 [a]	<0.05
	Trafficked	12.4 ± 0.054 [b]	13.9 ± 0.005 [b]	<0.05
F	Untrafficked	9.5 ± 0.004 [c]	16.7 ± 0.002 [c]	<0.05
	Trafficked	7.7 ± 0.002 [d]	13.3 ± 0.004 [b]	<0.05
C	Control	14.2 ± 0.029 [e]	14.2 ± 0.029 [a,c]	>0.05
p-value		<0.05	<0.05	

Note: Different letters after means within each treatment indicate significant differences by Tukey test ($p < 0.05$).

However, the trends are for a slow but clear recovery, which became visible two years after harvesting in the winch/trafficked treatment, and one year later in the forwarder/trafficked treatment (Figure 4). OM content was affected by the uncovering effect due to coppicing (i.e., mineralization due to the access of light) in the four years after harvesting and it was lower in the harvested but untrafficked areas than in the control ones (Table 9 and Figure 4). The regression models built (Table 10 and Figure 4) are all statistically significant with a medium to high explanatory value ($r^2_{adj} \sim 0.5$ for C/Untrafficked, and $r^2_{adj} \sim 0.7$) for W/Trafficked and F/Trafficked). During the fourth year post-harvesting untrafficked soil showed a complete recovery of soil OM content (Figure 4), while the OM content of soil trafficked by winching and forwarding was not back to the original values even five years after harvest: trends indicated that full recovery could be expected between the seventh and the eighth year.

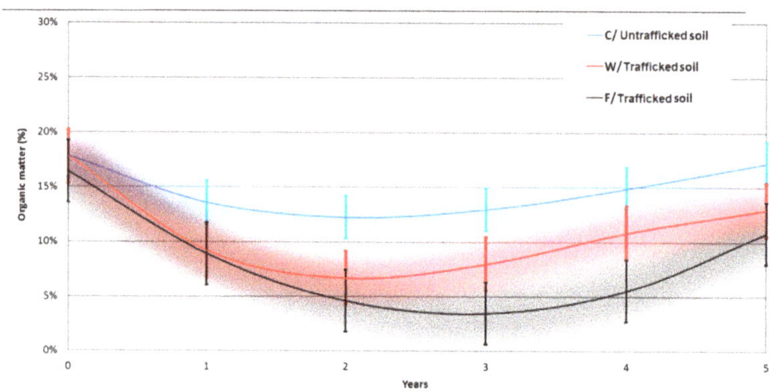

Figure 4. Graphical interpretation of the non-linear regression analysis for OM in relation to the time (in years) post-harvesting. The polynomial curves showed with a halo composed by ± regression standard estimate error. C/Untrafficked $r^2_{adj} = 0.538$, $F(3,176) = 35.558$, $p < 0.001$. W/Trafficked $r^2_{adj} = 0.695$, $F(3,176) = 68.479$, $p < 0.001$. F/Trafficked $r^2_{adj} = 0.719$, $F(3,176) = 76.958$, $p < 0.001$. W: winching; F: forwarding; C: control.

Table 10. Results of the non-linear regression analysis for Organic Matter content (dependent variable, %) in relation to the time (in years) post-harvesting. W: winching; F: forwarding; C: control.

Area/Soil Condition	Regression	Beta	Beta Std. Err.	B	B Std. Err.	t	p-Level
C/Untrafficked	Intercept	–	v	0.179	0.005	36.612	<0.001
	Years	−3.628	0.574	−0.060	0.010	−6.324	<0.001
	Years^2	5.915	1.485	0.019	0.005	3.983	<0.001
	Years^3	−2.238	0.983	−0.001	0.001	−2.276	<0.05
W/Trafficked	Intercept	–	–	0.178	0.006	28.953	<0.001
	Years	−4.840	0.466	−0.124	0.012	−10.377	<0.001
	Years^2	8.495	1.208	0.042	0.006	7.036	<0.001
	Years^3	−3.881	0.800	−0.004	0.001	−4.853	<0.001
F/Trafficked	Intercept	–	–	0.164	0.007	22.798	<0.001
	Years	−2.912	0.447	−0.091	0.014	−6.510	<0.001
	Years^2	2.660	1.158	0.016	0.007	2.297	<0.05
	Years^3	0.223	0.767	0.001	0.001	0.291	>0.05

3.2.5. Soil pH

Soil pH is an important soil characteristic, because its variations affect a number of pedological parameters and processes [52] (Picchio et al., 2019). In this study (Table 11, Table 12 and Figure 5), soil pH did not seem affected by either silvicultural treatment or logging technique.

Table 11. Results of the ANOVA and Tukey test for pH (average ± SD). Difference tested between trafficked, untrafficked, and control soil for the two mechanization levels (data showed in columns). Results of the *t*-test for pH, difference tested between two time periods (data showed in rows). W: winching; F: forwarding; C: control.

Area	Soil Condition	pH		*p*-Value
		1 Year	5 Years	
W	Untrafficked	6.8 ± 0.02 ^{a}	6.7 ± 0.05	>0.05
	Trafficked	6.7 ± 0.12 ^{a,b}	6.8 ± 0.105	>0.05
F	Untrafficked	6.5 ± 0.45 ^{b}	6.8 ± 0.05	>0.05
	Trafficked	7.0 ± 0.26 ^{a}	6.8 ± 0.01	>0.05
C	Control	6.7 ± 0.43 ^{a,b}	6.7 ± 0.43	>0.05
p-value		<0.05	>0.05	

Note: Different letters after means within each treatment indicate significant differences by Tukey test ($p < 0.05$).

Table 12. Results of the factorial ANOVA for pH. Difference tested between trafficked and untrafficked soil for the two mechanization levels in the five periods observed. W: winching; F: forwarding.

Variables	Sum of Square	Degree of Freedom	Mean of Square	F	*p*-Value
Year	4.73	5	0.95	37.8	<0.001
Untrafficked/W and F: Trafficked soil	0.4	2	0.2	8	<0.001
Year X Untrafficked/Wand F: Trafficked soil	5.25	10	0.52	20.9	<0.001
Error	6.31	252	0.03		

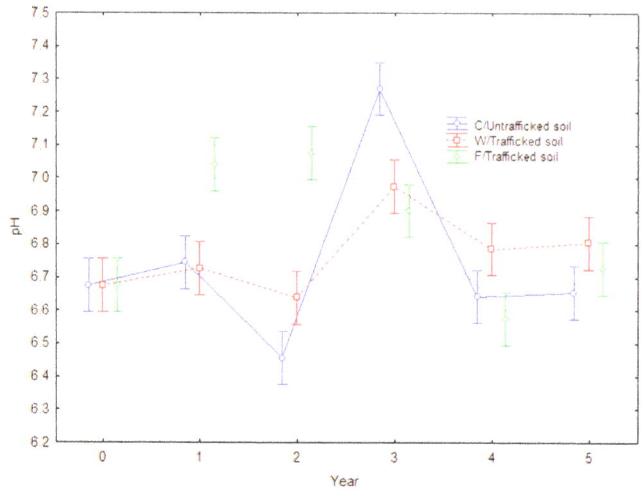

Figure 5. Graphical interpretation of the factorial ANOVA for pH (point: average value; bar: ±standard deviation). Difference tested between trafficked and untrafficked soil for the two mechanization levels in the five periods observed. W: winching; F: forwarding; C: control.

3.3. *Soil Biodiversity Aalysis*

The QBS-ar index (QBS-ar) showed statistically significant differences between the two mechanization levels, soil strata (trafficked and untrafficked), and five periods (data partially shown in Table 13, for the first and fifth year after coppicing). QBS-ar content was clearly lower in the trafficked areas than in the untrafficked ones in all the five observed periods, with the highest impact one year after coppicing. The effect was stronger for the F treatment than for the W treatment, but this trend reversed in the fifth year when the residual impact was stronger for the W treatment compared with the F treatment. Complete recovery was not achieved within five years for any of the two treatments, neither in the trafficked nor the untrafficked strata.

Table 13. Results of the Kruskall–Wallis and Duncan test for QBS-ar index. Difference tested between trafficked, untrafficked, and control soil for the two mechanization levels (data showed in columns). Results of the Kolmogorov–Smirnov test for QBS-ar. Difference tested among two time periods (data showed in rows). W: winching; F: forwarding; C: control.

Area	Soil Condition	QBS-ar		*p*-Value
		1 Year	5 Years	
W	Untrafficked	201 [a]	197 [a]	>0.05
	Trafficked	106 [b]	110 [b]	>0.05
F	Untrafficked	136 [c]	181 [a,c]	<0.05
	Trafficked	81 [d]	173 [c]	<0.05
C	Control	254 [e]	254 [d]	>0.05
p-value		<0.05	<0.05	

Note: Different letters after means within each treatment indicate significant differences by Tukey test ($p < 0.05$).

However, trend analysis showed that QBS-ar recovery started already in the second year after harvest for all treatments (Figure 6). The QBS-ar index was affected by the uncovering effect due to coppicing and during the five years after harvesting it was lower in the harvested areas than in the control ones, even in the absence of visible disturbance (Table 13 and Figure 6). The regression models (Table 14 and Figure 6) are all statistically significant and have a very high explanatory value (r2adj ~ 0.9). During the 5 years post-harvesting neither the untrafficked nor the trafficked strata showed a complete recovery of the QBS-ar index (Figure 6). However, the trends indicate that full recovery could be achieved between the sixth and the seventh year for untrafficked areas and between the eighth (forwarding) and the ninth year (winching) for trafficked areas.

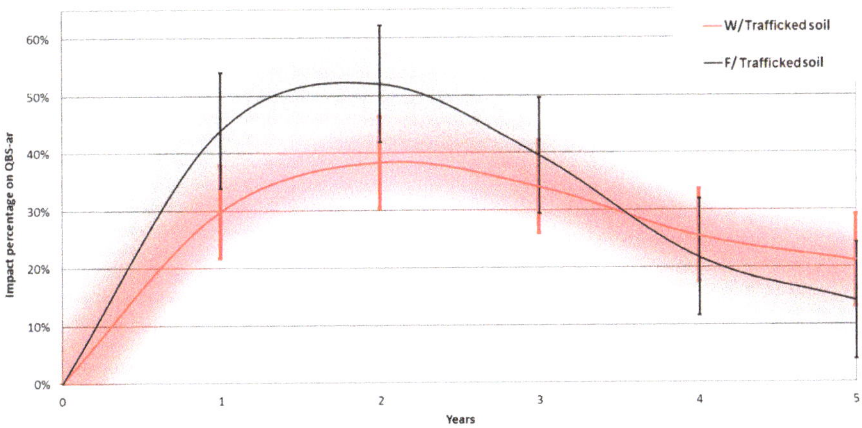

Figure 6. Graphical interpretation of the non-linear regression analysis for QBS-ar index percentage of impact (referred to the untrafficked soil) in relation to the time (in years) post-harvesting. The polynomial curves showed with a halo composed by ± regression standard estimate error. C/Untrafficked r^2_{adj} = 0.882, $F(3,176) = 225.15$, $p < 0.001$. W/Trafficked r^2_{adj} = 0.947, $F(3,176) = 533.85$, $p < 0.001$. W: winching; F: forwarding.

Table 14. Results of the non-linear regression analysis for QBS-ar index (dependent variable, expressed as the percentage of impact referred to the untrafficked soil) in relation to the time (in years) post-harvesting. W: winching; F: forwarding.

Area/Soil Condition	Regression	Beta	Beta Std. Err.	B	B Std. Err.	T	p-Level
W/Trafficked	Intercept	4.451	0.386	0.434	0.038	11.534	<0.001
	Years	−6.465	0.946	−0.149	0.022	−6.836	<0.001
	Years^2	2.816	0.596	0.014	0.003	4.725	<0.001
	Years^3	4.451	0.386	0.434	0.038	11.534	<0.001
F/Trafficked	Intercept	5.842	0.259	0.670	0.030	22.530	<0.001
	Years	−9.423	0.635	−0.256	0.017	−14.829	<0.001
	Years^2	4.304	0.401	0.026	0.002	10.747	<0.001
	Years^3	5.842	0.259	0.670	0.030	22.530	<0.001

4. Discussion

First of all, it is important to state upfront the possible limitations of this study so that any conclusions are interpreted with due caution, especially when it comes to generalization. Even if the analyses were done with statistical normalization and standardized methodologies, some residual limitations are still present. The main possible limitation is the reference to one specific stand and machine system. This is a common limitation of most studies of this type and while the stand was representative of a wider typology, it is clear that different results could have been obtained on different stands or on the same stand type growing on a different soil substrate. The same goes with machine selection, since both winching and forwarding can be applied with a wide range of different machines with different competence and with different levels of attention to minimizing undesirable environmental effects. Nevertheless, this study offers an interesting example of how different extraction techniques can impact the physical and biological characteristics of forest soils, and it is still suitable as an approximate general reference until more data will be available for other forest and machine types.

Comparison with previous studies supports cautions generalization. The proportion of the trafficked surface is only slightly higher than reported by Marchi et al. [4] and very close to that reported by Venanzi et al. [5], Picchio et al. [53], and Jourgholami et al. [54]. In fact, the highest value reported in the study for forwarding—31%—matches almost perfectly the 33% benchmark offered by Spinelli et al. [55] for the harvesting of coppice stands with traditional ground-based technology in Central Italy. In that regard, one must be aware that the proportion of total surface disturbed by forest operations is widely variable and is strongly dependent on the type of intervention, with the lightest harvesting operations—such as the selective removal of individual trees—impacting as little as 5% of the total surface [56] and the heaviest ones—the salvage of large windthrown areas—affecting over 50% of the total surface [57], and that independently of the harvesting technique. As a matter of fact, this study also highlights the role of silviculture, indicating that significant soil impacts occur even in the absence of machine traffic, by merely removing the forest canopy, which can be easily construed as the most traumatic event for the forest ecosystem.

Corroboration for the results of this study is also offered by previous studies of soil compaction consequent to harvesting. The 14% to 19% soil bulk density increase recorded here matches quite well the 12% increase recorded in Mediterranean pine forests by Kleibl et al. [58]. This value is twice as large as recorded by Magagnotti et al. [59] for dedicated forest skidders (increase in soil BD ca. 6%), but one must account for the different technology and for the very good floatation capacity of the dedicated forest equipment used in the quoted study. In any case, the post-impact soil BD values recorded in these studies and those obtained from the current experiment are almost the same, and range between 1.1 and 1.3 g cm^{-3}. This is very important because several studies indicate that root growth is impaired only when soil BD reaches higher values than here, and in the range of 1.7 to 1.8 g cm^{-3} [60,61]. Therefore, it is unlikely that the level of compaction recorded in this study may stunt stand growth, which is also confirmed by the fact that the soil started recovering relatively quickly, and in most cases the original soil properties were fully restored within five years. Incidentally, this result is even better than reported by Kleibl et al. [58] for Mediterranean pine stands also located in Central Italy, where full recovery was not achieved within the sixth year.

The study also showed that winching caused lighter soil disturbance compared with forwarding, and generally allowed for faster recovery, except in the case of QBS-ar. However, this is a contentious subject because the studies that have compared the soil impacts caused by tree-length and cut-to-length harvesting offer contrasting results: some support the findings of this research and indicate that TL causes lighter impacts [2], while others support the exact contrary [62,63]. The issue is likely one of machine selection and operational planning. In particular, one may argue that the forwarder used in this study may have been too big for the work at hand, and that the higher level of disturbance it caused could have been avoided if one had selected a lighter machine, like one of the many mini-forwarder available on the market and used in Italy, too [64]. In any case, the problem is that winching is a labor-intensive work technique with a much lower technical and financial performance compared with forwarding [65]. What is more, winching is a very tiresome job and none of the solution adopted to relieve operator's fatigue has been fully successful [66,67]. As a matter of fact, this work technique has almost disappeared from the coppice operations conducted in more industrialized countries like France [31,68]. Italian loggers are now looking with increasing interest to modern forwarder technology [69]. Finally, a decisive step towards mechanization is the best way to reduce fatalities in forest operations, which should be a strategic objective and a strong ethical obligation [70].

In decreasing order of disturbance, soil SR, QBS-ar index, OM content, and BD were influenced by coppicing.

Full recovery was observed between the fourth and fifth year for OM content and BD. Instead, soil PR and pH were not influenced at all by coppicing. This was shown by the comparison of untrafficked soil surfaces in the coppiced areas with the control areas (left unmanaged for the last two decades). These findings are marginally similar to what was found by Venanzi et al. [2–5] and Marchi et al. [4]). These soil parameters could be affected by weather events, but due to the quick canopy regeneration in coppice management these impacts were limited to a period of few years after harvesting (3–5 years).

As found also by many authors [2,4,5,28,29,32,48,71,72], in the short-term, the soil disturbance caused by forest operations showed significant differences between trafficked and untrafficked soil samples. In this case the highest impact being found for forwarding rather than winching, especially in the first year after harvest. However, a recovery trend clear emerged within the second or third year, depending on specific soil property and treatment.

Five years after logging, the two harvesting techniques showed different recovery trends. For the areas extracted by winching, BD and PR showed a complete recovery, while SR, OM, and QBS-ar index showed an important but incomplete recovery with percentages of residual impact varying between 24% and 44%. Conversely, in the areas extracted by forwarding, BD, PR, SR, OM, and QBS-ar index showed an important but incomplete recovery with residual impact varying between 4% and 144%. Consequently, as also suggested by Venanzi et al. [2], there is a need to discuss a possible limit on forest soil surface directly affected by the moving of machinery and logs, even if recovery could be considered relatively fast. The better results shown by winching was related to the limited movements of the tractor directly on the forest floor, since the tractor stationed on the road.

Differently from what was found by Venanzi et al. [2], soil organic matter content was not clearly influenced by coppicing and complete recovery was possible four years after logging. In general, the chemical and physical parameters observed were similar to those found in other studies [2,4,5] but in this forest typology the recovery of the soil disturbance specifically caused by coppicing (canopy removal) was recovered 3–4 years after harvesting.

As found also by Venanzi et al. [2], logging activities showed significant modification of OM content for both harvesting techniques. Recovery started 2–3 years after logging and it was expected to be complete after 7–8 years. The negative variation in OM content during the first years after coppicing may be linked to canopy removal, which means a lack of leaves contributing to litter formation and an increase in the respiratory activity of soil microorganisms. Quick recovery shown was linked to fast canopy regeneration and to the release of the twigs on the ground, both typical of coppice management.

Soil pH did not show any clear statistical relation with treatments or time. As found in other studies [2,4,5,12,71], soil pH generally shows low variation and is not clearly connected with soil disturbance.

As also showed in other research [2,4,5] QBS-ar index was negatively influenced by coppicing, and recovery was slower here than showed in other studies (expected 6–7 years after logging). That was consistent with the physical and chemical properties of the observed soil. This was shown by a comparison of the untrafficked soil surfaces in the coppiced areas with the control areas.

The QBS-ar index is often linked to physical and chemical soil parameters, therefore it is logical that significant differences in the QBS-ar index would be found between trafficked and untrafficked soil samples. The greatest impact was for winching areas (48%) and the lowest for forwarding areas (40%). Similar values were found in other similar studies [2,4,5,28,29,48]. Soil recovery was evident three years after coppicing and it showed different rates for the two harvesting techniques. Five years after harvesting, the forwarding areas showed an important recovery with only 4% residual impact, while the winching areas showed a modest recovery with still 40% residual impact. This result, in contrast to the other soil parameters, is closely linked to the different working capacities of the two machines used, which has resulted in the forwarder completing the harvesting in a much shorter time. Therefore, the disturbance inflicted on soil micro-arthropods lasted a much shorter time compared with the disturbance inflicted by the tractor with winch.

Although these findings underline the vulnerability of forest soil due to natural and/or human disturbance [73], in the case of coppice management the forest soil showed a significant recovery in a short period, highlighting that this silvicultural practice is sustainable and that coppice stands are quite resilient in the face of external disturbance.

Within such context, precision forestry could be an interesting approach to reduce impacts, through rationalized planning of logging operations [74]. The design and application of low impact logging methods [18–27] and sustainable forest operations (SFO) criteria, together with operator training is the mainstay of reduced impact logging, rather than the mechanization level.

Principal non-metric multidimensional scaling (NMDS) tests produced a two-dimensional ranking (Figure 7) that provided a significantly greater reduction in statistical stress than expected by chance ($\alpha = 0.05$). When considering BD, PR, SR, OM, pH, and QBS-ar index, the two axes explained 96.2% of the overall variance. These six variables showed the maximum correlation with the ordination axes. The variables BD, PR, SR, and pH illustrated the soil scenario on the weighted scale of axis 1 (Figure 7). The impact arrangement along axis 2 was dominated mainly by QBS-ar index and in part by OM content (Figure 7).

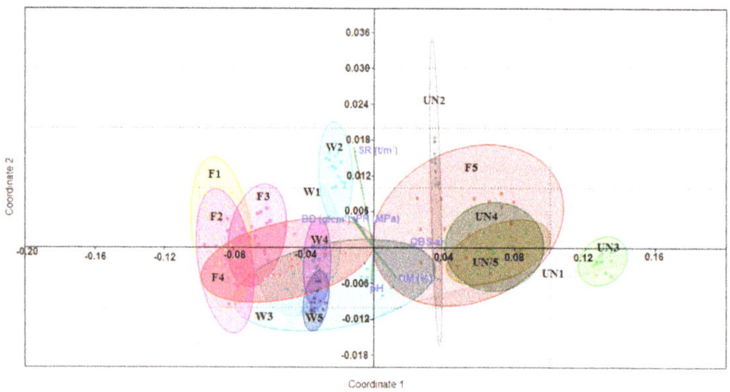

Figure 7. Non-metric multidimensional scaling (NMDS) analysis of the main indexes and indicators of the logging impact on soil (BD bulk density; PR: penetration resistance; SR: shear resistance; OM: organic matter content; QBS-ar: QBS-ar index). Difference tested between the three areas and the five time periods observed (UN: untrafficked soil; W: trafficked soil by winching; F: trafficked soil by forwarding; 1-5: years after coppicing).

The nMDS for the ten disturbed and five undisturbed scenarios showed a negative relationship between the time after harvesting and the impact levels. Therefore, five years after harvesting disturbed areas showed similar conditions to undisturbed areas. The two technology type scenarios (Figure 7) showed an initial differentiation with lower impact for the winching scenario but starting from the fourth year after coppicing they reached similar impact degrees.

Comparing the results of this work with those shown by Venanzi et al. [2], there are some differences to be attributed mainly to the different soil types and climatic conditions. These differences are reflected in terms of the need for longer recovery times in Mediterranean areas with greater aridity. The phytoclimatic zone reported in Venanzi et al. [2] is warm *Castanetum*, while in this study the zone is an intermediate *Lauretum* (according to Pavari phytoclimatic classification [75]). Thus, there seems to be a positive relationship between recovery time and actual or perceived aridity of the forest ecosystem. In Venanzi et al. [2] the recovery of soil impacts caused by coppicing as a management practice (i.e., canopy gap) was almost complete three years after harvesting, while from in this study the same level of recovery took almost five (6–7 years for complete recovery of the QBS-ar index). The recovery from logging disturbance showed a clear positive trend, but in Venanzi et al. [2] 4–5 years post-harvesting it was possible to confirm, statistically, a complete recovery while from this study the soil recovery was expected 8–9 years after coppicing.

5. Conclusions

Part of this applied research has the potential to translate into "forest harvesting best practices" in order to increase the knowledge for a sustainable management of coppice forests in the Mediterranean area, supporting the decision making of forest managers.

As found in other studies, the physical, chemical, and biological soil features were partially disturbed by the act of coppicing for itself, due to the sudden and drastic interruption of canopy cover. Machine traffic compounded such disturbance, through its mechanical action on the soil structure, resulting in a substantial alteration of the physical-mechanical soil components.

Between the two extraction techniques on test, winching caused the least disturbance while forwarding had stronger impacts. That was likely related to the small size of the trees being extracted (which minimized the impact of winching) and to the choice of a heavy forwarder model instead of a lighter one.

However, soil recovery was almost complete five years after harvesting without substantial differences between logging techniques.

Soil recovery after logging showed a statistical positive trend with similar results in the fifth year for both harvesting techniques, although five years after harvesting it was not possible to confirm that recovery was yet complete. However, full recovery is likely to be achieved approximately eight or nine years after harvest, regardless of the logging technique tested.

Similar studies are important for their potential contribution to updating the guidelines, criteria, and indicators for sustainable forest management, as proposed by Forest Europe and Reduced Impact Logging (RIL) in a perspective of SFOs application.

Author Contributions: Conceptualization, R.V., R.P., S.G., and R.S.; methodology, R.V. and R.P.; software, R.V. and R.P.; validation, R.V., R.P., and S.G.; formal analysis, R.V. and R.P.; investigation, R.V. and R.P.; data curation, R.V. and R.P.; writing—original draft preparation, R.V., R.P., S.G., and R.S.; writing—review and editing, R.V., R.P., S.G., and R.S.; supervision, R.V., R.P., S.G., and R.S. All authors read and agreed to the published version of the manuscript.

Funding: This research received no external funding.

Acknowledgments: This work was supported by the Italian Ministry for education, University and Research (MIUR) for financial support (Law 232/2016, Italian University Departments of excellence)—UNITUS-DAFNE WP3.

Conflicts of Interest: The authors declare no conflicts of interest. The funders had no role in the design of the study; in the collection, analyses, or interpretation of data; in the writing of the manuscript; or in the decision to publish the results.

References

1. Unrau, A.; Becker, G.; Spinelli, R.; Lazdina, D.; Magagnotti, N.; Nicolescu, V.N.; Buckley, P.; Bartlett, D.; Kofman, P.D. (Eds.) *Coppice Forests in Europe*; Freiburg i. Br.; Albert Ludwig University of Freiburg: Freiburg, Germany, 2018; 392p.
2. Venanzi, R.; Picchio, R.; Grigolato, S.; Latterini, F. Soil and forest regeneration after different extraction methods in coppice forests. *For. Ecol. Manag.* **2019**, *454*, 117666. [CrossRef]
3. Szabò, P. Driving forces of stability and change in woodland structure: A case-study from the Czech lowlands. *For. Ecol. Manag.* **2010**, *259*, 650–656. [CrossRef]
4. Marchi, E.; Picchio, R.; Mederski, P.S.; Vusić, D.; Perugini, M.; Venanzi, R. Impact of silvicultural treatment and forest operation on soil and regeneration in Mediterranean Turkey oak (*Quercus cerris* L.) coppice with standards. *Ecol. Eng.* **2016**, *95*, 475–484. [CrossRef]
5. Venanzi, R.; Picchio, R.; Piovesan, G. Silvicultural and logging impact on soil characteristics in Chestnut (*Castanea sativa* Mill.) Mediterranean coppice. *Ecol. Eng.* **2016**, *96*, 82–89. [CrossRef]
6. Matula, R.; Šrámek, M.; Kvasnica, J.; Uherková, B.; Slepička, J.; Matoušková, M.; Kutchartt, E.; Svátek, M. Pre-disturbance tree size, sprouting vigour and competition drive the survival and growth of resprouting trees. *For. Ecol. Manag.* **2019**, *446*, 71–79. [CrossRef]

7. Müllerová, J.; Hédl, R.; Szabó, P. Coppice abandonment and its implications for species diversity in forest vegetation. *For. Ecol. Manag.* **2015**, *343*, 88–100. [CrossRef]
8. Imamura, K.; Managi, S.; Saito, S.; Nakashizuka, T. Abandoned forest ecosystem: Implications for Japan's Oak Wilt disease. *J. For. Econ.* **2017**, *29*, 56–61. [CrossRef]
9. Picchio, R.; Spina, R.; Maesano, M.; Carbone, F.; Lo Monaco, A.; Marchi, E. Stumpage value in the short wood system for the conversion into high forest of a oak coppice. *For. Stud. China* **2011**, *13*, 252–262. [CrossRef]
10. Gimmi, U.; Bürgi, M.; Stuber, M. Reconstructing anthropogenic disturbance regimes in forest ecosystems: A case study from the Swiss Rhone valley. *Ecosystems* **2008**, *11*, 113–124. [CrossRef]
11. Olajuyigbe, S.; Tobin, B.; Saunders, M.; Nieuwenhuis, M. Forest thinning and soil respiration in a Sitka spruce forest in Ireland. *Agr. For. Meteorol.* **2012**, *157*, 86–95. [CrossRef]
12. Picchio, R.; Mercurio, R.; Venanzi, R.; Gratani, L.; Giallonardo, T.; Lo Monaco, A.; Frattaroli, A.R. Strip clear-cutting application and logging typologies for renaturalization of pine afforestation—A case study. *Forests* **2018**, *9*, 366. [CrossRef]
13. Klvac, R.; Vrána, P.; Jirousek, R. Possibilities of using the portable falling weight deflectometer to measure the bearing capacity and compaction of forest soils. *J. For. Sci.* **2010**, *56*, 130–136. [CrossRef]
14. Korb, J.E.; Fulé, P.Z.; Gideon, B. Different restoration thinning treatments affect level of soil disturbance in ponderosa pine forests of Northern Arizona, USA. *Ecol. Restor.* **2007**, *25*, 43–49. [CrossRef]
15. Williamson, J.R.; Neilsen, W.A. The influence of soil and forest site on rate and extent of soil compaction and profile disturbance of skid-trails during ground-based harvesting. *Can. J. For. Res.* **2000**, *30*, 1196–1205. [CrossRef]
16. Grigal, D.F. Effects of extensive forest management on soil productivity. *For. Ecol. Manag.* **2000**, *138*, 167–185. [CrossRef]
17. Vusić, D.; Šušnjar, M.; Marchi, E.; Spina, R.; Zečić, T.; Picchio, R. Skidding operations in thinning and shelterwood cut of mixed stands—Work productivity, energy inputs and emissions. *Ecol. Eng.* **2013**, *61*, 216–223. [CrossRef]
18. Picchio, R.; Pignatti, G.; Marchi, E.; Latterini, F.; Benanchi, M.; Foderi, C.; Venanzi, R.; Verani, S. The application of two approaches using GIS technology implementation in forest road network planning in an Italian mountain setting. *Forests* **2018**, *9*, 277. [CrossRef]
19. Spinelli, R.; Lombardini, C.; Marchi, E.; Aminti, G. A low-investment technology for the simplified processing of energy wood from coppice forests. *Eur. J. For. Res.* **2019**, *138*, 31–41. [CrossRef]
20. Tolosana, E.; Spinelli, R.; Aminti, G.; Laina, R.; López-Vicens, I. Productivity, efficiency and environmental effects of whole-tree harvesting in Spanish coppice stands using a drive-to-tree disc saw Feller-Buncher. *Croat. J. For. Eng.* **2018**, *39*, 163–172.
21. Picchio, R.; Neri, F.; Maesano, M.; Savelli, S.; Sirna, A.; Blasi, S.; Baldini, S.; Marchi, E. Growth effects of thinning damage in a Corsican pine (*Pinus laricio* Poiret) stand in central Italy. *Forest Ecol. Manag.* **2011**, *262*, 237–243. [CrossRef]
22. Laschi, A.; Marchi, E.; González-García, S. Forest operations in coppice: Environmental assessment of two different logging methods. *Sci. Total Environ.* **2016**, *562*, 493–503. [CrossRef] [PubMed]
23. Bertolotto, P.; Calienno, L.; Conforti, M.; D'Andrea, E.; Lo Monaco, A.; Magnani, E.; Marinšek, A.; Venanzi, R. Assessing indicators of forest ecosystem health. *Ann. Silvi. Res.* **2016**, *40*, 64–69.
24. Picchio, R.; Spina, R.; Calienno, L.; Venanzi, R.; Lo Monaco, A. Forest operations for implementing silvicultural treatments for multiple purposes. *Ital. J. Agron.* **2016**, *11*, 156–161.
25. Sohrabi, H.; Jourgholami, M.; Tavankar, F.; Venanzi, R.; Picchio, R. Post-harvest evaluation of soil physical properties and natural regeneration growth in steep-slope terrains. *Forests* **2019**, *10*, 1034. [CrossRef]
26. Mederski, P. A comparison of harvesting productivity and costs in thinning operations with and without midfield. *For. Ecol. Manag.* **2006**, *224*, 286–296. [CrossRef]
27. Picchio, R.; Latterini, F.; Mederski, P.S.; Venanzi, R.; Karaszewski, Z.; Bembenek, M.; Croce, M. Comparing accuracy of three methods based on the GIS environment for determining winching areas. *Electronics* **2019**, *8*, 53. [CrossRef]
28. Cambi, M.; Certini, G.; Fabiano, F.; Foderi, C.; Laschi, A.; Picchio, R. Impact of wheeled and tracked tractors on soil physical properties in a mixed conifer stand. *IForest* **2016**, *9*, 89–94. [CrossRef]
29. Picchio, R.; Mederski, P.S.; Tavankar, F. How and how much, do harvesting activities affect forest soil, regeneration and stands? *Curr. For. Rep.* **2020**. [CrossRef]

30. Marchi, E.; Picchio, R.; Spinelli, R.; Verani, S.; Venanzi, R.; Certini, G. Environmental impact assessment of different logging methods in pine forests thinning. *Ecol. Eng.* **2014**, *70*, 429–436. [CrossRef]
31. Picchio, R.; Magagnotti, N.; Sirna, A.; Spinelli, R. Improved winching technique to reduce logging damage. *Ecol. Eng.* **2012**, *47*, 83–86. [CrossRef]
32. Picchio, R.; Venanzi, R.; Tavankar, F.; Luchenti, I.; Bodaghi, A.I.; Latterini, F.; Nikooy, M.; Di Marzio, N.; Naghdi, R. Changes in soil parameters of forests after windstorms and timber extraction. *Eur. J. For. Res.* **2019**, *138*, 875–888. [CrossRef]
33. Proto, A.R.; Macrì, G.; Visser, R.; Harrill, H.; Russo, D.; Zimbalatti, G. A case study on the productivity of forwarder extraction in small-scale Southern Italian forests. *Small-Scale For.* **2018**, *17*, 71–87. [CrossRef]
34. Proto, A.R.; Zimbalatti, G. Firewood cable extraction in the southern Mediterranean area of Italy. *For. Sci. Techn.* **2016**, *12*, 16–23. [CrossRef]
35. Sánchez-Moreno, S.; Minoshima, H.; Ferris, H.; Jackson, L.E. Linking soil properties and nematode community composition: Effects of soil management on soil food webs. *Nematology* **2006**, *8*, 703–715.
36. Spinelli, R.; Pari, L.; Aminti, G.; Magagnotti, N.; Giovannelli, A. Mortality, re-sprouting vigor and physiology of coppice stumps after mechanized cutting. *Ann. For. Sci.* **2017**, *74*, 5. [CrossRef]
37. Buckley, G. *Ecology and Management of Coppiced Woodlands*; Chapman & Hall: London, UK, 1992; 336p.
38. Joys, A.C.; Fuller, R.J.; Dolman, P.M. Influences of deer browsing, coppice history, and standard trees on the growth and development of vegetation structure in coppiced woods in lowland England. *For. Ecol. Manag.* **2004**, *202*, 23–37. [CrossRef]
39. Benes, J.; Cizek, O.; Dovala, J.; Konvicka, M. Intensive game keeping, coppicing and butterflies: The story of Milovicky Wood, Czech Republic. *For. Ecol. Manag.* **2006**, *237*, 353–365. [CrossRef]
40. Bugalho, M.N.; Caldeira, M.C.; Pereira, J.S.; Aronson, J.; Pausas, J.G. Mediterranean cork oak savannas require human use to sustain biodiversity and ecosystem services. *Front. Ecol. Environ.* **2011**, *9*, 278–286. [CrossRef]
41. Hédl, R.; Kopecky´, M.; Komárek, J. Half a century of succession in a temperate oakwood: From species-rich community to mesic forest. *Divers. Distrib.* **2010**, *16*, 267–276. [CrossRef]
42. Spitzer, L.; Konvicka, M.; Tropek, R.; Tuf, I.H.; Tufová, J. Does closure of traditionally managed open woodlands threaten epigeic invertebrates? Effects of coppicing and high deer densities. *Biol. Conserv.* **2008**, *141*, 827–837. [CrossRef]
43. Vodka, Š.; Cizek, L. The effects of edge-interior and understorey-canopy gradients on the distribution of saproxylic beetles in a temperate lowland forest. *For. Ecol. Manag.* **2013**, *304*, 33–41. [CrossRef]
44. Abbas, D.; Di Fulvio, F.; Spinelli, R. European and United States perspectives on forest operations in environmentally sensitive areas. *Scand. J. For. Res.* **2018**, *33*, 188–201. [CrossRef]
45. Marchi, E.; Chung, W.; Visser, R.; Abbas, D.; Nordfjell, T.; Mederski, P.S.; McEwan, A.; Brink, M.; Laschi, A. Sustainable Forest Operations (SFO): A new paradigm in a changing world and climate. *Sci. Total Environ.* **2018**, *634*, 1385–1397. [CrossRef]
46. Saxton, K.E.; Rawls, W.J.; Romberger, J.S.; Papendick, R.I. Estimating generalized soil-water characteristics from texture. *Soil Sci. Soc. Am. J.* **1986**, *50*, 1031–1036. [CrossRef]
47. Corona, P.; Ascoli, D.; Barbati, A.; Bovio, G.; Colangelo, G.; Elia, M.; Garfì, V.; Iovino, F.; Lafortezza, R.; Leone, V.; et al. Integrated forest management to prevent wildfires under Mediterranean environments. *Ann. Silvi. Res.* **2015**, *39*, 1–22.
48. Picchio, R.; Neri, F.; Petrini, E.; Verani, S.; Marchi, E.; Certini, G. Machinery-induced soil compaction in thinning two pine stands in central Italy. *For. Ecol. Manag.* **2012**, *285*, 38–43. [CrossRef]
49. Gee, G.W.; Bauder, J.W. Particle-size analysis. In *Methods of Soil Analysis. Part I. Agronomy Monographs, 9*, 2nd ed.; Klute, A., Ed.; ASA and SSSA: Madison, WI, USA, 1986; pp. 383–411.
50. Spinelli, R.; Magagnotti, N.; Aminti, G.; De Francesco, F.; Lombardini, C. The effect of harvesting method on biomass retention and operational efficiency in low-value mountain forests. *Eur. J. For. Res.* **2016**, *135*, 755–764. [CrossRef]
51. Rigal, S.; Devictor, V.; Dakos, V. A method for classifying and comparing non-linear trajectories of ecological variables. *Ecol. Indic.* **2020**, *112*, 106113. [CrossRef]
52. Picchio, R.; Tavankar, F.; Nikooy, M.; Pignatti, G.; Venanzi, R.; Lo Monaco, A. Morphology, growth and architecture response of beech (*Fagus orientalis* Lipsky) and maple tree (*Acer velutinum* Boiss.) seedlings to soil compaction stress caused by mechanized logging operations. *Forests* **2019**, *10*, 771. [CrossRef]

53. Picchio, R.; Tavankar, F.; Bonyad, A.; Mederski, P.S.; Venanzi, R.; Nikooy, M. Detailed analysis of residual stand damage due to winching on steep terrains. *Small-scale For.* **2019**, *18*, 255–277. [CrossRef]
54. Jourgholami, M.; Majnounian, B.; Abari, M.E. Effects of tree-length timber skidding on soil compaction in the skid trail in Hyrcanian forests. *For. Syst.* **2014**, *23*, 288–293. [CrossRef]
55. Spinelli, R.; Magagnotti, N.; Nati, C. Benchmarking the impact of traditional small-scale logging systems used in Mediterranean forestry. *For. Ecol. Manag.* **2010**, *260*, 1997–2001. [CrossRef]
56. Spinelli, R.; Lombardini, C.; Magagnotti, N. Salvaging windthrown trees with animal and machine systems in protected areas. *Ecol. Eng.* **2013**, *53*, 61–67. [CrossRef]
57. Magagnotti, N.; Picchi, G.; Spinelli, R. A versatile machine system for salvaging small-scale forest windthrow. *Biosyst. Eng.* **2013**, *115*, 381–388. [CrossRef]
58. Kleibl, M.; Klvač, R.; Lombardini, C.; Porhaly, J.; Spinelli, R. Soil compaction and recovery after mechanized final felling of Italian coastal pine plantations. *Croat. J. For. Eng.* **2014**, *35*, 63–71.
59. Magagnotti, N.; Spinelli, R.; Güldner, O.; Erler, J. Site impact after motor-manual and mechanised thinning in Mediterranean pine plantations. *Biosyst. Eng.* **2012**, *113*, 140–147. [CrossRef]
60. Ampoorter, E.; Goris, R.; Cornelis, W.; Verheyen, K. Impact of mechanized logging on compaction status of sandy forest soils. *For. Ecol. Manag.* **2007**, *241*, 162–174. [CrossRef]
61. Heilman, P. Root penetration of Douglas-fir seedlings into compacted soil. *For. Sci.* **1981**, *27*, 660–666.
62. Cudzik, A.; Brennensthul, M.; Białczyk, W.; Czarnecki, J. Damage to Soil and Residual Trees Caused by Different Logging Systems Applied to Late Thinning. *Croat. J. For. Eng.* **2017**, *38*, 83–95.
63. Dudek, T.; Sosnowski, J. Evaluation of the environmental impact of selected timber extraction technologies used in mountain forests. *Sylwan* **2011**, *155*, 413–420.
64. Spinelli, R.; Magagnotti, N. Performance and cost of a new mini-forwarder for use in thinning operations. *J. For. Res.* **2010**, *15*, 358–364. [CrossRef]
65. Spinelli, R.; Magagnotti, N. The effects of introducing modern technology on the financial, labour and energy performance of forest operations in the Italian Alps. *Forest Policy Econ.* **2011**, *13*, 520–524. [CrossRef]
66. Spinelli, R.; Aalmo, G.O.; Magagnotti, N. The effect of a slack-pulling device in reducing operator physiological workload during log winching operations. *Ergonomics* **2015**, *58*, 781–790. [CrossRef] [PubMed]
67. Magagnotti, N.; Aalmo, G.O.; Brown, M.; Spinelli, R. A new device for reducing winching cost and worker effort in steep terrain operations. *Scand. J. For. Res.* **2016**, *31*, 602–610. [CrossRef]
68. Spinelli, R.; Cacot, E.; Mihelic, M.; Nestorovski, L.; Mederski, P.; Tolosana, E. Techniques and productivity of coppice harvesting operations in Europe: A meta-analysis of available data. *Ann. For. Sci.* **2016**, *73*, 1125–1139. [CrossRef]
69. Ferrari, E.; Spinelli, R.; Cavallo, E.; Magagnotti, N. Attitudes towards mechanized Cut-to-Length technology among logging contractors in Northern Italy. *Scand. J. For. Res.* **2012**, *27*, 800–806. [CrossRef]
70. Bell, J. Changes in logging injury rates associated with use of feller-bunchers in West Virginia. *J. Saf. Res.* **2002**, *33*, 436–471. [CrossRef]
71. Cambi, M.; Hoshika, Y.; Mariotti, B.; Paoletti, E.; Picchio, R.; Venanzi, R.; Marchi, E. Compaction by a forest machine affects soil quality and Quercus robur L. seedling performance in an experimental field. *For. Ecol. Manag.* **2017**, *384*, 406–414. [CrossRef]
72. Tavankar, F.; Bonyad, A.E.; Nikooy, M.; Picchio, R.; Venanzi, R.; Calienno, L. Damages to soil and tree species by cable-skidding in caspian forests of Iran. *For. Syst.* **2017**, *26*, e009. [CrossRef]
73. Vossbrink, J.; Horn, R. Modern forestry vehicles and their impact on soil physical properties. *Eur. J. For. Res.* **2004**, *123*, 259–267. [CrossRef]
74. Picchio, R.; Proto, A.R.; Civitarese, V.; Di Marzio, N.; Latterini, F. Recent contributions of some fields of the electronics in development of forest operations technologies. *Electronics* **2019**, *8*, 1465. [CrossRef]
75. Pavari, A. *Studio Preliminare Sulla Coltura di Specie Forestali Esotiche in Italia*; Parte Generale, Volume I; Annali del Regio Istituto Superiore Forestale Nazionale; Tipografia Ricci: Florence, Italy, 1916; 221p.

sustainability

MDPI

Article

Experimental Study Based on Game Theory on the Private, Voluntary Supply Mechanisms of Goods for Forestry Infrastructure from the Perspective of Quasi-Public Goods

Liying Zhang, Chengliang Wu and Yang Zhang *

School of Economics and Management, Beijing Forestry University, Beijing 100083, China; zhangliying@bjfu.edu.cn (L.Z.); wubjfu@163.com (C.W.)

* Correspondence: zhangyang052012@aliyun.com

Received: 13 February 2020; Accepted: 31 March 2020; Published: 2 April 2020

Abstract: The existing research on forestry infrastructure has focused on suggestions from other areas of forestry research: that forestry infrastructure should be completed and improved. However, research on forestry infrastructure is relatively rare. In the real world, there are various problems with creating forestry infrastructure, such as complex approval procedures for facility construction, irrational facility layout, insufficient funding for facility construction, and conflicts between the nature of land used for facility construction and the nature of forest land. This paper uses game theory to analyze the behavior of forest infrastructure goods suppliers. Relevant parameters related to forest area infrastructure were designed, including communication, environmental certainty, information feedback, and reward and punishment mechanisms, and experimental economics methods were used to simulate accurate behavior regarding the supply of goods. Then, the key factors that affect the provision of quasi-public goods for forestry infrastructure were studied. At the end of the paper, some targeted suggestions that distinguish rural infrastructure from general infrastructure are given.

Keywords: forestry infrastructure; supply mechanism; game theory; experimental economics

1. Introduction

Forestry infrastructure refers to engineering facilities for forestry public services that provide economic, social, and ecological benefits, such as paths for cycling, tool storerooms, and transfer facilities. Forestry infrastructure helps with the sustainable development of the forestry industry. Research on the private and voluntary supply of goods for forestry infrastructure is important to ensure long-term supplies. Sound forestry infrastructure can promote the sustainable development of forestry production.

Up to now, the research on forestry infrastructure supplies has mainly focused on analyses of, e.g., insufficient supplies, low quality, and a shortage of funds [1–8]. Some scholars have also studied forest roads. Road location and design is a complex engineering problem involving economic and environmental requirements [9]. For instance, Enache et al. [10] carried out research on how the extraction distance and correction factors can be computed and used for assessing forest road options in a more efficient and effective manner, utilizing process automation in geographic information systems. Hayati et al. [11] have considered skidding costs, road construction, maintenance costs, and harvesting volume to confirm the optimum forest road network density and evaluate the quantity and quality of existing forest road networks. Laschi et al. [12] have developed a decision support system to assist managers in the process of forest road network planning, exploiting multicriteria analyses, an analytic hierarchy process, and geographic information systems. Parsakhoo and Mostafa [13] have summarized the results of a road network analysis (RNA) and have evaluated the shortest path (to save travel time) in the city of Gorgan's public road network in Iran. Picchio et al. [14] have tested accuracy in estimating areas accessible for winching along skid trails: they used three geographic information systems, namely the correct distance method (CDM), the real distance buffer method 12 (RDBM12), and the real distance buffer method 10 (RDBM10). However, research on forestry infrastructure supplies has not been sufficient.

Forestry infrastructure is getting more and more attention. Many studies have mentioned the importance of forestry infrastructure. In terms of forestry production activities, many scholars believe that a sound forestry infrastructure can promote the development of a forestry economy and enhance the forest tenure reform system (the forest tenure reform system distinguishes between forest ownership, forest use rights, and forest land use rights; its purpose is to analyze property rights clearly) and forest operations [5,15–18]. Gumus and Turk [19] have analyzed timber extraction by farm tractors, developing a new skid trail pattern design using linear programming (LP) and geographic information systems.

The development of an under-forest economy (an under-forest economy is based on the ecological environment of forest land, using forest land resources, and carrying out multiple forest and agricultural operations, animal husbandry, and other projects within a forest) and disaster prevention in forest areas are inexorably linked to a sound forestry infrastructure [20–25]. Sound forestry infrastructure is not only conducive to accelerating the circulation of forest products, but also can improve the efficiency of the use of forestry industry information resources [26,27]. In terms of environmental protection, one study has shown that the more extreme climatic changes are, the more infrastructure support is required [28]. As for environmental planning, some scholars believe that a sound forestry infrastructure is integral to ecological restoration after a disaster [29,30]. Therefore, many scholars have suggested that infrastructure construction and investment should be strengthened to improve forestry foundation facilities [5,31,32].

Forestry infrastructure is a part of "general goods", which can be divided into private goods, club goods, crowded goods, or pure public goods according to a four-point method. Of these, club goods and crowded goods can also be called quasi-public goods. In reality, the supply of quasi-public goods comes from diverse sources, and problems with adequate supply are common. Therefore, an ample supply of quasi-public goods can greatly improve economic outcomes. There have been many studies on the supply of quasi-public goods. In terms of research on the effective supply of quasi-public goods, most scholars have written that the private supply of quasi-public goods is more efficient than a government supply alone [33–36]. In terms of methods for supplying quasi-public goods, scholars have written that except for the government's supply, individuals, enterprises, and organizations should also be involved [37,38]. This paper studies supply mechanisms for goods for forestry infrastructure from the perspective of quasi-public goods.

In terms of research on the supply of quasi-public goods, most scholars have used game theory [39–43]. They have shown that it is feasible to analyze the supply mechanism of quasi-public goods utilizing game theory. This paper draws on existing research and uses game theory to analyze the supply mechanisms of quasi-public goods for forestry infrastructure. Most papers on the supply of quasi-public goods, on the contrary, have been limited to a theoretical level and have lacked empirical research.

One of the main focuses of experimental economics methods is the study of public goods [44]. With regard to experiments on the supply of quasi-public goods, some scholars have found that environmental uncertainty will affect the level of cooperation [45]. Jane and Rick [46] have studied the impact of information feedback on cooperation, indicating that information feedback can promote the supply of public goods. However, Weimann [47] and Croson [48] did not find a role for information feedback in their experiment. Some scholars have studied the impact of information feedback methods on the supply of public goods [49–52]. In addition, other scholars have applied punishment mechanisms to the study of the impact of information feedback on the supply of public goods [53,54]. Traditional methods of data collection have generally included literature surveys and questionnaires, which can lead to static data and may only represent a single moment in time. Experimental economic methods are solutions to these problems. Therefore, this paper uses experimental economics methods to verify the results of a game theory analysis. We selected four control variables for the experiments. Using the above-mentioned research, we combined the characteristics of the supply of goods for forestry infrastructure: communication, information feedback, environment determination, and reward and punishment mechanisms.

To sum up, this paper uses game theory to study the supply mechanisms of quasi-public goods for forestry infrastructure and then uses experimental economics to verify the theoretical results. Using game theory analysis, we include noncooperative games and cooperative games, and determine a game payment matrix according to the current situation of forestry production. In the empirical research section, we use experimental economics, selecting four control variables: whether to communicate, whether the environment is determined, whether the information is feedback information, and whether there is a reward and punishment mechanism. The research object is quasi-public goods for forestry infrastructure, where we simulate the supply process of quasi-public goods for forestry infrastructure. Finally, we verify the main factors that affect cooperative games.

2. Materials and Methods

2.1. Game Theory Analysis

Game theory is a mathematical theory and method for studying the phenomena of struggle or competition. The basic concepts include people, actions, information, strategies, benefits, equilibrium, and results. These can be divided into different categories according to different classification methods. Among them, there are noncooperative games and cooperative games (depending on the relationship between individual interests and collective interests) [55,56]. Using this game classification, we first studied the noncooperative game equilibrium results, and then studied the cooperative game equilibrium results. In conclusion, the cooperative game equilibrium results were better than the results from the noncooperative games. A detailed analysis follows.

2.1.1. Noncooperative Games

Noncooperative game theory can be defined as game theory with complete rules. Thus, it can be described with three features. First, the rules are complete. Second, the ultimate decision units are the individual players. Third, commitments are not available, unless allowed for by the rules of the game [56].

• Forestry Infrastructure Is Not Provided When Income Is the Same

Suppose there are two rational people in a forest area, subject A and subject B, with the same income. Due to the difference in the quantity of forest land resources owned by forest farmers and the different requirements for forestry infrastructure, different forest farmers also receive different benefits from quasi-public goods for forestry infrastructure. When these two people cooperate to provide goods, the cost of each person's burden varies according to the degree of demand. Those with a high demand have a higher cost, and those with low demand have lower costs. Suppose subject A has a large quantity of forest land resources and a high demand for quasi-public goods for forestry infrastructure. The benefit of providing forestry infrastructure is $R1$, and the cost of the burden is $C1$. Subject B has a small quantity of forest land resources and has a low demand for public-goods-based forestry infrastructure. The benefit of providing forestry infrastructure is $R2$, and the cost of the burden is $C2$. The cost of forestry infrastructure construction is $2C$, of which $C1 + C2 = 2C$. If both of them refuse to provide goods, the return is 0 [57]. The payment matrix is shown in Table 1.

Table 1. Payment matrix when income is the same.

Subject A \ Subject B	Offers	Does Not Offer
Offers	R1−C1, R2−C2	R1−2C, R2
Does not offer	R1, R2−2C	0, 0

The prisoner's dilemma game is a standard example of a game that is analyzed in game theory that shows why two completely rational individuals might not cooperate, even if it appears that it is in their best interests to do so. Two members of a criminal gang are arrested and imprisoned. Each prisoner is in solitary confinement with no means of communicating with the other. The prosecutors lack sufficient evidence to convict the pair on the principal charge, but they have enough to convict both on a lesser charge. Simultaneously, the prosecutors offer each prisoner a bargain. Each prisoner is given the opportunity either to betray the other by testifying that the other committed the crime, or to cooperate with the other by remaining silent. Here, when $2C > R1 > 0$ and $2C > R2 > 0$, the game is a prisoner's dilemma game, and the Nash equilibrium is (not provide, not provide). The reason is that the construction costs of some quasi-public goods for forestry infrastructure are relatively high, but an individual's income from forestry infrastructure is very low. If individuals provide money alone, the fund can run a deficit. At that time, rational participants will choose not to provide. (Not provide, not provide) is the Nash equilibrium, because not providing is the optimal strategy for everyone. No one has the motivation to change his/her strategy. A private, voluntary supply of forestry infrastructure will thus not exist [57].

The chicken game, also known as the hawk–dove game or the snowdrift game, is a game theory model of conflict for two players. The principle of the game is that while the outcome is ideal for one player to yield (to avoid the worst outcome if neither yields), individuals try to avoid this out of pride, not wanting to look like a "chicken". Thus, each player taunts the other to increase the shame in yielding. However, when one player yields, the conflict is avoided, and the game is for the most part over. Here, when $R1 > 2C > 0$ and $R2 > 2C > 0$, the game is a chicken game, and the Nash equilibrium is (provide, not provide) and (not provide, provide). In this scenario, according to a rational person's assumptions, each forest farmer wants to benefit from the actions of the others (hitchhiking). As both enjoy the same income, neither has the incentive to choose or provide first. At this time, the game is also deadlocked, and the private, voluntary supply of forestry infrastructure will not exist [57].

In summary, when two rational people in a forest area share the same income, if they do not choose to cooperate, they will not voluntarily provide goods for forest infrastructure.

• An Insufficient Supply of Forestry Infrastructure with Different Incomes

Suppose there is forest farmer A and forest farmer B in the same forest area. They have different incomes. Forest farmer A has a high income and forest farmer B has a low income. Each of them has two strategies for providing forestry infrastructure: providing or not providing.

Because high-income people usually provide forestry infrastructure in pursuit of some social benefit, such as reputation or fame, and low-income people only hope to meet their production needs, high-income earners benefit more from forestry infrastructure construction. In addition, high-income earners have enough money to provide goods for forestry infrastructure, but low-income earners do not. From the perspective of opportunity costs, the opportunity cost of providing forestry infrastructure is far lower for high-income earners.

Suppose the cost of building forestry infrastructure is $2C$. In this scenario, both forest farmers choose to provide, the income of high-income earners is R1, and the cost is C1. The income of low-income earners is $R2$, and the cost is $C2$. At this point, $R1$ and $R2$ represent economic benefits, or basic production needs. When the high-income individual chooses to provide and the low-income individual chooses not to provide, the high-income individual earns $R11$, the cost is $2C$, and the low-income individual earns *R*2. At this time, R11 is the sum of the economic benefits and social benefits of the high-income individual who provides forestry infrastructure, so *R*11 > *R*1. When the high-income individual chooses not to provide, and the low-income individual chooses to provide, the income of the high-income individual is *R*1, and the income of the low-income individual is *R*2. Because the opportunity cost of providing forestry infrastructure is higher for low-income individuals, the total cost is (2*C* + *C*0). When neither the high-income individual nor the low-income individual provide, their benefits are zero, and *R*1 > *R*2, *R*2 < (2*C* + *C*0), *R*11 > *R*1 > 2*C*. This situation is a typical rational pigs' scenario [57]. With rational pigs, there are two pigs in a pigsty, a big pig and a small pig. The pigsty has a pedal on one side. When a pig steps on the pedal, a small amount of food will fall into the feeding tray on the other side of the pigsty, away from the pedal. When one pig steps on the pedal, the other pig has a chance to eat the food first. When a small pig steps on the pedal, the big pig will eat all the food before the small pig steps on the pedal and then runs to the feeding tray. When a big pig steps on the pedal, the small pig cannot eat all the food before the big pig steps on the pedal and then runs to the feeding tray. The specific payment matrix is shown in Table 2.

Table 2. Rational pigs.

Subject B / Subject A	Offers	Does Not Offer
Offers	R1−C1, R2−C2	R11−2C, R2
Does not offer	R1, R2−(2C + C0)	0, 0

In Table 2, subject A represents a high-income earner and subject B represents a low-income earner. The Nash equilibrium of this game is (provide, not provide), i.e., the high-income earner pays for the forestry infrastructure, and the low-income earner hitchhikes. This happens because the high-income earner does not mind letting the low-income earner hitchhike, and forestry infrastructure is provided. The high-income earner does not mind paying more, even if he/she bears the full cost.

2.1.2. Cooperative Games

Since rules are only broadly defined, the individual decision problems of players cannot be directly analyzed. Cooperative game theory avoids these difficulties by emphasizing coalitions of players. The study of coalitions implicitly assumes that such coalitions contain a presumption that players commit themselves, either by explicitly signing enforceable contracts or by transferring their decision-making powers. Hence, cooperative game theory is characterized by three features. First, rules are kept implicit. Second, the emphasis is on coalitions. Third, commitments are available [56].

- A Feasibility Analysis of Cooperative Games

When forest farmers cooperate to supply forestry infrastructure, the costs borne by individual forest farmers must be reduced, regardless of whether forest farmers have the same demand for forestry infrastructure. At this point, for a single forest farmer, consumer surplus will increase. The supply of forestry infrastructure increases for the entire forest area, achieving Pareto optimality. Assuming that forest farmers have the same demand for forestry infrastructure, a supply–demand model can be used to study the feasibility of a cooperative supply of goods for forestry infrastructure [57]. The specific analysis is shown in Figure 1.

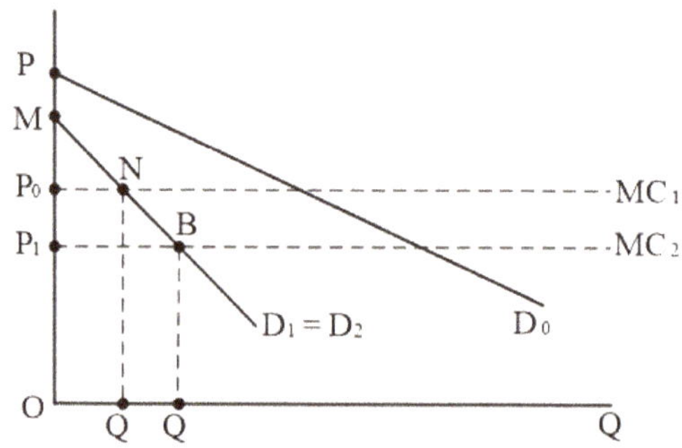

Figure 1. Economic analysis of the private joint provision of forestry infrastructure.

In Figure 1, D_1 and D_2 represent the demand curve for forestry infrastructure for forest farmer A and forest farmer B, respectively. Because the demand for forestry infrastructure is the same, the two-day demand curves overlap. MC_1 indicates the cost when forest farmers provide forestry infrastructure separately. MC_2 indicates the supply cost borne by individual forest farmers when they cooperate with each other. Cooperation reduces the supply cost of individual forest farmers, so $MC_2 < MC_1$. It can be seen from Figure 1 that after cooperation, the supply of goods for forestry infrastructure increases from Q_0 to Q_1, and consumer surplus increases from $S_{\triangle MNP0}$ to $S_{\triangle MBP1}$. Through cooperation, the consumer surplus of individual forest farmers increases, and the supply of goods for forestry infrastructure is closer to the collective optimal quantity, achieving Pareto optimality.

• Conditions for Realizing a Cooperative Game of Private, Voluntary Supply

The "club" size of community: The "club" size of community must meet two conditions. One is that the "club" will exist steadily for a period of time to ensure that members of the community can repeat the game, and the other is that reward and punishment mechanisms can be implemented within the "club" to restrict free rides by community members. Medium-sized rural villages meet the above-mentioned "club" conditions [57].

When the game is repeated, members of medium-sized rural areas are relatively stable, and they are more familiar with each other and can monitor each other. Because this is a repeated game, participants in the community not only consider the benefits that can be obtained at the current stage, but also comprehensively consider the benefits that can be obtained in the next game, as well as the costs they are going to face. Therefore, the strategic choice over repeated games and the strategic choice during one game are different. At the same time, under the constraints of reward and punishment mechanisms, members of the community are more likely to choose cooperation strategies [57].

In the collection of quasi-public goods for forestry infrastructure in medium-sized rural areas, the members are relatively stable and can repeat the game [57]. Therefore, quasi-public goods for forestry infrastructure can be provided cooperatively in medium-sized rural areas.

2.2. Experimental Design

In order to study the main factors that affect cooperative games, this paper uses experimental economics methods. On the basis of the literature and real-life situations, four control variables were selected: whether there was communication, whether the environment was determined [58], whether the information was feedback information [58], and whether there was a reward and punishment mechanism [59]. A total of 12 experiments were designed, including 6 experiments in the nonexchange group and 6 experiments in the exchange group. The specific experimental design is shown in Table 3. The realization of the experimental design relied on Z-Tree software. This paper only shows a part of the experimental process and experimental procedures. For related information, see Appendix A.

Table 3. Specific design of private cooperative supply experiments.

Group Members Could Not Communicate	Group Members Could Communicate
Experiment 1 (NC): NF × U (10 rounds)	Experiment 1 (C): NF × U (10 rounds)
Experiment 2 (NC): NF × NU (10 rounds)	Experiment 2 (C): NF × NU (10 rounds)
Experiment 3 (NC): F × U × NP (10 rounds)	Experiment 3 (C): F × U × NP (10 rounds)
Experiment 4 (NC): F × U × P (10 rounds)	Experiment 4 (C): F × U × P (10 rounds)
Experiment 5 (NC): F × NU × NP (10 rounds)	Experiment 5 (C): F × NU × NP (10 rounds)
Experiment 6 (NC): F × NU × P (10 rounds)	Experiment 6 (C): F × NU × P (10 rounds)

Note: NC indicates that players could not communicate with each other, and C indicates that they could communicate with each other; F indicates that there was information feedback, and NF indicates that there was no information feedback; U indicates that the environment was determined, and NU indicates that the environment was uncertain; P indicates that there were reward and punishment measures, and NP indicates that there were no reward and punishment measures.

2.2.1. Experimental Design Ideas

A voluntary donation mechanism was used to design the experiments. A total of n subjects were divided into m groups, and each group of (n/m) individuals played 10 rounds of games. Before the start of each round, everyone had an initial fund of e. They could choose to keep the funds in their own private accounts or to invest the funds in a public project. The funds could not be carried into the next round. If a member's fund in the private account in round t was x_{it} $(0 \leq x_{it} \leq e)$, and the investment amount in the public project was g_{it} $(0 \leq g_{it} \leq e; x_{it} + g_{it} = e)$, then the public project's total investment was $G_t = \sum g_{it}$ in round t. Assume that the return on investment in a public project is β_t $(0 \leq \beta_t < 1)$. When there is no environmental uncertainty, $\beta_1 = \beta_2 = \ldots \ldots = \beta_T = \beta$. When "environmental uncertainty" exists, β_t will be an independent and identically distributed random variable. If the distribution is set to a uniform distribution, that is, $\beta_t \in [\beta, \beta]$, and the expected β_t $(E\beta_t)$ is β, that is, $E\beta_t = \beta$, then the gain i in round t is as follows: $\pi = x_{it} + \beta_t G_t = e - g_{it} + \beta_t \sum g_{it}$ [58].

Four variables were selected: whether the group members could communicate, whether the information was feedback information, whether the environment was determined, and whether there was a reward and punishment mechanism. A total of 12 experiments were conducted. The specific experimental arrangement is shown in Table 3 below.

In each experiment, the group members were partners, that is, during the 10 rounds of experiments, the members of each group were fixed.

In the experiments in which members could communicate with each other, at the beginning of each experiment, the group members could discuss how to invest and decide how much to donate.

In the experiments with information feedback, after the end of each period, each group member was given the investment amount and investment income of others, in addition to his/her own investment amount, investment income, and β-value from that round. In the experiment without information feedback, after the end of each round of experiments, each group member only knew his/her own investment amount and investment income for the round.

In the experiments with a certain environment, at the beginning of each experiment, the group members knew the return rate β of the public account, which was relatively stable in reality. With environmental uncertainty, the investors did not know the return rate β before the start of each round of experiments, when they needed to estimate the return rate β and then decide how much to invest based on the estimated value of β. After the end of each round of experiments, they re-estimated the return rate β according to their own returns from the previous round and the influence of the social environment.

Rewards and punishments only existed in the experiments with information feedback. In the experiments with rewards and punishments, after each round of experiments, each group member in the group could decide to reward or punish the other members in accordance with the feedback information. After the information feedback was over and before the next round of experiments began, everyone in the group received 10 additional chips and had the opportunity to reward or punish others (note: they could not punish or reward themselves). When a person was rewarded or punished with one chip, the person's gain increased or decreased by three chips. The maximum number of chips that each person could use for each target member was 5, which meant that each member could increase or decrease the profit of a target member by 15 at most [59].

2.2.2. Choice of Experimental Subjects

Experimental economics uses real people in society as experimental objects. In order to ensure the scientific nature of experimental economics, the selected subjects only have small differences or have differences that have nothing to do with or have little influence on the purpose of the experiment. Undergraduates and graduate students were chosen as candidates because of their relatively simple living environments, their relatively few differences, and their relatively strong ability to learn and understand. Moreover, using undergraduate and graduate students in economic experiments that require little or no practical experience can lead to better and more realistic effects. The recruitment of subjects was voluntary. Recruitment information that included the purpose and requirements of the experiment and the possible benefits and costs of participating in the experiment were also released to the public [60].

2.2.3. Statistical Analyses

The statistical analyses of the personal characteristics of the experimental subjects are shown in Table 4. Variables such as gender, age, income, and trustworthiness of the experimental subjects were counted.

Table 4. Statistical analyses of experimental objects.

Variables	Mean	Standard Deviation	Min	Median	Max
Gender	0.100	0.300	0.000	0.000	1.000
Ethnic group	0.950	0.218	0.000	1.000	1.000
Communist or not	0.050	0.218	0.000	0.000	1.000
Has taken out a loan or not	0.200	0.400	0.000	0.000	1.000
Part-time or not	0.500	0.500	0.000	0.500	1.000
Participated in the experiment or not	0.150	0.357	0.000	0.000	1.000
Average household income per month	3.250	0.829	1.000	3.000	4.000
Evaluation of self-reliability	3.800	0.600	2.000	4.000	5.000
Evaluation of stranger's credibility	2.700	0.781	1.000	3.000	4.000

Note: gender (0 = female; 1 = male); ethnic group (0 = ethnic minority; 1 = ethnic Han); communist or not (0 = no; 1 = yes); Has taken out a loan or not (0 = no; 1 = yes); part-time or not (0 = no; 1 = yes); participated in the experiment or not (0 = no; 1 = yes); average household income per month (1 = 0–3000 RMB; 2 = 3001–6000 RMB; 3 = 6001–9000 RMB; 4 = 9000 RMB or more); evaluation of self-reliability(5 = very trustworthy; 4 = trustworthy; 3 = fair; 2 = untrustworthy; 1 = quite untrustworthy); evaluation of stranger's credibility (5 = very trusting; 4 = trusting; 3 = fair; 2 = untrusting; 1 = quite untrusting).

2.3. *Methods*

2.3.1. Mann–Whitney *U* Test

The Mann–Whitney *U* test, also called the "Mann–Whitney rank sum test", was proposed by Mann and Whitney in 1947. The purpose was to test whether the means of two samples were significantly different. We used this method to analyze whether the experimental results of these two groups of experiments were different when a certain experimental variable changed. When the P value was less than 0.05, this indicated that the means of the two independent samples were significantly different.

2.3.2. Wilcoxon Signed-Rank Test

The Wilcoxon signed-rank test, or Wilcoxon symbol test, was proposed by Wilcoxon (F. Wilcoxon) in 1945. We used this method to test whether the differences between two types of data in the same group were significant. For example, a Wilcoxon signed-rank test was used to test whether the difference between "average donations after consultation" and "the average of actual donations" was significant (in experiment 1, with communication). When the *P* value was less than 0.05, this indicated that the means of the two independent samples were significantly different.

2.3.3. Multiple Regression Analysis

We used multiple regression to analyze the main factors that affect donations under different experimental conditions.

When *P* was less than 0.05, H0 could be rejected (there was no relationship between the two variables). Therefore, the *P*-value of the variable was less than 0.05, which indicated the explanatory variable had a significant effect on the dependent variable. Here, * means significant at the 0.05 probability level, and ** means significant at the 0.01 probability level.

3. Results

3.1. Statistical Analyses of Donation Amounts and Income under Different Experimental Conditions

In this paper, the donation amounts from the 12 groups of experiments were calculated using the mean, standard deviation, minimum value, median value, and maximum value. Basic statistical information on donation amounts from the different experiments is shown in Table 5.

Table 5. Statistics of donation amounts under different experimental conditions.

Experiments	Variable	Mean	Standard Deviation	Min	Median	Max
Experiment 1 (NC) (10 rounds)	C1	2.520	2.106	0.000	2.000	7.200
Experiment 2 (NC) (10 rounds)	C2	6.275	4.483	0.000	6.000	20.000
Experiment 3 (NC) (10 rounds)	C3	3.800	1.646	0.000	4.000	6.000
Experiment 4 (NC) (10 rounds)	C4	14.925	3.085	5.000	15.000	20.000
Experiment 5 (NC) (10 rounds)	C5	4.790	1.934	1.000	5.000	10.000
Experiment 6 (NC) (10 rounds)	C6	13.125	3.116	5.000	13.500	20.000
Experiment 1 (C) (10 rounds)	C11	7.300	6.604	0.000	4.800	20.000
Experiment 2 (C) (10 rounds)	C22	5.975	6.207	0.000	5.000	20.000
Experiment 3 (C) (10 rounds)	C33	7.975	6.191	0.000	10.000	20.000
Experiment 4 (C) (10 rounds)	C44	14.675	2.611	9.000	15.000	20.000
Experiment 5 (C) (10 rounds)	C55	5.160	1.960	1.000	5.000	10.000
Experiment 6 (C) (10 rounds)	C66	10.075	4.424	5.000	8.500	20.000

Note: *NC* indicates that players could not communicate with each other, and *C* indicates that they could communicate with each other; *F* indicates that there was information feedback, and *NF* indicates that there was no information feedback; U indicates that the environment was determined, and *NU* indicates that the environment was uncertain; *P* indicates that there were reward and punishment measures, and *NP* that there were no reward and punishment measures. C1, C2, C3, C4, C5, C6, C11, C22, C33, C44, C55, and C66 represent the number of experimental coins donated under different experimental conditions. Before each round of experiments began, each member had 50 initial coins.

It is shown in Table 4 that the averages of Experiments 4 and 6 were the highest, and the average value of the communication group was higher than that of the noncommunication group, which shows that communication as well as reward and punishment mechanisms could improve the donation level as a whole.

Figure 2 shows the key factors affecting the supply of goods for forestry infrastructure, summarizing the results of five variables out of the six groups of experiments for comparative analyses: the "mean value of donations without communication", "mean value of donations with communication", "mean value of planned donations after communication", "mean value of earnings without communication", and "mean value of earnings with communication". In Figure 2, C11, C21, C31, C41, C51, and C61 represent the "mean values of donations without communication"; C12, C22, C32, C42, C52, and C62 represent the "mean values of donations with communication"; C13, C23, C33, C43, C53, and C6 represent the "mean values of planned investment after communication"; *R*11, *R*21, *R*31, *R*41, *R*51, and *R*61 represent the " mean values of earnings without communication"; and *R*12, *R*22, *R*32, *R*42, *R*52, *R*62 represent the "mean values of earnings with communication".

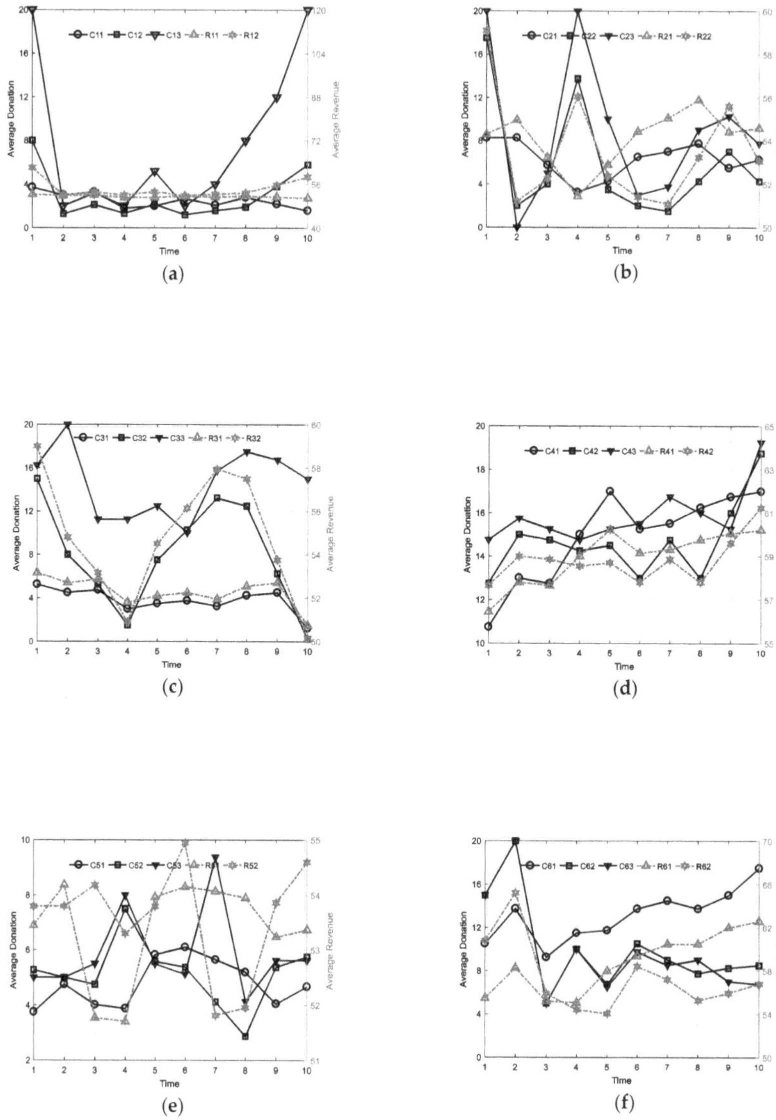

Figure 2. The average donation amount for each period in the experiments with and without communication: (**a**) the statistics for average donations and average returns in Experiment 1 (*C* & *NC*); (**b**) statistics for average donations and average returns in Experiment 2 (*C* & *NC*); (**c**) statistics for average donations and average returns in Experiment 3 (*C* & *NC*); (**d**) statistics for average donations and average returns in Experiment 4 (*C* & *NC*); (**e**) statistics for average donations and average returns in Experiment 5 (*C* & *NC*); (**f**) statistics for average donations and average returns in Experiment 6 (*C* & *NC*). "Mean values of donations without communication", "mean values of donations with communication", "mean values of planned investment after communication", "mean values of earnings without communication", and "mean values of earnings with communication" represent the number of experimental coins under different experimental conditions. Before each round of experiments began, each member had 50 initial coins. "Time" is the number of rounds in the experiment, and 1–10 represents rounds 1–10. Each experiment had 10 rounds.

The analysis of Figure 2 is as follows. Donations in Experiment 2 (*C* & *NC*) were slightly higher than in Experiment 1 (*C* & *NC*), which indicated that an uncertain environment could increase the donation level when there was no information feedback. However, the fluctuation in Experiment 2 (*C* & *NC*) was relatively large because of the uncertain environment. Experiment 3 (*C* & *NC*) had higher values, while Experiment 5 (*C* & *NC*) had greater fluctuations due to the uncertainty of the environment, indicating that when there was information feedback, a certain environment could improve the donation amounts. As for Experiment 4 and Experiment 6 (*C* & *NC*), Experiment 4 (*C* & *NC*) had slightly higher values than did Experiment 6 (*C* & *NC*), but the overall donation levels of Experiment 4 (*C* & *NC*) and Experiment 6 (*C* & *NC*) were higher than in the other experimental groups. In addition, Experiment 6 (*C* & *NC*) had large fluctuations due to the uncertain environment, which shows that a reward and punishment mechanism could improve the level of donations.

3.2. *Impact of Communication on Donations*

3.2.1. An Analysis of Hitchhiking Using Different Experimental Mechanisms

In order to study the free-riding situation due to different experimental mechanisms, a Wilcoxon signed-rank test was performed for "donations determined after consultation" and "actual donations" for the six groups of experiments.

Table 6 indicates the results of a Wilcoxon signed-rank test that was used to check whether there was a difference between "donations determined after consultation" and "actual donations" in the six groups of experiments. The "median of the average of donations after consultation" and the "median of the average of actual donations" were calculated. "Median difference" refers to the median difference between "donations determined after consultation" and "actual donations", and the "*P*-value" represents the significance level of the difference. When the *P*-value was less than 0.05, the difference was significant.

Table 6. Wilcoxon signed-rank test results of the mean values of donations determined after consultation and actual donations in the experimental groups that were allowed to communicate.

Groups	Median of the Average of Donations after Consultation	Median of the Average of Actual Donations	Median Difference	*P*-Values
Experiment 1 (C)	4.600	5.050	−0.133	0.674
Experiment 2 (C)	8.375	4.083	2.875	0.012
Experiment 3 (C)	15.375	7.750	5.667	0.007
Experiment 4 (C)	15.4375	14.5833	0.7500	0.021
Experiment 5 (C)	5.500	5.308	0.125	0.192
Experiment 6 (C)	8.750	8.750	−0.200	0.207

Table 6 shows the results when there was a reward and punishment mechanism (Experiment 4 (C) and Experiment 6 (C)). The actual and negotiated results of Experiment 4 (C) were significantly different, but those of Experiment 6 (C) were not significant. This shows that with a reward and punishment mechanism and a certain environment, the actual result was lower than the negotiated result. When the environment was uncertain, the actual and negotiated results were basically the same. The reason is that when the environment was determined, members could reasonably estimate their costs and benefits, and the actual donation amount was slightly lower than the negotiated result (but matched their needs). The *P*-value of Experiment 2 (C) was less than 0.05, which indicates that the difference in Experiment 2 (C) between the average value of the actual donation amount and the average value of the donation amount determined after consultation was significant. However, the *P*-value of Experiment 1 (C) was not significant. This shows that members were more likely to choose free-riding when there was no information feedback and the environment was uncertain. The difference between the average value of the actual donation amount and the average value of the donation amount determined after consultation was significant in Experiment 3 (C), while in Experiment 5 (C), it was not. This shows that if information feedback was provided, it was easier for members to choose to be a free-rider when the environment was determined.

3.2.2. Mann–Whitney *U* Test for the Effect of Communication

A Mann–Whitney *U* test was used to analyze whether communication could increase group members' supply of goods for forestry infrastructure. The test results are shown in Table 7.

Table 7. Mann–Whitney *U* test for the influence of communication on the supply of goods for forestry infrastructure.

Contrast Groups	*P*-Values	Contrast Groups	*P*-Values
Experiment 1 (NC) and Experiment 1 (C)	0.001	Experiment 4 (NC) and Experiment 4 (C)	0.363
Experiment 2 (NC) and Experiment 2 (C)	0.161	Experiment 5 (NC) and Experiment 5 (C)	0.406
Experiment 3 (NC) and Experiment 3 (C)	0.021	Experiment 6 (NC) and Experiment 6 (C)	0.028

An analysis of Table 7 is given below. It could be concluded that the difference between Experiment 1 (*NC*) and Experiment 1 (C) was significant, while the difference between Experiment 2 (*NC*) and Experiment 2 (C) was not significant. This shows that when there was no information feedback and the environment was determined, communication could increase donations. The difference between Experiment 3 (*NC*) and Experiment 3 (C) was significant, while the difference between Experiment 5 (*NC*) and Experiment 5 (C) was not. This shows that when there was information feedback, but no reward and punishment mechanism, communication could increase donations in a determined environment. It could be concluded that the difference between Experiment 6 (*NC*) and Experiment 6 (C) was significant, but the difference between Experiment 4 (*NC*) and Experiment 4 (C) was not, indicating that when there was a reward and punishment mechanism and an uncertain environment, communication could increase donations.

When there is no reward or punishment, everyone may choose to hitchhike. If the environment is uncertain during communication, members will still choose to donate less or even to not donate because they cannot accurately estimate the specific benefits. When the environment is determined, members know the specific benefits. In this situation, through communication, everyone can make the best choice according to their needs, which shows that communication works. When there are rewards and punishments, members want to donate more because they are afraid of being punished or because they want further reward. If, in addition, the environment is determined, members will make rational decisions based on their needs, that is, they will choose the minimum amount required. At this point, communication does not work. However, if the environment is uncertain, everyone in the communication process will know how much reward or punishment they will get. At this time, members will choose to invest more (within their ability to give), and communication plays a role.

3.3. *Impact of Environmental Certainty on Donations*

3.3.1. Statistical Analyses of Environmental Certainty

In this paper, environmental certainty mainly refers to whether the return on investment in forestry infrastructure is known before investment. The returns of Experiment 2 (*C* & *NC*), Experiment 5 (*C* & *NC*), and Experiment 6 (*C* & *NC*) were uncertain. In order to analyze the members' estimations of the rate of return in an uncertain environment, we used Wilcoxon signed-rank tests in Experiment 2 (*C* & *NC*), Experiment 5 (*C* & *NC*), and Experiment 6 (*C* & *NC*). The results of the test are shown in Table 8. At the same time, we compared the true value and estimated value of the return rate β in the three groups of experiments. The results are shown in Figure 2.

Table 8. Wilcoxon signed-rank test results of the true value and estimated value of the return rate β.

Groups	Median of Mean of True Values of β	Median of Mean of Estimated Values of β	Median Difference	*P*-Values
Experiment 2 (NC)	0.420	0.407	0.010	0.439
Experiment 5 (NC)	0.430	0.468	−0.040	0.021
Experiment 6 (NC)	0.420	0.426	−0.020	0.058
Experiment 2 (C)	0.420	0.410	−0.015	0.306
Experiment 5 (C)	0.430	0.430	0.013	0.959
Experiment 6 (C)	0.430	0.408	0.030	0.105

Table 8 includes a Wilcoxon signed-rank test that was used to check whether there was a difference between "the true value of the return rate β" and "the estimated value of the return rate β". The "mean of true values of β", the "mean of estimated values of β", the "median of the mean of true values of β", and the "median of the mean of estimated values of β" were each calculated. "Median difference" refers to the median of the difference between "the true value of the return rate β" and "the estimated value of the return rate β". "*P*-value" represents the level of significance. When the "*P*-value" was less than 0.05, the difference was significant.

As can be seen from Table 8, when there was communication, there was no significant difference between the estimated value and the actual value of the return rate β. The reason is that after consultation, the disadvantage of information asymmetry was obviated and the accuracy of the estimation of the return rate β improved, which made the estimated value too close to the true value. This indirectly proved the power of cooperation. Without communication, only the *P*-values of Experiment 5 (*NC*)—given by the Wilcoxon signed-rank test—were significant. Combining this with the information in Figure 3, it could be concluded that the estimated return rate β was higher than the true value, indicating that when there was no communication, the return rate β was overestimated.

In analyzing Figure 3, it could be concluded that with communication, information feedback as well as reward and punishment mechanisms had little effect on the estimated value of the return rate β, which fluctuated around the true value and was close to the true value. This was consistent with the results of the Wilcoxon signed-rank test. In the absence of communication, when there was information feedback as well as a reward and punishment mechanism (Experiment 6 (*NC*) and Experiment 6 (C)), the actual value of the return rate β was not significantly different from the estimated value. The reason is that on the one hand, information feedback allowed for members to have a clear understanding of the previous period's return rate, which helped in their estimation of the rate for the next period; on the other hand, the existence of a reward and punishment mechanism allowed members to make an objective estimation by comprehensively considering the economic benefits and social benefits. When there was no information feedback (Experiment 2 (*NC*) and Experiment 2 (C)), the estimated rate of return β was lower than the actual value, because members would avoid risks when making decisions without knowing the return rate β of the previous year or without being able to communicate. In Experiment 5 (*NC*) and Experiment 5 (C), the estimated value of the return rate β was higher than the actual value, which indicates that information feedback as well as reward and punishment led to overestimation.

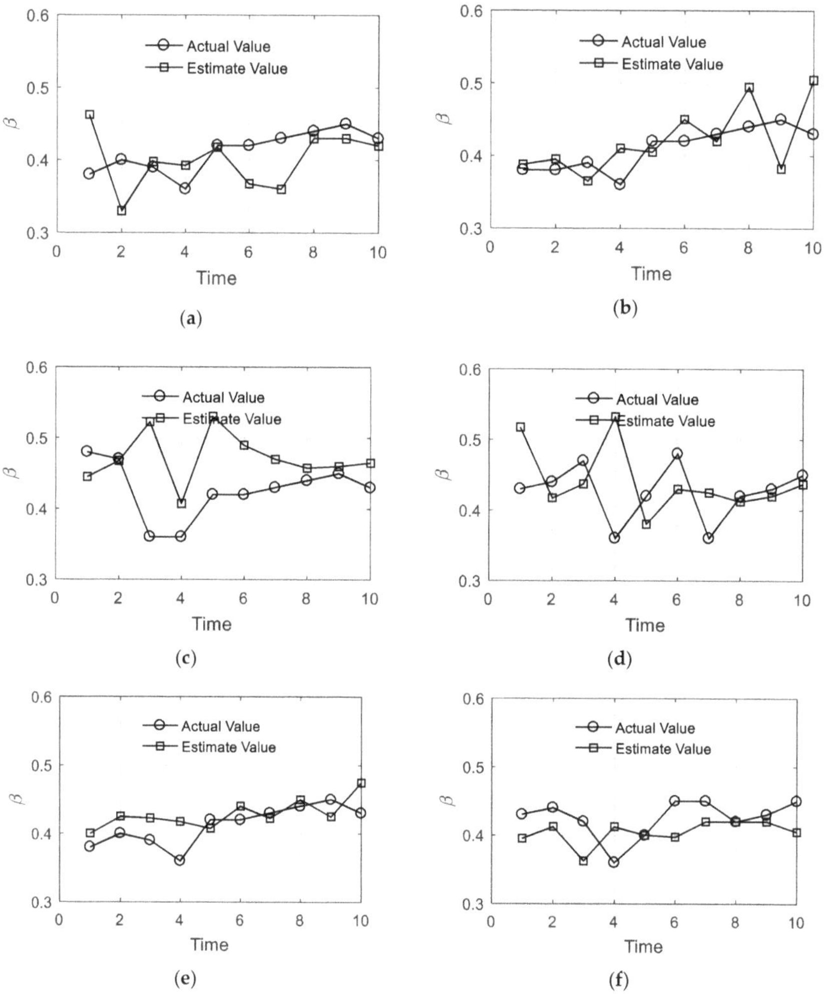

Figure 3. The average estimated and actual value of the rate of return β in each period: (**a**) under the conditions of Experiment 2 (*NC*), when the group members could not communicate (statistics on the estimated and true value of the rate of return β); (**b**) under the conditions of Experiment 2 (*C*), when the group members could communicate (statistics on the estimated and true value of the rate of return β); (**c**) under the conditions of Experiment 5 (*NC*), when the group members could not communicate (statistics on the estimated and true value of the rate of return β); (**d**) under the conditions of Experiment 5 (*C*), when the group members could communicate (statistics on the estimated and true value of the rate of return β); (**e**) under the conditions of Experiment 6 (*NC*), when the group members could not communicate (statistics on the estimated and true value of the rate of return β); (**f**) under the conditions of Experiment 6 (*C*), when the group members could communicate (statistics on the estimated and true value of the rate of return β). "Actual value" and "estimated value" represent the number of experimental coins under different experimental conditions. Before each round of experiments began, each member had 50 initial coins. "Time" is the number of rounds in the experiment, and 1–10 represents rounds 1–10. Each experiment contains 10 rounds.

3.3.2. Mann–Whitney *U* Test on the Effect of Environmental Certainty

We used a Mann–Whitney *U* test to analyze whether a determined environment—compared to environmental uncertainty—could improve the supply of goods given by members of a group for forestry infrastructure. The test results are shown in Table 9.

Table 9. Mann–Whitney *U* test of the impact of environmental certainty on the supply level of good for forestry infrastructure.

Contrast Groups	*P*-Values	Contrast Groups	*P*-Values
Experiment 1 (NC) and Experiment 2 (NC)	0.000	Experiment 1 (C) and Experiment 2 (C)	0.384
Experiment 3 (NC) and Experiment 5 (NC)	0.054	Experiment 3 (C) and Experiment 5 (C)	0.089
Experiment 4 (NC) and Experiment 6 (NC)	0.088	Experiment 4 (C) and Experiment 6 (C)	0.000

In the experimental groups in which there was no communication, the *P*-value of Experiment 1 (*NC*) and Experiment 2 (*NC*), given by the Mann–Whitney *U* test, were significant. Through a comparison of the donation amounts (with no communication) in Experiment 1 (*NC*) and Experiment 2 (*NC*) (i.e., C11 and C21 in Figure 2), it could be found that C21 was higher than C11, indicating that when there was no reward and punishment mechanism and no information feedback, the uncertainty of the environment increased the supply levels of goods for forestry infrastructure. The reasons were as follows. Without communication between members, a single member had limited information, and environmental certainty was important in making decisions. When the environment was uncertain, a single member tended to overestimate the return rate and invest more.

In the experimental groups in which there was communication, the *P*-value of Experiment 4 (*C*) and Experiment 6 (*C*), given by the Mann–Whitney *U* test, were significant. Through a comparison of the donation amounts (with no communication) in Experiment 4 (*C*) and Experiment 6 (*C*) (i.e., C41 and C61 in Figure 2), it was found that C41 was higher than C61. This shows that when there was information feedback and reward and punishment mechanisms, a determined environment could increase the donation amount. The results of Experiment 1 (*C*) and Experiment 2 (*C*) were not significant, indicating that without information feedback, but with communication, environmental certainty did not work. The reasons are shown below. When there was no communication, information was relatively occluded, and a determined environment was an important basis for decision-making among members. When there was communication, members could obtain comprehensive information about the forestry infrastructure being built. When members made decisions, environmental certainty was important. However, it was not the only important consideration. In this way, the role of environmental certainty weakened. The test results of Experiment 3 (*C*) and Experiment 5 (*C*) were not significant, indicating that environmental certainty did not work when there was information feedback and no reward and punishment mechanism.

3.4. *Impact of Information Feedback on Donations*

A Mann–Whitney *U* test was used to compare the donations in Experiment 1 (*C* & *NC*) and Experiment 3 (*C* & *NC*), and Experiment 2 (*C* & *NC*) and Experiment 5 (*C* & *NC*), to study the impact of information feedback on the supply levels of goods for forestry infrastructure The specific results are shown in Table 10.

Table 10. Mann–Whitney *U* test on the impact of information feedback on the supply levels of goods for forestry infrastructure.

Contrast Groups	*P*-Values	Contrast Groups	*P*-Values
Experiment 1 (NC) and Experiment 3 (NC)	0.009	Experiment 1 (C) and Experiment 3 (C)	0.496
Experiment 2 (NC) and Experiment 5 (NC)	0.028	Experiments 2(C) and Experiment 5 (C)	0.256

According to Table 10, when there was communication, information feedback had no significant impact on the supply levels of goods for forestry infrastructure. When there was no communication, the Mann–Whitney U-test results for Experiment 1 (*NC*) and Experiment 3 (*NC*) and Experiment 2 (*NC*) and Experiment 5 (*NC*) were significant. Combining this with Figure 2, it can be concluded that donations in Experiment 3 (*NC*) were higher than in Experiment 1 (*NC*) (C31 was higher than C11) and that donations in Experiment 2 (*NC*) were higher than in Experiment 5 (*NC*) (C21 was higher than C51), indicating that when the environment was determined, information feedback increased donations. When the environment was uncertain, a lack of information feedback increased donations.

3.5. Impact of Reward and Punishment on Donations

3.5.1. Statistical Analyses of Reward and Punishment

In order to indicate the impact of reward and punishment in different experimental scenarios, Figure 4 summarizes the four variables from Experiments 4 (*C* & *NC*) and 6 (*C* & *NC*): "average donation amounts for rewards", "average donation amount for punishment", "average reward received", and "average penalty received". In Figure 4, "*A*1, *A*2, *A*3, and *A*4" mean "average donation amounts for rewards"; "*P*1, *P*2, *P*3, and *P*4" mean "average donation amounts for punishment"; "*A*11, *A*21, *A*31, and *A*41" mean "average reward received"; and "*P*11, *P*21, *P*31, and *P*41" mean "average penalty received".

In analyzing Figure 4, it can be seen that if there was no communication, the donations used for rewards declined, regardless of whether the environment was determined or not. Donations used for punishment fluctuated at around 1 in Experiment 4 (*C* & *NC*), while Experiment 6 (*C* & *NC*) showed an upward trend. In Experiment 4 (*C* & *NC*), rewards received had a downward trend, while in Experiment 6 (*C* & *NC*), they fluctuated steadily at around 1. In Experiment 4 (*C* & *NC*), penalties obtained fluctuated smoothly around 1, while in Experiment 6 (*C* & *NC*), they had an upward trend. In general, because there was no communication before making decisions, the intensity of reward and punishment after decision-making was relatively low, and the fluctuations in trends were relatively stable.

If there was communication, donations used for rewards increased in Experiment 4 (*C* & *NC*), while in Experiment 6 (*C* & *NC*), they decreased. Donations used for punishment were at 0 in Experiment 4 (*C* & *NC*), while in Experiment 6 (*C* & *NC*) they had an upward trend. In Experiment 4 (*C* & *NC*), rewards received had an upward trend, while in Experiment 6 (*C* & *NC*), they declined steadily. In Experiment 4 (*C* & *NC*), penalties obtained were at 0, while in Experiment 6 (*C* & *NC*), they had an upward trend. In general, with communication, the amount of money donated for reward and the amount of money donated for punishment were higher than the amounts given when there was no communication, and donations for reward or punishment fluctuated greatly.

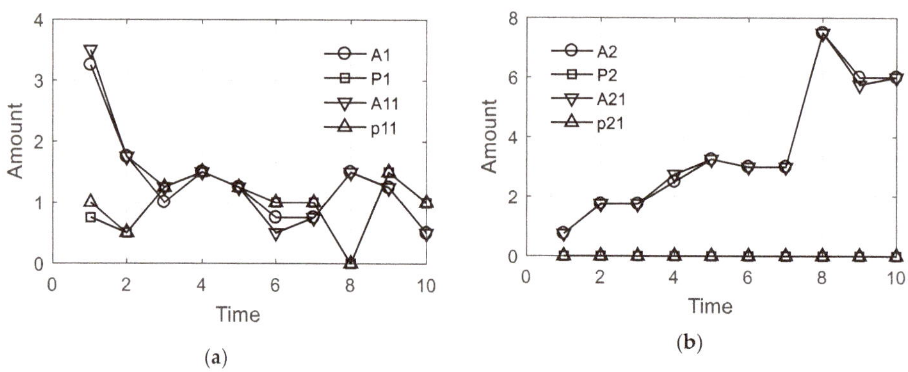

Figure 4. *Cont.*

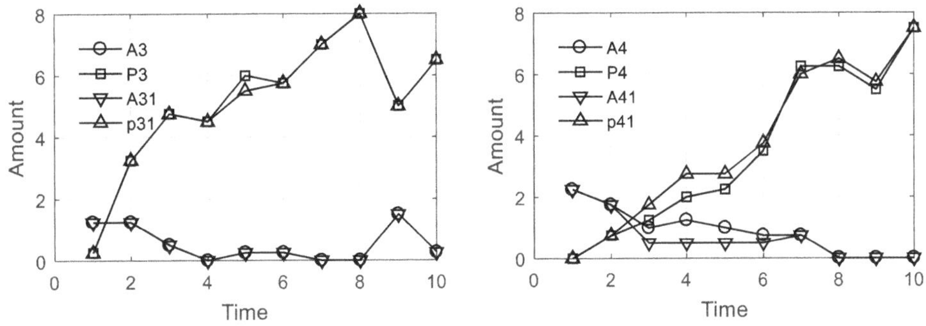

(c) (d)

Figure 4. Summary of reward and punishment in the groups with reward and punishment: (**a**) under the conditions of Experiment 4 (*C* & *NC*), when the group members could not communicate (statistics on rewards and punishments meted out to other people or statistics on rewards and punishments received by members); (**b**) under the conditions of Experiment 4 (*C* & *NC*), when the group members could communicate (statistics on rewards and punishments meted out to other people or statistics on rewards and punishments received by members); (**c**) under the conditions of Experiment 6 (*C* & *NC*), when the group members could not communicate (statistics on rewards and punishments meted out to other people or statistics on rewards and punishments received by members); and (**d**) under the conditions of Experiment 6 (*C* & *NC*), when the group members could communicate (statistics on rewards and punishments meted out to other people or statistics on rewards and punishments received by members). "Average donation amount for rewards", "average donation amount for punishment", "average reward received", and "average penalty received" represent the number of experimental coins under different experimental conditions. Before each round of experiments began, each member had 50 initial coins. "Time" is the number of rounds in the experiment, and 1–10 represents rounds 1–10. Each experiment had 10 rounds.

3.5.2. Mann–Whitney *U* Test of Reward and Punishment Mechanisms

In order to analyze whether a reward and punishment mechanism significantly affected the supply level of goods for forestry infrastructure, a Mann–Whitney *U* test was used to compare the "mean value of donations without communication" to the "mean value of donations with communication" in Experiment 3 (*C* & *NC*) and Experiment 4 (*C* & *NC*) and in Experiment 5 (*C* & *NC*) and Experiment 6 (*C* & *NC*). The test results are shown in Table 11.

Table 11. Mann–Whitney *U* test on the influence of reward and punishment on the supply level of goods for forestry infrastructure.

Contrast Groups	*P*-Values	Contrast Groups	*P*-Values
Experiment 3 (NC) and Experiment 4 (NC)	0.000	Experiment 3 (C) and Experiment 4 (C)	0.003
Experiment 5 (NC) and Experiment 6 (NC)	0.000	Experiment 5 (C) and Experiment 6 (C)	0.001

In analyzing Table 11, it can be concluded that a reward and punishment mechanism could significantly affect the supply level of goods for forestry infrastructure. In Figure 2, it can be seen that the average donation amount in Experiment 6 (*NC*) was significantly higher than that in Experiment 5 (*NC*) and that the average donation amount in Experiment 6 (*C*) was significantly higher than that in Experiment 5 (*C*), which indicates that setting up a reward and punishment mechanism could significantly increase the supply level of goods for forestry infrastructure. The reason is that, after a reward and punishment mechanism was set up, members would consider not only their own economic interests but also other social benefits such as reputation and fame while making decisions. Therefore, members would choose to increase their donations to obtain more benefits. Moreover, in the long term, the members of the community were fixed and could not avoid being punished due to the punishment mechanism. Therefore, members would choose to increase their donations in order to avoid punishment by other members.

3.6. Research on the Influencing Factors on Donation Amounts in Different Experimental Situations

In order to study the influencing factors—the direction and degree of influence over the actual donation amounts due to different experimental mechanisms—a regression analysis on the data from the six groups of experiments were conducted.

3.6.1. Experiment 1 (C & NC)

Taking "actual donation amount without communication (*ADNC*)" and "actual donation amount with communication (*ADC*)" as dependent variables, and taking "planned donation (*PD*)", "return from the previous round (*R*)", and "degree of trust in strangers (*S*)" as independent variables, a regression was implemented. The results are shown in Table 12.

Table 12. Regression analysis results for Experiment 1 (C & NC).

Independent Variables	Dependent Variables	
	ADNC	*ADC*
PD	-	0.3806 ** (0.0779)
R	−2.0149 ** (0.2975)	0.0670 (0.1720)
S	−0.2046 (0.7246)	−0.1938 (0.9113)
Constant	110.5538 ** (15.8225)	−0.4205 (9.7951)
F	23.11 **	8.13 **
R^2	0.5834	0.4326

Note: * significant at the 0.05 probability level; ** significant at the 0.01 probability level. *ADNC*: "actual donation amount without communication"; *ADC*: "actual donation amount with communication".

Table 12 shows that at a 99% confidence level, the last round of earnings had a significant impact on the actual donation amount when there was no communication. For every unit increase in income in the previous round, the actual donation amount decreased by 2.0149 units when there was no communication. In this case, the increase in income in the previous round was due to reduced investment compared to other members and the extra income received through free-riding. Therefore, when income increased, less was invested. At a 99% confidence level, the planned donation amount had a significant impact on the actual donation amount when there was communication. For every unit increase in the planned donation amount, the actual donation amount increased by 0.3806 units when there was communication. The *F*-statistics were all significant.

3.6.2. Experiment 2 (C & NC)

The regression results for the dependent variables "actual donation amount without communication (*ADNC*)" and "actual donation amount with communication (*ADC*)", as well as the independent variables "planned donation (*PD*)", "estimated value of β (*EV*)", "return from the previous round (*R*)", the "actual value of β in the previous round (*AV*)", and "degree of trust in strangers (*S*)" are shown in Table 13.

Table 13 shows that at a 95% confidence level, the estimated value of β had a significant impact on the actual donation amount when there was no communication. For each unit increase in the estimated value of β, the actual donation amount increased by 21.7732 units when there was no communication. That is, the higher the estimated rate of return β was, the more confidence there was in investment prospects, and the more money was donated. At a 99% confidence level, the planned donation amount had a significant impact on the actual donation amount when there was communication. For every unit increase in the planned donation amount, the actual donation amount increased by 0.5395 units when there was communication. The *F*-statistics were all significant.

Table 13. Regression analysis results for Experiment 2 (C & NC).

Independent Variables	Dependent Variables	
	ADNC	*ADC*
PD	-	0.5395 ** (0.1226)
EV	21.7732 * (8.4576)	5.9781 (10.3440)
R	0.1414 (0.1747)	0.0170 (0.1591)
AV	−2.7987 (25.2626)	−2.5214 (25.4660)
S	0.5917 (1.5337)	−0.5469 (1.5920)
Constant	−11.3111 (13.5503)	−0.1530 (15.7398)
F	2.84 *	4.64 **
R^2	0.2681	0.4363

Note: * significant at the 0.05 probability level; ** significant at the 0.01 probability level. *ADNC*: "actual donation amount without communication"; *ADC*: "actual donation amount with communication".

3.6.3. Experiment 3 (*C* & *NC*)

Taking "actual donation amount without communication (*ADNC*)" and "actual donation amount with communication (*ADC*)" as the dependent variables and "planned donation amount (*PD*)", "return from last round (*R*)", and "degree of trust in strangers (*S*)" as the independent variables, a regression was implemented, and the results are shown in Table 14.

Table 14. Regression analysis results for Experiment 3 (C & NC).

Independent Variables	Dependent Variables	
	ADNC	*ADC*
PD	-	0.3806 ** (0.0779)
R	0.8662 ** (0.0449)	0.0670 (0.1720)
S	−0.3889 ** (0.0743)	−0.1938 (0.9113)
Constant	−41.4704 ** (2.3666)	−0.4205 (9.7951)
F	199.73 **	8.13 **
R^2	0.9237	0.4326

Note: * significant at the 0.05 probability level; ** significant at the 0.01 probability level. *ADNC*: "actual donation amount without communication"; *ADC*: "actual donation amount with communication".

Table 14 shows that at a confidence level of 99%, the benefits received in the previous round and the degree of trust in strangers had a significant impact on the actual donation amount when there was no communication. For each additional unit of income received in the previous round, the actual donation amount increased by 0.8662 units when there was no communication. The increase in income could be attributed to an increase in the amount in the public account, which was caused by the overinvestment of group members. Therefore, there was a positive correlation between the benefits received in the previous period and the actual donation amount when there was no communication. For each additional unit in the "degree of trust in strangers (*S*)" variable, the actual donation amount when there was no communication decreased by 0.3889 units. At a 99% confidence level, the planned donation amount had a significant impact on the actual donation amount when there was communication: for every unit increase in the planned donation amount, the actual donation amount increased by 0.3806 units. The *F*-statistics were significant.

3.6.4. Experiment 4 (*C* & *NC*)

Taking the "actual donation amount without communication (*ADNC*)" and the "actual donation amount with communication (*ADC*)" as the dependent variables and the "planned donation amount

(*PD*)", "return from the previous round (*R*)", "reward received from the previous round (RP)", "penalty received from the previous round (*PP*)", "revenue after the previous round of reward and punishment (*RRP*)" and "degree of trust in strangers (*S*)" as the independent variables, a regression was implemented, the results of which are shown in Table 15.

Table 15. Regression analysis results for Experiment 4 (C & NC).

Independent Variables	Dependent Variables	
	ADNC	*ADC*
PD	-	0.9277 ** (0.1710)
R	−0.0372 (0.3619)	0.0014 (0.1615)
RP	−1.1947 (0.9250)	−0.3030 (0.5093)
PP	1.7193 (1.0083)	-
RRP	0.5720 * (0.2781)	0.0875 (0.0825)
S	0.8822 (0.5812)	0.0614 (0.3912)
Constant	−23.5713 (15.4177)	−6.2810 (12.5589)
F	1.93	6.60 **
R^2	0.2434	0.5238

Note: * significant at the 0.05 probability level; ** significant at the 0.01 probability level. *ADNC*: "actual donation amount without communication"; *ADC*: "actual donation amount with communication".

Table 15 shows that at a 95% confidence level, earnings after the previous round of reward and punishment had a significant impact on the actual donation amount when there was no communication. For each additional unit of earnings after the previous round of reward and punishment, the actual donation amount increased by 0.5720 units when there was no communication. At a 99% confidence level, the planned donation amount had a significant impact on the actual donation amount when there was communication. For every unit increase in the planned donation amount, the actual donation amount increased by 0.9277 units when there was communication.

3.6.5. Experiment 5 (C & NC)

Taking "the actual donation amount without communication (*ADNC*)" and "the actual donation amount with communication (*ADC*)" as the dependent variables and the "planned donation amount (*PD*)", "estimated value of β (*EV*)", "return from the previous round (*R*)", "actual value of β in the previous round (*AV*)", and "degree of trust in strangers (*S*)" as the independent variables, a regression was implemented. The results are shown in Table 16.

Table 16. Regression analysis results for Experiment 5 (C & NC).

Independent Variables	Dependent Variables	
	ADNC	*ADC*
PD	-	0.1941 (0.1338)
EV	1.7577 (3.1826)	9.4234 * (4.5536)
R	−0.6062 ** (0.1638)	−0.1654 (0.1859)
AV	5.7869 (7.3403)	2.8021 (9.3125)
S	−0.3296 (0.3630)	−0.4663 (0.4228)
Constant	34.7118 ** (7.7161)	8.6229 (8.8104)
F	5.34 **	3.49 *
R^2	0.4081	0.3678

Note: * significant at the 0.05 probability level; ** significant at the 0.01 probability level. *ADNC*: "actual donation amount without communication"; *ADC*: "actual donation amount with communication".

Table 16 shows that at a 99% confidence level, the earnings from the previous round had a significant impact on the actual amount of donations when there was no communication. For each unit increase in the previous round's income, the actual amount of money donated decreased by 0.6062 units when there was no communication. If there was communication, the estimated value of β had a significant effect on the actual donation amount at a 95% confidence level. For each additional unit of the estimated value of β, the actual donation amount increased by 9.4234 units when there was communication. The *F*-statistics were significant.

3.6.6. Experiment 6 (*C* & *NC*)

Taking the "actual donation amount without communication (*ADNC*)" and the "actual donation amount with communication (*ADC*)" as the dependent variables and the "planned donation amount (*PD*)", "return from the previous round (*R*)", "reward received from the previous round (*RP*)", "penalty received from the previous round (*PP*)", "returns after one round of reward and punishment (*RRP*)", "estimated value of β (*EV*)", "actual value of β in the previous round (*AV*)", and "degree of trust in strangers (*S*)" as the independent variables, a regression was implemented, the results of which are shown in Table 17.

Table 17. Regression analysis results for Experiment 6 (C & NC).

Independent Variables	Dependent Variables	
	ADNC	*ADC*
PD	-	0.8344 ** (0.0670)
R	−0.02810 (0.3152)	−0.0039 (0.1300)
RP	−0.03906 (1.2356)	0.2757 (0.6025)
PP	0.3405 (0.6826)	−0.4356 (0.3273)
RRP	0.1126 (0.1970)	−0.0865 (0.0952)
EV	15.9091 * (6.9543)	6.3467 (9.8157)
AV	65.7716 * (27.6774)	2.8950 (13.3299)
S	0.5950 (1.0832)	0.1638 (0.5813)
Constant	−12.5920 (11.7948)	3.9333 (7.2328)
F	2.43 *	28.50 **
R^2	0.3780	0.8941

Note: * significant at the 0.05 probability level; ** significant at the 0.01 probability level. *ADNC*: "actual donation amount without communication"; *ADC*: "actual donation amount with communication".

Table 17 shows that at a 95% confidence level, the estimated value of β had a significant effect on the actual donation amount when there was no communication. For each additional unit in the estimated β value, the actual donation amount increased by 15.9091 units when there was no communication. At a 95% confidence level, the actual value of β in the previous round had a significant impact on the actual donation amount when there was no communication. For each increase in the actual value of β in the previous round, the actual donation amount increased by 65.7716 units when there was no communication. At a 99% confidence level, the planned donation amount had a significant impact on the actual donation amount when there was communication. For every unit increase in the planned donation amount, the actual donation increased by 0.8344 units when there was communication. The *F*-statistics were significant.

4. Discussion and Conclusions

Two main issues were studied in this paper: (1) Which is better, cooperation or noncooperation? (2) What are the key factors affecting the private, voluntary supply of goods for forestry infrastructure? The following was found: (1) With private cooperation, the supply was higher than with private noncooperation. (2) Communication, environmental determination, information feedback, and reward and punishment mechanisms affected the quantity of private, voluntary supplies. (3) Communication could increase supply. (4) Reward and punishment mechanisms could also increase supply. (5) The impact of environmental certainty on supply depended on reward and punishment mechanisms. When there was no reward and punishment mechanism, environmental uncertainty increased the supply of goods for forestry infrastructure. When incentives and penalties were in place, a defined environment increased the supply of goods for forestry infrastructure. (6) The impact of information feedback on supply depended on environmental certainty. When the environment was determined, information feedback increased donations. When the environment was uncertain, no information feedback increased donations.

Choi and Ahn have written that rewards and punishments can increase supply [59]. This is consistent with our findings. Zhou and other scholars have indicated that environmental uncertainty could increase donations [58]. Zhou's experimental design considered leaders and did not consider reward and punishment mechanisms, which is not exactly what our experimental design did. Zhang found that information feedback could affect free-riding behavior [61]. Our research did not consider the method of information feedback, which is a disadvantage of our experimental design.

The conclusions of our experiments are instructive for the private, voluntary provision of quasi-public goods for forestry infrastructure. On the one hand, low-cost quasi-public goods can be provided by private, voluntary foresters in a forest. Meanwhile, cooperation between forest farmers should be promoted as far as possible. On the other hand, reward and punishment mechanisms ought to be more public, and members should communicate more.

Author Contributions: Conceptualization, C.W.; methodology, Y.Z.; data curation, L.Z.; writing—original draft preparation, L.Z.; writing—review and editing, Y.Z. and C.W. All authors have read and agreed to the published version of the manuscript.

Funding: This research was supported by the Fundamental Research Funds for the Central Universities, grant number 2017ZY61 and the National Natural Science Foundation of China, grant number 71573019.

Conflicts of Interest: The authors declare that there are no conflicts of interest.

Appendix A. Software for Experimental Design Implementation: Z-Tree

1. Z-tree user interface

1.1. Interface when entering Z-tree

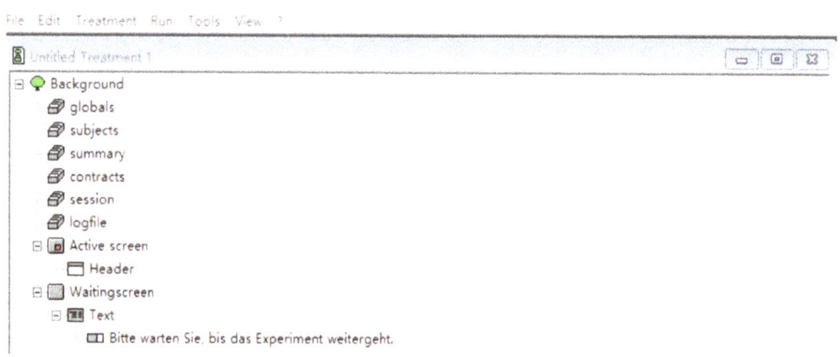

1.2. Interface when the Z-leaf is connected to the Z-tree

1.3. Design of experimental procedures

1.3.1. Program display of Experiment 1

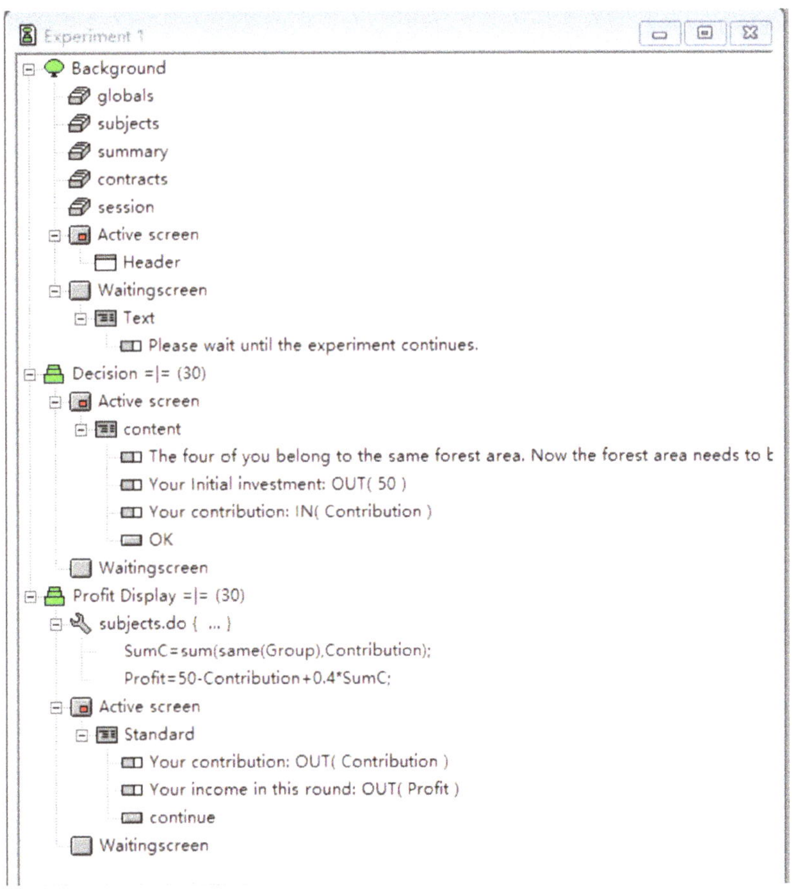

1.3.2. Experimental operation interface of Experiment 1

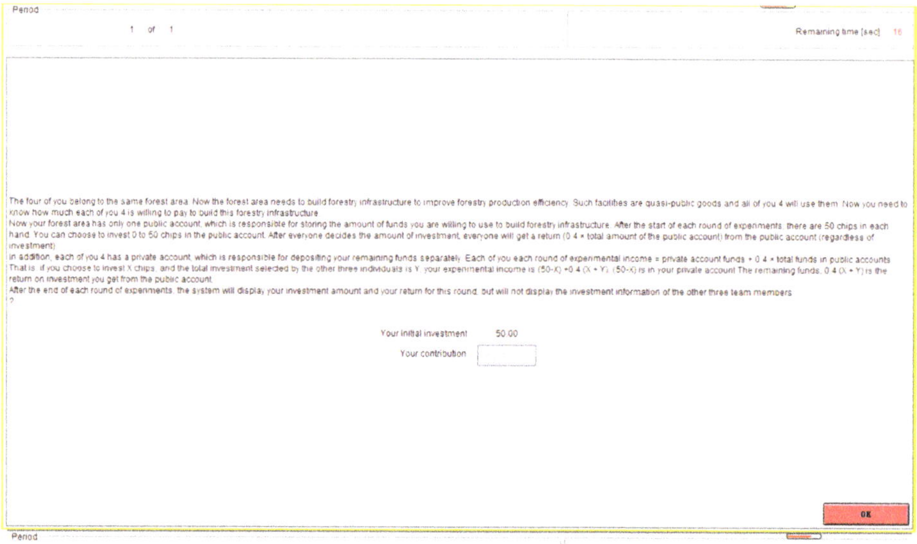

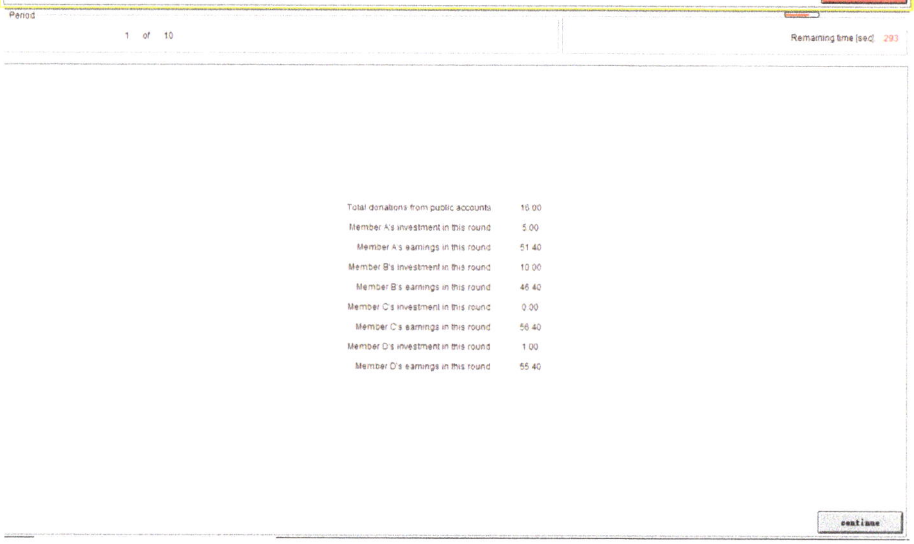

1.3.3. Description of Experiment 1

• Explanation of the experiment

You are welcome to participate in this experiment and we thank you for your cooperation.

The four of you belong to the same forest area. Now, the forest area needs to build forestry infrastructure to improve forestry production efficiency. Such facilities are quasi-public goods, and all four of you will use them. Now you need to know how much each of you four is willing to pay to build this forestry infrastructure.

Your forest area only has one public account, which is where the funds you are willing to use to build forestry infrastructure will be stored. After the start of each round of experiments, each person will have 50 chips. You can choose to invest 0 to 20 chips in the public account. After everyone has decided on the amount of his/her investment, everyone will get a return (0.4 × the total amount in the public account) from the public account (whether or not the investment is made), where 0.4 is the response rate. That is, if you choose to invest X chips, and the total investment amount chosen by the other three individuals is Y, then 0.4 $(X + Y)$ is the investment return you will get from the public account. In addition, each of the four of you has a private account, which is where your remaining funds will be deposited, namely (50 - X). After the end of each round of experiments, the system will display your investment amount and your return for this round, but will not display the investment information of the other three team members.

After the end of this round of experiments, the next round of experiments will be carried out. In each round of experiments, you will get 50 chips from the experimenter. We will carry out 10 rounds of experiments. After the 10 rounds of experiments are over, we will pay you cash for participating in this experiment. Your cash income = 0.1 × (appearance fee + experiment income). The entrance fee for the 10 rounds of experiments is fixed at 30, and it is paid only once. The experiment income = $(50 - X) + 0.4(X + Y)$.

During the experiment, it is strictly prohibited to communicate with the other participants, and your mobile phone must be turned on vibrate. If you have any questions, please raise your hand and we will answer your questions individually. Your compliance with the rules is very important, as otherwise we must exclude you from the experiment without giving you any compensation.

• Test questions

Before you start the experiment, you need to complete three test questions to make sure you fully understand the experimental rules. You can then make optimal decisions that maximize your revenue.

(1) If three other people—A, B, and C—each invest 10, 10, and 10 chips in the public account, and you also invest 10 chips in the public account, the return on investment of the public account is 0.4. Therefore, the remaining funds in the private accounts of each of you are (), (), (), and (); the gains from the public accounts are (), (), (), and (); and the experimental benefits are (), (), (), and ().

(2) If three other people —A, B, and C—each invest 10, 10, and 10 chips in the public account, and you invest 0 chips in the public account, the return on investment of the public account is 0.4. Therefore, the remaining funds in the private accounts of each of you are (), (), (), and (); the gains from the public accounts are (), (), (), and (); and the experimental benefits are (), (), (), and ().

(3) You invest 0 chips in the public account, and the other three persons (A, B, and C) each invest 5, 5, and 20 chips in the public account. The return on investment of the public account is 0.4. Therefore, the remaining funds in the private accounts of each of you are (), (), (), and (); the gains from the public accounts are (), (), (), and (); and the experimental benefits are (), (), (), and ().

References

1. Hagadone, T.A.; Grala, R.D. Business clusters in Mississippi's forest products industry. *For. Policy Econ.* **2012**, *20*, 16–24. [CrossRef]
2. Czajkowski, M.; Bartczak, A.; Giergiczny, M.; Navrud, S.; Zylicz, T. Providing preference-based support for forest ecosystem service management. *For. Policy Econ.* **2014**, *39*, 1–12. [CrossRef]
3. Atewamba, C.; Boimah, M. Policy forum: Potential options for greening the concessionary forestry business model in rural Africa. *For. Policy Econ.* **2017**, *85*, 46–51. [CrossRef]
4. Xu, H.W. On the application of forestry engineering cost management information. *China For. Prod. Ind.* **2015**, *42*, 56–57. [CrossRef]
5. Zhang, F.Y.; Du, J.; Kang, X.L.; Zhu, S.B. Empirical research on influencing factors and action mechanism of farmer's forestry income. *For. Econ.* **2017**, *39*, 15–22.
6. Flatberg, T.; Nørstebø, V.S.; Bjørkelo, K.; Astrup, R.; Søvde, N.E. A mathematical model for infrastructure investments in the forest sector of coastal Norway. *For. Policy Econ.* **2018**, *92*, 202–209. [CrossRef]
7. Svetlana, P.; Jussi, H.; Esa, V. Policy forum: Challenges of forest governance: Biomass export from Leningrad Oblast, north-west of Russia. *For. Policy Econ.* **2018**, *95*, 13–17.

8. Falcone, P.M.; Tani, A.; Tartiu, V.E.; Imbriani, C. Towards a sustainable forest-based bioeconomy in Italy: Findings from a SWOT analysis. *For. Policy Econ.* **2020**, *110*, 101910. [CrossRef]
9. Akay, A.E.; Sessions, J. Roading and transport operations. In *Encyclopedia of Forest Sciences*; Burley, J., Evans, J., Youngquist, J., Eds.; Elsevier Academic Press: Amsterdam, The Netherlands, 2004; pp. 259–269.
10. Enache, A.; Pentek, T.; Ciobanu, V.D.; Stampfer, K. GIS based methods for computing the mean extraction distance and its correction factors in Romanian mountain forests. *Šumar List.* **2015**, *139*, 35–46.
11. Hayati, E.; Majnounian, B.; Abdi, E. Qualitative evaluation and optimization of forest road network to minimize total costs and environmental impacts. *Forests* **2012**, *5*, 121–125. [CrossRef]
12. Laschi, A.; Neri, F.; Brachetti Montorselli, N.; Marchi, E. A methodological approach exploiting modern techniques for forest road network planning. *Croat. J. For. Eng.* **2016**, *37*, 319–331.
13. Parsakhoo, A.; Mostafa, M. Road network analysis for timber transportation from a harvesting site to mills (Case study: Gorgan county–Iran). *J. For. Sci.* **2015**, *61*, 520–525. [CrossRef]
14. Picchio, R.; Latterini, F.; Mederski, P.S.; Venanzi, R.; Karaszewski, Z.; Bembenek, M.; Croce, M. Comparing accuracy of three methods based on the GIS environment for determining winching areas. *Electronics* **2019**, *8*, 53. [CrossRef]
15. Feng, Q.L.; Qin, F.D.; Liao, H.J.; Zhao, F.H. An empirical study on the influencing factors of the quality of poverty alleviation work of the forestry industry in Guangxi. *For. Econ.* **2018**, *40*, 26–32.
16. Su, S.P.; Wu, J.Y.; Gan, J.B. Comparative analysis of total factor productivity change among family forestry operators since forest tenure reform in Fujian, Zhejiang and Jiangxi provinces. *Resour. Sci.* **2015**, *37*, 112–124.
17. He, D.H.; Zhu, D.L. The collective forest tenure reform implementation and performance evaluation: 2014 monitoring observations of collective forest tenure reform. *For. Econ.* **2015**, *37*, 13–27.
18. Picchio, R.; Spina, R.; Calienno, L.; Venanzi, R.; Lo Monaco, A. Forest operations for implementing silvicultural treatments for multiple purposes. *Ital. J. Agron.* **2016**, *11*, 156–161.
19. Gumus, S.; Turk, Y. A new skid trail pattern design for farm tractors using linear programing and geographical information systems. *Forests* **2016**, *7*, 306. [CrossRef]
20. Zhu, H.Y.; Zhao, J.Y. Analysis on Chinese herbal medicine planting subsidies pilots. *For. Econ.* **2015**, *37*, 61–64.
21. Xue, Y.Z. Study on forest zone disaster-reduction indicator system construction and countermeasures in western China. *Issues For. Econ.* **2015**, *35*, 133–137.
22. Han, J.H.; Lin, H.J.; Zhang, J.Z. Analysis of people's livelihood forestry developing in state forest zone: A study case in Chifeng forestry. *For. Econ.* **2015**, *37*, 75–79.
23. Chen, S.Z.; Bai, X.P. Support system for forestry-related disaster prevention and mitigation policy in Japan and its implications. *World For. Res.* **2014**, *27*, 67–74.
24. Song, W.Z.; Wang, A.J. Analysis of impacts and causes of ice and snow disasters: Taking bamboo forest in Zhejiang as an example. *For. Resour. Manag.* **2008**, *37*, 45–49.
25. Corona, P.; Ascoli, D.; Barbati, A.; Bovio, G.; Colangelo, G.; Elia, M.; Garfì, V.; Iovino, F.; Lafortezza, R.; Leone, V.; et al. Integrated forest management to prevent wildfires under Mediterranean environments. *Ann. Silvic. Res.* **2015**, *39*, 1–22.
26. Liao, B.; Gao, X.P.; Liao, W.M.; Zhang, G.L. A Comparative analysis on the structure of forestry fixed assets investment in Jiangxi. *Issues For. Econ.* **2014**, *34*, 154–159.
27. Sun, W.; Cao, S.S.; Pu, Z.; Fu, L.Y. Cloud infrastructure of forest resources information. *J. Northwest For. Univ.* **2014**, *29*, 71–79.
28. Pinkard, E.; Battaglia, M.; Bruce, J.; Matthews, S.; Callister, A.N.; Hetherington, S.; Last, I.; Mathieson, S.; Mitchell, C.; Mohammed, C.; et al. A history of forestry management responses to climatic variability and their current relevance for developing climate change adaptation strategies. *Forestry* **2015**, *88*, 155–171. [CrossRef]
29. Wang, Z.X.; Yan, H.W.; Mo, M. Some thoughts on forestry recovery and reconstruction after rain and snow freezing in the south of China. *For. Resour. Manag.* **2008**, *37*, 1–5.
30. Wu, B. Some Thoughts on forestry and ecological recovery and reconstruction in the southern rain and snow freezing areas. *Sci. Silvae Sin.* **2008**, *54*, 2–4.
31. Kong, F.B.; Liao, W.M. Forestry development effects of forestry marketization process in China: Based on statistical data of 31 Provinces (cities and autonomous regions) during 2002 to 2011. *Sci. Silvae Sin.* **2013**, *49*, 126–135.

32. Xiao, Z.; Chen, Y.G.; Zhou, Y. Dynamic relationship between public service investment and forestry economic growth: A study based on OLS regression initial model. *Issues For. Econ.* **2012**, *32*, 177–184.
33. Gordon, T. The economics of public choice/perspectives on public choice: A handbook. *Econ. J.* **1998**, *108*, 1886–1888.
34. Krauze, A. Let the public sector go free. *New Statesman* **2000**, *129*, 5.
35. Shen, M.H.; Xie, H.M. On the issue of public goods and its solution: A review on the public goods theory. *J. Zhejiang Univ. (Hum. Soc. Sci.)* **2009**, *39*, 133–144.
36. Li, S.S. An analysis of the disadvantages and countermeasures of the government's provision of public goods. *Party Sch. Le Shan Munic. Comm. C.P.C.* **2011**, *13*, 45–47.
37. Henry, H. Ownership of the firm. *J. Law Econ. Organ.* **1988**, *4*, 267.
38. Savas, E.S. *Privatization and Public-Private Partnerships*; Renmin University of China Press: Beijing, China, 2002; pp. 57–321.
39. Fu, J.L.; Cui, H.; Huang, X.H. Farmers' voluntary provision of public goods in rural communities: Analysis based on the theory of reputation. *Econ. Surv.* **2007**, *24*, 106–109.
40. Hong, B.G.; Zhang, Y.B. Rent-seeking game analysis on quasi-public goods supplied by private in rural areas: Rural passenger transport system as an example. *J. Xiangtan Univ. (Philos. Soc. Sci.)* **2009**, *33*, 44–48.
41. Xiao, W.; Zhu, Y.Z. Game analysis of the supply of rural public goods on the basis of contracts: Taking the rural areas of Hunan province as an example. *Chin. Rural. Econ.* **2010**, *29*, 26–36.
42. Liu, B.X.; Liu, W.L. Game analysis of private supply mechanism of quasi-public goods: Taking Chinese transportation infrastructure investment as an example. *China Soft Sci.* **2007**, *22*, 145–151.
43. Liu, C.L.; Xu, Y.Q. The application of game model of public goods in information resources sharing in China. *Inf. Sci.* **2007**, *28*, 50–53.
44. Dong, Z.Y. *Experimental Economics*; Peking University Press: Beijing, China, 2008; pp. 8–11.
45. Rami, S.; David, V.B.; Amnon, R. Provision of step-level public goods with uncertain provision threshold and continuous contribution. *Group Decis. Negot.* **2001**, *10*, 253–274.
46. Jane, S.; Rick, K.W. Levels of information and contributions to public goods. *Soc. Forces* **1991**, *70*, 107–124.
47. Weimann, J. Individual behavior in a free riding experiment. *J. Public Econ.* **1994**, *54*, 185–200. [CrossRef]
48. Croson, R.T.A. Feedback in voluntary contribution mechanisms: An experiment in team production. *Res. Exp. Econ.* **2001**, *8*, 85–97.
49. Bigoni, M.; Suetens, S. Feedback and dynamics in public good experiments. *J. Econ. Behav. Organ.* **2012**, *82*, 86–95. [CrossRef]
50. Irlenbusch, B.; Ter Meer, J. Fooling the nice guys: Explaining receiver credulity in a public good game with lying and punishment. *J. Econ. Behav. Organ.* **2013**, *93*, 321–327. [CrossRef]
51. Huang, G.B.; Zhou, Y.A. Review of leadership research in experiment economics. *Econ. Theory Bus. Manag.* **2014**, *34*, 48–59.
52. Zhou, Y.; Zhang, Q.L.; Fu, L.; Yang, Y.Z.; Sun, G.Y. Effect of information disclosure mechanism on free riding behavior: Evidence from laboratory. *J. Manag. Sci. China* **2014**, *17*, 86–94.
53. Nikos, N. Feedback, Punishment and cooperation in public good experiments. *Games Econ. Behav.* **2010**, *68*, 689–702.
54. Faillo, M.; Grieco, D.; Zarri, L. Legitimate punishment, feedback and the enforcement of cooperation. *Games Econ. Behav.* **2013**, *77*, 271–283. [CrossRef]
55. Xie, S.Y. *Economic Game Theory*; Fudan University Press: Shanghai, China, 2017; pp. 23–24.
56. Klaus, R. *Foundations of Non-Cooperative Game Theory*; Oxford University Press: Oxford, UK, 2003; pp. 7–8.
57. Guo, X.J. Research on the Problem about the Private Provision of Quasi-Public Goods in China from the Perspective of Game Theory. Master's Thesis, Shanxi University of Finance & Economics, Shanxi, China, 2016.
58. Zhou, Y.A.; Huang, G.B.; He, H.R.; Liu, M.W. Can leaders really serve as role models: A study based on public goods game experiments. *Manag. World* **2014**, *30*, 75–90.
59. Choi, J.-K.; Ahn, T.K. Strategic reward and altruistic punishment support cooperation in a public goods game experiment. *J. Econ. Psychol.* **2013**, *35*, 17–30. [CrossRef]

60. Hai, J.J. Research on Cooperative Behavior in Public Goods Game Experiment under Leadership-chasing Random System. Ph.D. Thesis, Shan Dong University, Shandong, China, 2018.
61. Zhang, J.B. Free Rider Behavior of Public Goods Based on Experimental Economics. Master's Thesis, Beijing University of Posts and Telecommunications, Beijing, China, 2014.

sustainability

MDPI

Article

Rut Depth Evaluation of a Triple-Bogie System for Forwarders—Field Trials with TLS Data Support

Michael Starke [1,*], Cédric Derron [1], Felix Heubaum [2] and Martin Ziesak [1]

1 School of Agricultural, Forest and Food Sciences (HAFL), Bern University of Applied Sciences, 3052 Zollikofen, Switzerland; Cedric.Derron@students.bfh.ch (C.D.); Martin.Ziesak@bfh.ch (M.Z.)
2 Staatsbetrieb Sachsenforst, Maschinenstation Königstein, 01824 Königstein, Germany; Felix.Heubaum@smul.sachsen.de
* Correspondence: Michael.Starke@bfh.ch

Received: 19 June 2020; Accepted: 7 August 2020; Published: 10 August 2020

Abstract: In 2019, the machine manufacturer HSM presented a forwarder prototype for timber hauling in cut-to-length processes fitted with a new 10-wheel triple-bogie (TB) setup approach aimed at promoting sustainable forest management by reducing the ecological impact of forest operations, especially under soft-soil working conditions. The purpose of our study was to assess the resulting soil-protection effect emerging from additional wheel-contact surface area. For this, the rut development under known cumulative weight, related to the soil conditions of shear strength and moisture content, was recorded for later comparison. Additional terrestrial laser scanning (TLS) was used to generate a multi-temporal digital terrain model (DTM) in order to enhance the data sample, assess data quality, and facilitate visualization of the impact of local disturbance factors. In all TB configurations, a rut depth of 10 cm (5.8–7.2 cm) was not exceeded after the hauling of a reference amount of 90 m^3 of timber (average soil shear strength reference of 67 kPa, volumetric water content (VMC) 43%). Compared to a reference dataset, all observed configurations ranked in the lowest-impact machine categories on related soil stability classes, and the configuration without bogie tracks revealed the highest machine weight to weight distribution trade-off potential.

Keywords: forest operation; forwarder; cut-to-length; soft soil; soil protection; rut depth; TLS data

1. Introduction

Since the introduction of larger machines in forestry, their effect on the environment has been a continuously debated topic. Soil protection is a particular point of concern [1]. With increased awareness and defined restrictions, the surrounding conditions for forest work have become more important, as global warming forces forestry contractors to work under suboptimal conditions outside the frost period [2]. This has also started to affect regions that historically could rely on frozen soils in certain time periods during the year [3,4]. Normally, workers would interrupt harvesting during unsuitable weather conditions but with the increasing absence of good working conditions, amplified by high machine costs, they must continue their work for economic reasons [5]. In order to continue supplying timber with a tolerable ecological impact in the context of sustainable forest management (SFM), different machine designs, that are further described below, are gaining interest, in addition to such operational solutions as the intense use of brush mats to reduce soil impact [6].

When machine configurations without bogie tracks are discussed, the enlargement of the soil contact area is generally divided into three main machine configuration-related solutions with different soil protection effects: the number of tires on the machine (effect 55%), the air pressure in the tires (35%), and the tire width (10%) [7]. The use of a low tire-pressure machine setup, along with the number of tires, is a particularly suitable approach [8–10]. However, the resulting change in machine weight must always be kept in mind as a trade-off parameter [2].

Other basic trends to achieve a high soil-protection effect are the use of bogie tracks [11] or a switch to a tracked undercarriage of the machine. A possible extension of the bogie axle length should also be noted here [12,13]. Specially designed machines for soft soil conditions such as the Elliator [14] or the Ponsse "OnTrack" prototype [15] are existing examples of tracked forwarder setups.

With the now common switch from 6- to 8-wheeled machines, the addition of another pair of wheels seems a logical development. Hence, also 10-wheeled forwarder setups are not a new development per se [2], and several variants have already been tested with proven positive effects.

One such market-available solution is provided by Ponsse with its "10w setup". Based on the Wisent, Elk, and Buffalo forwarder model types, the larger carriage designed models have the option to vertically adjust a third pair of wheels at the rear axle of the machine, whereas the wheel of the lighter model remains in a fixed position [16]. One three-axle bogie, or triple-bogie (TB), solution considers the trade-off between the energy efficiency of the machine and the friction, in terms of machine-soil interaction, resulting in the invention of a new rocker design of the bogie axle that only uses the maximum contact surface of the additional pair of wheels at higher torque requirements and thus only when needed [17,18].

Another solution has recently been developed by "Hohenloher Spezial Maschinenbau GmbH" (HSM) with its TB system. This 10-wheeled forwarder setup is based on a HSM 208F 10 t machine [19] and provides the option to be used either with specially designed bogie tracks for extremely soft soil conditions or with low pressurized "BigFoot" tires with three actively driven wheels.

To compare the resulting soil protection effect of different machinery, an assessment method is needed. There are various approaches to evaluating the impact of a forest machine on sensitive soils [20]. The first methodological option consists of calculation-based estimations, e.g., by determining the nominal ground pressure (NGP) of a forest machine [21]. Furthermore, more accurate calculations such as the mean maximum pressure (MMP) [22] of a certain machine can be included in order to derive the soil impact from calculated machine characteristics. Dynamic local tire-soil effects make this approach not only very difficult to implement, but also potentially inaccurate under certain conditions (ibid.), with the necessary input data sometimes hard to acquire [20].

Another possibility is to measure the impact of the machine in describing the soil characteristics, e.g., with the bulk-density change over time during the harvest operation [22–25]. The machine impact can thereby be distinguished between shearing effects caused by traction loss and direct compaction [26], both of which lead to ruts in the driving lanes. As the bulk density cannot be visually observed in the field and does not explain technical accessibility, guidelines for practical use focus on the visually detectable rut depth and its shape, related to previously described soil parameters [27].

Related research uses various methods of describing and measuring this rut expression. For fast or larger area observations, it can be separated by classes, where the rut depths or the moved soil volumes have been determined visually [28,29], with photogrammetric methods (such as an unmanned aerial vehicle (UAV)-based evaluation method [30]), or have been measured with a traditional terrestrial or newer portable laser scanners [26,31,32]. To achieve a higher resolution for the comparison of different machine types, other studies have manually recorded the rutting development [33] and checked its correlation to the cumulative weight consisting of machine weight, load weight, equipment weight, and the number of transits [2,34–36]. Because manual measurements are very labor-intensive [31], combined approaches have been developed using close-range photogrammetry and multi-temporal digital terrain models (DTM) to measure the rut developments with higher resolution [8,37,38]. This method has already been successfully used with additional consideration of the number of machines passes with known impact weight, thus offering additional possibilities of data recording and evaluation [8,38,39].

In such evaluations, it is important to record the related soil parameters as influencing factors [20,40]. Soil parameters that can be used to rank soil stability are shear strength (τ) [2,38,40], penetration resistance, or E-module [41]. Shear-strength classification may have higher replicability and objectivity compared to the cone penetrometer test for determining penetration resistance [42].

A remaining problem with soil characterization using the cone penetrometer is the occurrence of skeletal content, which can influence the reliability of the data [35]. A similar effect may arise during manual measurements with a lower number of sample points caused by the root system of the trees [36]. Nevertheless, both measuring setups are regarded as common and reliable techniques to characterize physical soil parameters [40] and are also used in combination [23]. The related (volumetric) soil moisture content and the Unified Soil Classification System (USCS) or similar soil type description is valuable additional information as they both directly influence the soil characteristics [20].

Based on the previously mentioned methodological elements, applications for comparing the rut depth development of multiple different machines do exist [36,42], which can be used to rate our own results. In one mentioned application [36], the rutting development of different forwarder bogie-track setups were manually recorded on flat terrain during the hauling process under predominantly soft soil conditions. For this, a Valmet 860.4 base machine was equipped with different bogie-track types. To illustrate the overall impact on the skid trails, the cumulative weights (consisting of machine weight, load weight, and the weight of the bogie tracks) of each variant were recorded. Furthermore, the respective soil conditions were documented in order to classify soil conditions by shear strength in three classes [36] (Table 1).

Table 1. "Sachsenforst" reference soil stability classes for the Leipzig region (Wermsdorfer Wald) for mean soil characteristic values [36].

Soil Stability Class	Description	Mean Shear Strength τ [kPa]	Mean Volumetric Soil Moisture Content (VMC) [%]
SC1	Less sensitive soils	124	39
SC2	Sensitive soils	91	46
SC3	Very sensitive soils	67	48

The soil moisture ranges were then used to cluster the data as a further common influence factor [20,33,41,43,44].

This following study investigates the newly developed HSM TB system configurations with regards to their rutting effects on predominantly very sensitive soil conditions. It is therefore compared with existing 8-wheel and bogie-track setups from direct measurements and additional reference data available for similar soil conditions. Terrestrial laser scanning (TLS) data will be used to visualize small-scale influences that may distort rutting effects and to directly support the selected manual measurement setup, including data-quality assessment and additionally recorded samples.

2. Materials and Methods

The study was performed in Saxony, Germany, near the cities of Tharandt (50.94° N, 13.49° E) and Schönheide (50.47° N, 12.51° E), in two field trials and in combination with a pre-test that was made to test the functionality of the machine setup and to assess the applicability of the study layout [45]. In the second campaign (Schönheide), a further developed machine setup was finalized to be tested. All measurements were taken in springtime to guarantee high soil water saturation. The selection of the testing areas was based on an assumed soil stability classified as B3 of the Saxonian soil protection guideline: "Trafficability is severely limited by weather conditions" [46], which makes it comparable with the requirements of the additional reference data [36]. The measurement points (Figure 1) were selected inside forest stands on different, existing skid trails that had initially no visible soil disturbance from previous timber hauling or harvesting operations that had been carried out several years earlier. To observe a rut deepening effect, soil stress was then provoked by multiple passes over the area of interest with a loaded machine in a straight line while the changes were recorded (Table 2). For the layout documentation, the measuring points were named according to the tested machine configuration (1–4), the recording type when multiple methods were used (a: TLS; b: manual), the recording cycle (A, B), and the consecutive number of measuring points per setup (e.g., 1a_A1).

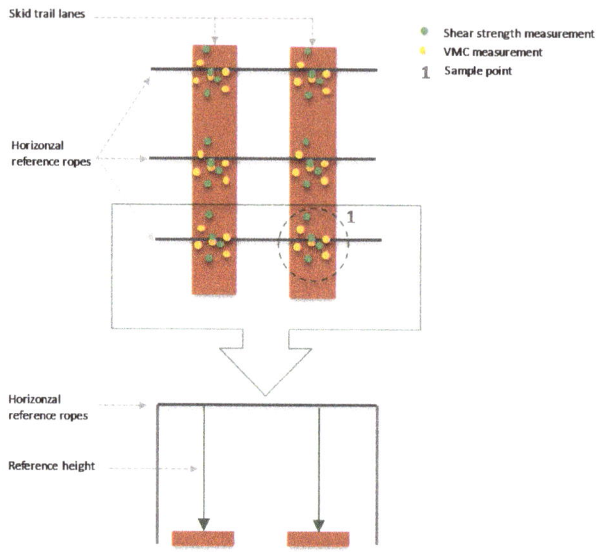

Figure 1. Measuring point, consisting of max. eight sample points for the manual measurements in horizontal (top) and vertical (bottom) schematic view (changed after [33,42,44,47]).

Table 2. Overview of the data recording sites for the rut-depth evaluation.

	Tharandt (Saxony, GER)					Schönheide (Saxony, GER)				
Recording Date	3–6 April 2018					5–11 April 2019				
Manual Measurements	X					X				
Terrestrial Laser Scanning (TLS) Measurements	X					-				
Soil Type [48]	Stagnosol					Gleysol, Cambisol				
Soil Compartments of the Predominant Soil Type [49]		Sand [%]	Silt [%]	Clay [%]	Skeletal Comp. [%]		Sand [%]	Silt [%]	Clay [%]	Skeletal Comp. [%]
Soil Horizon [max. depth]	I [12 cm]	20	40	40	10.5	I [8 cm]	64.6	26.8	9.6	23.3
	-	-	-	-	-	II [15 cm]	57.9	32.5	9.6	31.8
	II [30 cm]	8.8	48.6	42.6	17.9	III [27 cm]	55.8	33.6	10.6	23.9
	-	-	-	-	-	IV [51 cm]	55.0	35.0	10.0	35.6
	III [60 cm]	21.7	54.6	23.7	80.0	V [75 cm]	56.3	32.4	11.4	46.1

In advance of the trials, the selected sample points were cleared of organic litter to reveal the mineral soil. Shear strength and moisture content were then determined with six measurements in a 1 m^2 area around the sample point on the assumed driving lane directly in advance of the measurements (Figure 1), which defines the minimum, longitudinal sample point spacing. The data were collected manually with a shear-strength meter (varying depth of mainly 15 cm) and a time-domain reflectometry (TDR) probe to determine soil moisture as volumetric water content (VMC) with 20 cm-long probe forks.

To measure the rut depth, the vertical distance between the horizontal rope (Figure 1) was taken after each pass of the machine for at least 10 passes. The deepest point in the middle of the lane (inner 2.5 cm range) was measured for each sample point during the "Tharandt" measurement campaign and at the highest point in the middle of the lane during the second measurement campaign "Schönheide". This change of measurement principle promised smaller disturbances for the second measurement setup.

To create a multi-temporal DTM that is further used to enhance the data collected by manual measurements, a FARO Focus 3D X330 terrestrial laser scanner (TLS) was positioned outside the lane of the skid trail at four measurement points during the "Tharandt" campaign. With this setup, a scan of each measuring point was made after every second pass of the machine in the frequency [1,3,5,7,9,11] for the measurements 1a and [2,4,6,8,10] for the measurements 2a (see also Table 3).

Table 3. Overview of the collected data according to different machine configuration setups.

Machine Configuration Number	Configuration Description	No. of Measuring Points (TLS Datasets)	Related Machine Weight *with Tracks* and (Load) [t]	No. of Passes for 90 m³ (Cumulative Weight)	Location
1 (1a TLS, 1b manual data)	Triple-bogie incl. tracks	2 (2)	21.9 + *8.8* (+13)	1a:7 (305.9), 1b:6 (262.2)	Tharandt (Saxony, GER)
2 (2a TLS, 2b manual data)	Triple-bogie high tire air pressure (3 bar) (BigFoot)	2 (2)	21.9 (+13)	6 (209.4)	Tharandt (Saxony, GER)
3	Triple-bogie low tire air pressure (BigFoot)	7	21.9 (+10.8)	7 (228.9)	Schönheide (Saxony, GER)
4	8-wheeled reference machine (HSM 208F 710 mm-high tire pressure (3.5 bar))	3	16.9 (+10.8)	7 (193.9)	Schönheide (Saxony, GER)

The sample point validity of the evaluated manual data series was defined by an upper limit of a mean shear strength of 80 kPa out of six measurements, to focus on the average soil sensitivity range of 67 kPa on the fully manually recorded measuring points [36]. For the TLS measurements, the higher shear-strength data series of the directly related measurement points remained in the study as the samples are drawn in a continuous manner, cannot be separated, and are thus part of the measurement principle.

To compare the TLS data with the manual measurements, the changes in rut geometry over time were calculated. The first scan without impact was regarded as the initial situation. For each subsequent scan, the absolute difference from the initial situation was calculated in z-direction with a 6-kNN algorithm for the scan points using the cloud distance functionality of "CloudCompare" open-source software (Version 2.11 beta (Anoia)) and stored as a scalar field.

The data samples for evaluation were then taken as a longitudinally oriented segment for both lanes on the skid trail, covering the deepest point of the lane. As the visibility and the point density of the laser scan lose resolution with increasing distance from the scanning center, the sample length (Figure 2) was visually estimated to be reliable on a 7.8 m length of observation. The sample width of 20 cm ensured a positioning in the main center of the lane, thus allowing lateral machine movement without influencing the measurement point.

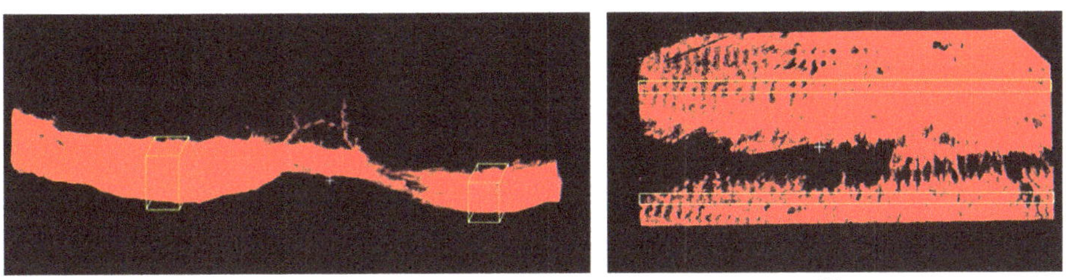

(**a**) (**b**)

Figure 2. Schematic presentation of samples taken from the TLS datasets in (**a**) cross-section and (**b**) bird's eye view: measurement point length 7.8 m and width 0.2 m.

The technical equipment investigated is a TB axle designed by HSM as a modular system to be used in exchange with a conventional bogie axle.

The total length of the bogie axle is 4528 mm (Figure 3) and is connected to the machine at the angle point of the center of the middle wheel and is thus mounted without suspension and not as a real bogie functionality setup. All wheels are actively driven and thus allow the use without bogie tracks and high traction values at the same time. The machine is designed to be equipped with "BigFoot" tires (Firestone 54 × 37.00 − 25, 940 mm [50]), but with the option to switch to lower tire widths.

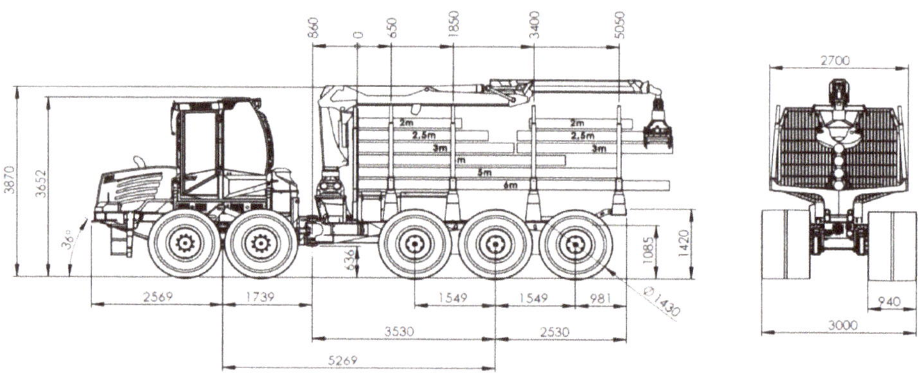

Figure 3. Schematic presentation of the Forwarder2020 base machine with the TB-BigFoot configuration [19].

Without bogie tracks, the TB axle setup has a determined weight of 21.9 t, including other prototype components, such as the enhanced hydraulic setup for energy recovery [51], that were installed on the machine. The definite weight of the series machine is therefore expected to be lower than the prototypes tested in this study and is more likely to approach the listed and communicated manufacturer's weight of 16 t [52].

An additional component of the machine is the specially produced bogie track for soft soil conditions. Clark TXL 150 TerraXlite bogie tracks were used with a special design to fit BigFoot tires. They were used in a symmetric design on the TB axle (width: 1345 mm, weight: 5740 kg) and in an asymmetric design on the front axle (width: 1175 mm, weight: 3058 kg).

The measurements were carried out with different machine setups. The TB prototype was tested in three different configurations: with the use of bogie tracks, with the standard tire pressure (also suitable for the use of tracks), and with a lowered tire pressure to achieve further soil protection effects. Table 3 shows all configurations, the number of collected datasets, and the measuring location.

The low tire pressure setup tested in this study was that specified in the user's manual as the lowest tire pressure configuration authorized for this machine setup without specially designed rims. One manufacturer's concern in this setup was related to the middle tire of the axle. As load peaks were expected when the machine passes over an obstacle without suspending the middle wheel, they advised that the pressure should stay above 2.1 bar. The final tire pressure setup is shown in Figure 4.

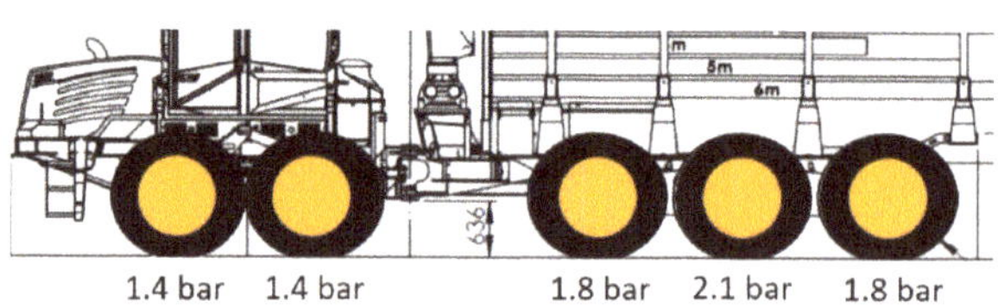

Figure 4. Tire pressure setup of the low-pressure measurements.

To compare the machine's behavior with conventional setups, an HSM 208F 10 t 8-wheeled forwarder, equipped with standard 710/45-26.5 tires with 3.5 bar tire pressure, was also observed. The base weight of this machine was 16.9 t [53].

In addition to the configuration setup, the machine carried a common load, defined as "just below" the stanchion height of the machine when equally loaded. The respective weight was measured by weighing small bundles of logs that were choked with a sling and lifted with the forwards crane to which a mobile crane scale was attached. The mobile crane scales that were used have an accuracy of at least 10 kg per load under static conditions. The loads used were thus 13,010 kg (15.7 m^3, 46 logs) for the Tharandt dataset and 10,750 kg (12.9 m^3, 36 logs) for the Schönheide dataset.

All datasets were evaluated with the statistic software R [54]. Due to the high number of samples in some data series, the normality assumption was assumed as given. For lower n-cases, a visual assessment followed by a Shapiro-Wilk test was performed. For the data series comparison of the different measuring setups, the data series were described by the measured shear strength and soil moisture, and a comparison of the rut depth at referenced 90 m^3 hauled timber was conducted. To prove the significance of different groups, an exact Wilcoxon Rank-Sum test without *p*-value correction was made.

3. Results

3.1. Data Quality Assessment of the Manual Rutting Development Measurements with TLS Data Enhancement

Figure 5 visualizes the TLS data for the optical recognition of small-scale influence parameters and the homogeneity of the data points. In Section 2a (Figure 5), the shadowing effect of the sidewise scanner setup is visible. The depth-value histogram of the 11th pass of the machine still shows a left skewed data series, supporting the theory that there is seemingly still no influence of the data quality on the validity of the data, and no higher value recordings are missing.

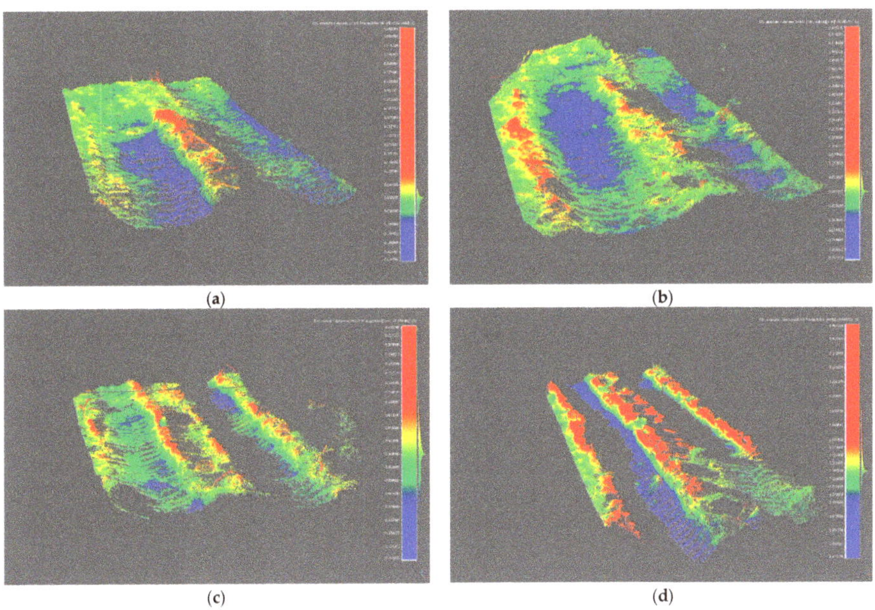

Figure 5. Visualization of the TLS measurement points after 11 passes, showing the absolute difference in *z*-direction (blue: 10 cm deepening, red: 10 cm warping color thresholds): 1a ((**a**) 1a_A1, (**b**) 1a_A4), 2a ((**c**) 2a_B1, (**d**) 2a_B4).

Within series 1a, the deepening effect changes locally in both visualizations but seems to have a higher expression in the 1a_A4 data series. The 2a_B4 series also shows a region of higher soil stability but is combined with extreme rut depths in the same lane and so displays a high inhomogeneity of the measurement point.

Figure 6 illustrates the TLS and manual data quality in direct comparison for two measuring points. Negative values are caused by local warping effects and objects that moved on the scan surface area. As in Figure 5, the variation of the two measuring points under the same soil types with comparable soil conditions is visible. The manual datasets and the TLS results thereby show the same tendencies towards deepening on the related measurement points but with a higher linearity in the TLS data.

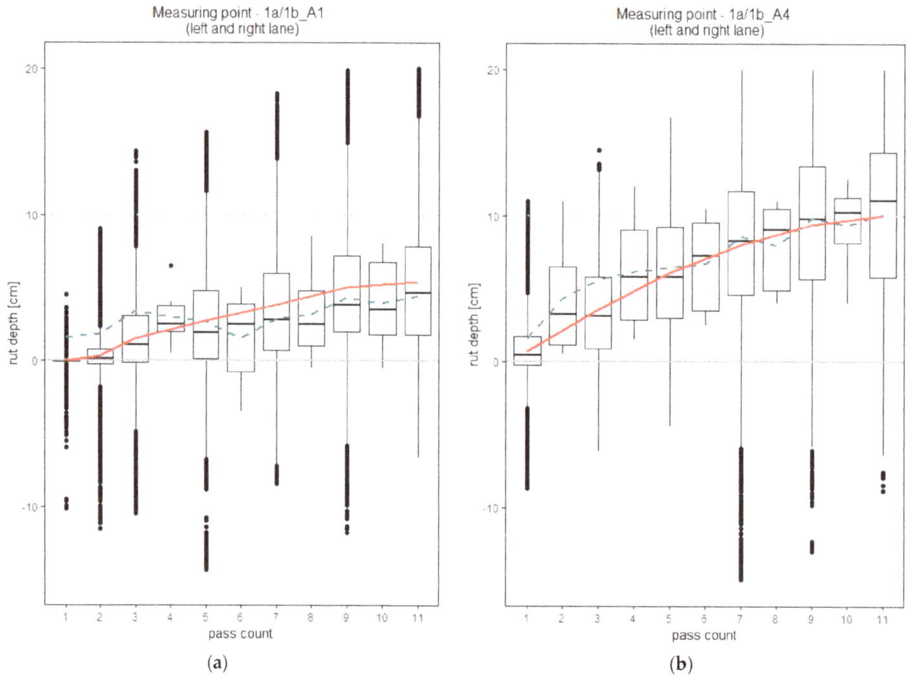

Figure 6. TLS results (red line, pass no: 1,2,3,5,7,9,11) of the measuring points (**a**) A1 (n = 12, τ = 61 kPa, sd_τ = 15 kPa, VMC = 43%, sd_{VMC} = 1.1%) and (**b**) A4 (n = 12, τ = 58 kPa, sd_τ = 24 kPa, VMC = 46%, sd_{VMC} = 5.6%), showing the related rutting effect on point 1a_A1 ($\Delta z_{mean\ pass\ 7}$ = 3.79 cm, $sd_{pass\ 7}$ = 4.21 cm, $SEM_{pass\ 7}$ = 0.01 cm (0.3%), 1a_A4 ($\Delta z_{mean\ pass\ 7}$ = 8.01 cm, $sd_{pass\ 7}$ = 4.56 cm, $SEM_{pass\ 7}$ = 0.01 cm (0.2%)) and the associated manual data (blue, dashed line, pass no: 1–11, 1b_A1 (n = 6 × 11, $\Delta z_{mean\ pass\ 7}$ = 2.83 cm, $sd_{pass\ 7}$ = 3.13 cm, $SEM_{pass\ 7}$ = 1.28 cm (45%)) 1b_A4 (n = 6 × 11, $\Delta z_{mean\ pass\ 7}$ = 8.58 cm, $sd_{pass\ 7}$ = 4.02 cm, $SE_{\ pass\ 7}$ = 1.64 cm (19%))).

Table 4 provides an overview of all data collected. At 76 kPa, the soil strength value of the 2a machine configuration lies 20% above the next comparable data series and is characterized by a relatively high mean shear strength value. The related dataset of manual measurements (2b) shows a lower mean value of 53 kPa and so focuses more on softer soil conditions. All other measurement point characterizations lie within the targeted range of soil classification.

Table 4. Data description of the measuring point characteristics (TSL points and related soil characteristics) characteristics in combination with machine configurations.

Machine Configuration Number	Measuring Point Name or Count of Points	Number of Sample Points	Number of Observations TLS	Related Shear Strength [kPa] Incl. (n, sd)	Related Soil Volume Moisture Content [%] Incl. (n, sd)
1a	A1, A4	12	2.03 m	63 (68, 24)	43.2 (70, 5.8)
1b	4 (count)	16	-	58 (98, 20)	45.3 (94, 5.7)
2a	B1; B4	12	0.57 m	76 (72, 27)	44.3 (71, 7.7)
2b	3 (count)	6	-	53 (36, 18)	48.8 (36, 5.4)
3	7 (count)	29	-	55 (168, 20)	38.3 (168, 8.9)
4	3 (count)	23	-	54 (138, 22)	34.2 (138, 9.0)

The results of the mean rut-depth development after each pass of the machine, and so the configuration characteristics and the machine behavior on the different machine setups (as described in Table 3), are visualized in Figure 7; Figure 8. Figure 7 shows the rut-depth development with the use of bogie tracks and without bogie tracks set with 3-bar tire pressure for the Tharandt measurement site. In Figure 8, the datasets collected in Schönheide are visualized. These series show the values of manual measurements of the TB setup related to the lower tire pressure adopted and the results of the 8-wheeled reference machine. Concerning the data quality of these manual measurement series, no issues are visible.

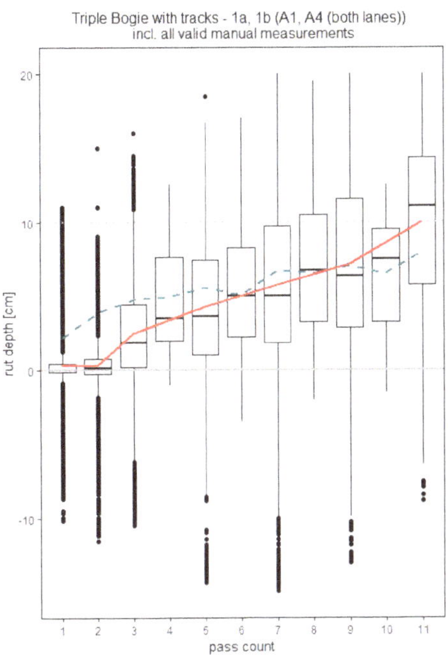

(a)

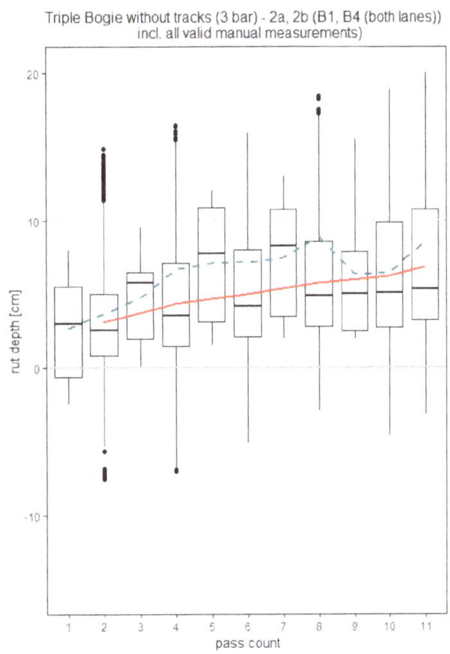

(b)

Figure 7. Rut-depth development data of TLS and manual measurements for each pass of the triple-bogie configuration (**a**) with bogie tracks and (**b**) without bogie tracks; the mean of the series values are visualized as a red line for the TLS data (row 1a left (n = 2.03 m), row 2a (right, n = 0.57 m)) and for all valid manual measurements as a dashed blue line (row 1b ((**a**), n = 16 × 11) and row 2b ((**b**), n = 6 × 11)).

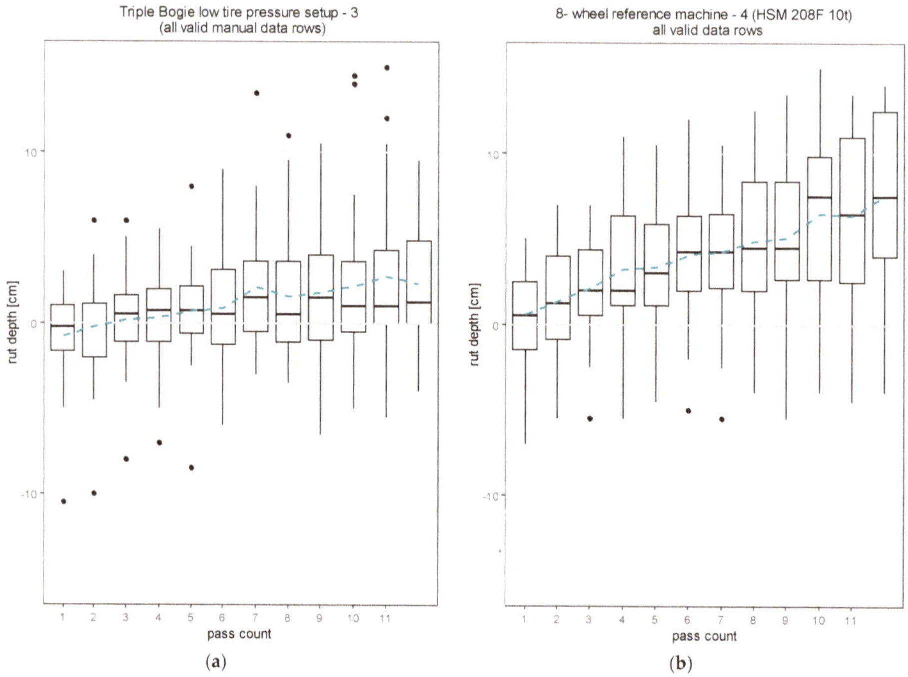

Figure 8. Rut-depth development data of manual measurements for each pass of the (**a**) triple-bogie configuration with lowered tire air pressure and (**b**) the results of the reference machine; the mean of the series values are visualized as a dashed blue line (row 3 ((**a**), n = 29 × 11) and row 4 ((**b**), n = 23 × 11)).

3.2. Comparison of Machine Setups

To enable a comparison of the different machine setups, the rutting effect is shown for all setups at a reference point of 90 m^3 of timber hauled (Figure 9). As the loads and the machine configurations differ between the setups, the soil-stress level varies according to the reference weights listed in Table 3. The TLS dataset reference was changed due to the scanning layout being performed every second time. Therefore, the closest number of passes was missing and was substituted by the next higher value, representing 91 t of load.

The normality assumption was not rejected for all tested manual datasets (Shapiro-Wilk test $p < 0.05$), but the according qqPlots showed additional anomalies. The following group comparison (Wilcoxon Rank-Sum) shows that the groups differ significantly ($p < 0.05$). Table 5 shows the p-values of the pairwise comparison of all valid data series. In contrast to the Tharandt machine configurations, where no directly significant difference of rut expression on the same testing area is given, the Schönheide data series shows significant differences on the same testing area.

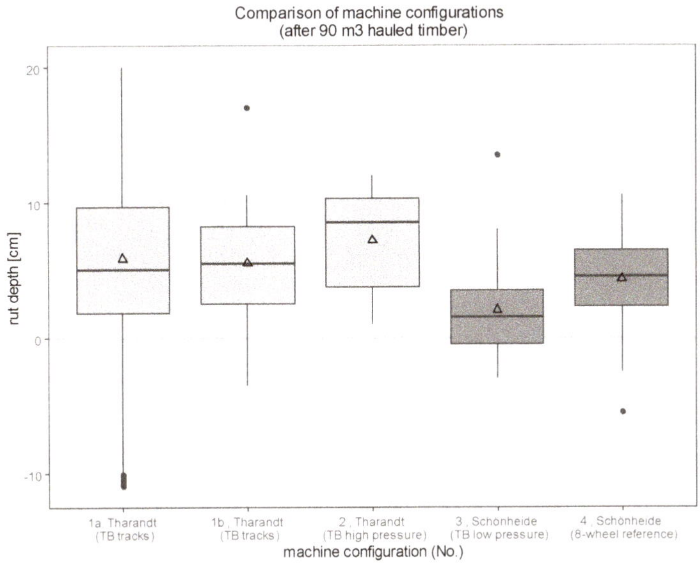

Figure 9. Machine setup comparison incl. the visualization of the mean value (triangle) and the median (horizontal line) of the tracked machine (TLS data: 1a (Δz_{mean} = 5.8 cm for 7 passes)), manual data: 1b (Δz_{mean} = 5.5 cm), 2: Triple-bogie (TB) with 3 bar tire air pressure (Δz_{mean} = 7.2 cm), 3: Triple-bogie with low tire pressure setup (Δz_{mean} = 2.1 cm), 4: 8-wheeled reference machine (Δz_{mean} = 4.3 cm), clustered by soil types (location 1, blue: τ_{mean} = 53–63 kPa, VMC_{mean} = 43–48%; location 2, green: τ_{mean} = 54–55 kPa, VMC_{mean} = 34–38%).

Table 5. *p*-Value table output from the exact Wilcoxon Rank-Sum test for group comparison (p = 0.05) for the different machine setups: 1a: Triple-bogie (TB) with tracks (TLS data), 1b: Triple-bogie with tracks (manual data), 2: Triple-bogie 3-bar tire pressure, 3: Triple-bogie adapted tire pressure, 4: 8-wheel reference machine. The grey background clusters the different soil types for direct comparison without measuring-site classification.

	1a (TB Tracks, TLS)	1b (TB Tracks, Man.)	2 (TB High Pressure)	3 (TB Low Pressure)
1b (TB tracks, man.)	0.911	-	-	-
2 (TB high pressure)	0.471	0.416	-	-
3 (TB low pressure)	0.000	0.010	0.021	-
4 (8-wheeled, 710)	0.275	0.520	0.205	0.024

4. Discussion

A rut-depth limit of 10 cm was introduced as a basic benchmark for rating the soil impact similar to the fundamental soil-impact warning threshold for forest operations used in other studies [55]. Within this threshold, soil impairments (soil life, decomposing organic matter, germination capacity of tree seeds, rooting capacity, and soil structure for drainage) start to occur, but in a reversible manner of potential natural regeneration. This is necessary as basic guidance for a potential sustainable forest operation. Above this value, irreversible destruction of soil functionalities can be assumed [27]. In this regard, all examined machine setups fulfilled this requirement for a reference amount of 90 m^3 timber hauled, indicating that they are suitable for use on related soil conditions.

To enable a direct comparison of the recorded absolute values with those of the reference data [36], our own reference machine results (Configuration 4) must be considered. A rut depth of more than 33.7 cm is expected for the standard setup without bogie tracks [36]. However, Configuration

4 is characterized by a low mean rut depth of only 4.3 cm. As the shear-strength values of all recorded data series lie within the defined limits of the reference data [36] for "very sensitive soil conditions (SC3)", soil moisture is considered separately to estimate its influence in case of equal shear-strength classes. With a 29% lower moisture content compared to the valid range of the reference soil stability class, the dataset of the reference machine is particularly affected by this unexpected ratio of shear strength to moisture content. It therefore does not fit the expected requirements of the SC3 site classification in this case. This dataset thus seems to underestimate the results between the different measurement sites. Even though moisture content should influence the shear-strength value directly [41], it must be separately taken into consideration. Within the framework used in the reference study, the "less-sensitive soils (SC1)" class was chosen for the Schönheide datasets because of its tendency towards lower soil moisture content. The related higher shear strength of the reference soils in the new class suggests that these classes are related to different soil types. Nevertheless, the systems can still be compared on this basis as the TB system classification is characterized by higher technical requirements to avoid soil disturbances than the reference classification.

In this new reference range, the low-pressure TB setup, as the only related TB configuration, is ranked equally with the best reference machine setup, represented by the 8-wheel Clark Terra-X TXL 150 bogie track equipment with a mean rut depth of 2.2 cm after 90 m^3 reference volume.

Furthermore, a significantly better result of the TB low-tire pressure setup compared to the 8-wheeled reference machine ($p < 0.05$) was observed at this measuring site, which underlines the consistency of the expected results.

In contrast to the already discussed measurements, those made in Tharandt are completely classified within the boundaries of the target range of the SC3 soil stability class. The results do not show any significant variations amongst the different setups of the tracked and high tire pressure machine configurations.

However, influences or disturbances of the data gathered in Tharandt become evident in the non-linear rut development of the manually gathered dataset (Figure 6; Figure 7). The manual data is therefore compared with the additionally collected TLS data to obtain clarity about the behavior. The occurred effects of the manually gathered data are not visible in the TLS data and thus seem to be either related to the manual measurement principle or are very locally expressed.

Another observation, visible in both the manual and the TLS data, is the difference in mean rut-depth development between the comparable measuring points 1a_A1 and 1a_A4. In the additional visual assessment of the multi-temporal DTMs (Figure 5), in the first (sample 1a_A1), second (sample 1a_A4), and especially the fourth image (sample 2a_B4), disturbances in the soil as areas with a higher soil stability appear at the beginning and the end of the right track. This also shows a related deepening effect behind the disturbance point and serves as an example of how the manual measurements can be both positively and negatively influenced by small- to medium-scale factors that are part of the machine behavior under real working conditions. The different expression of these structural changes at the 1a_A4 measuring point compared to the 1a_A1 point might be the reason for the relatively higher or lower values. As these influences are not obviously visible before starting measurements and are difficult to quantify without documentation, other studies have relied on open-area study sites like agricultural or similar areas [2,39,56] where these effects can be avoided. With a small number of sample points, as for the manual measurements, these expressions lead to a high variation in the data with a respective high standard error, especially between multiple data series. With the additional TLS data, the impact of one data point can be reduced, thus helping to raise the quality of the data and reduce the number of measuring points necessary. In this case, the standard error of means concerning one whole measurement point (1a/b_A1 with six sample points) was here reduced from 45% to 0.3%. The comparison of the manual data with the TLS measurements in Figure 5, the results in Figure 6, and also in view of the multiple frequency peaks underline the fact that, while one TLS measuring point tends to represent one unknown influenced situation, it does so very precisely.

With just two repetitions of scans at two sites, the TLS data on its own therefore promises more objective results on representing a variety of different measurement point characteristics.

After discussing the potential influence parameters of the different measurements sites, a direct comparison of the Tharandt results with the previous results of the reference study [36] can only be made with care, as not all parameters of the reference study are known. Assuming comparable basic conditions in terms of the frequency and intensity of disturbance factors, the results of the tracked-bogie setup (Δz_{mean} = 5.8 cm/5.5 cm) and the setup without bogie tracks and high tire pressure (Δz_{mean} = 7.2 cm) would be most closely ranked with the best results of the reference study. This is represented by the Clark Terra-X TXL 150 bogie-track equipment on the 8-wheeled reference machines with a mean rut development of Δz_{mean} = 4.2 cm after 90 m^3 of timber hauled. It must be mentioned that the relatively poor results of the high-pressure TB setup are still ranked within the range of all listed 8-wheeled bogie-track setups. The TB configuration without bogie tracks is at least an alternative to the use of 8-wheeled bogie track configurations as the differences in results are very small, even considering high tire pressure and in relation to track setups. This would confirm the previous study's conclusions that 10-wheeled configurations would benefit from less heavy configurations [2]. The lowered tire pressure setup shows a further significant improvement of the non-tracked setups ($p < 0.05$). This conclusion must be taken with care, because the respective data series, as previously discussed, are related to different soil conditions. Although the coherent soil-conservation effect of lower tire pressure setups was observed [8,10] and was already known from different studies [8,9,57], this configuration should receive further attention in follow-up studies in direct comparison with other machines on similar soils and thus direct comparable testing conditions.

5. Conclusions

In this study, the rutting effect of the new HSM triple-bogie system on sensitive soil conditions was tested. The results show that an at least equally high soil-protection effect is achieved on flat terrain compared to standard 8-wheeled machines with bogie tracks. Even when no additional bogie tracks are used on the TB system setup, the results are very promising. In particular, this setup shows additional potential when optimizing/lowering tire pressure. Due to the diverging soil conditions and the resulting higher soil stability classes, this effect should receive further attention, especially on soft soil conditions.

The TLS data was very valuable in this study as it helped minimize the effort of manual data collection, increased the data quality, and showed additional potential for assessing selected data samples with the support of multi-temporal digital terrain models. Indeed, apart from measuring soil characteristics, manual data recording could be fully replaced by the presented longitudinal sample drawing method.

Author Contributions: Conceptualization, M.S., F.H., and M.Z.; methodology, M.S. and F.H.; validation, M.S., C.D., and F.H.; formal analysis, M.S. and C.D.; investigation, M.S., C.D., and F.H.; resources, M.S., C.D., F.H., and M.Z.; data curation, M.S. and C.D.; writing—original draft preparation, M.S., C.D., F.H., and M.Z.; writing—review and editing, M.S., C.D., F.H., and M.Z. visualization, M.S.; supervision, M.S., C.D., F.H., and M.Z.; project administration, M.Z.; funding acquisition, M.Z. All authors have read and agreed to the published version of the manuscript.

Funding: This project has received funding from the European Union's Horizon 2020 research and innovation program under grant agreement No 727883.

Acknowledgments: We acknowledge the Forwarder2020 project partners' contribution in data acquisition and are further grateful to Anton Kunneke for his technical support in data preparation and to Karl Stampfer for his support in structuring the article. We appreciate the constructive comments of the three anonymous reviewers.

Conflicts of Interest: The authors declare no conflict of interest. The funders had no role in the design of the study; in the collection, analyses, or interpretation of data; in the writing of the manuscript, or in the decision to publish the results.

References

1. Frey, B.; Kremer, J.; Rüdt, A.; Sciacca, S.; Matthies, D.; Lüscher, P. Compaction of forest soils with heavy logging machinery affects soil bacterial community structure. *Eur. J. Soil Biol.* **2009**, *45*, 312–320. [CrossRef]
2. Ala-Ilomäki, J.; Högnäs, T.; Lamminen, S.; Sirén, M. Equipping a Conventional Wheeled Forwarder for Peatland Operations. *Int. J. For. Eng.* **2011**, *22*, 7–13. [CrossRef]
3. Contosta, A.R.; Casson, N.J.; Garlick, S.; Nelson, S.J.; Ayres, M.P.; Burakowski, E.A.; Campbell, J.; Creed, I.; Eimers, C.; Evans, C.; et al. Northern forest winters have lost cold, snowy conditions that are important for ecosystems and human communities. *Ecol. Appl.* **2019**, *29*, e01974. [CrossRef] [PubMed]
4. Soja, A.J.; Tchebakova, N.M.; French, N.H.F.; Flannigan, M.D.; Shugart, H.H.; Stocks, B.J.; Sukhinin, A.I.; Parfenova, E.I.; Chapin, F.S.; Stackhouse, P.W. Climate-induced boreal forest change: Predictions versus current observations. *Glob. Planet. Chang.* **2007**, *56*, 274–296. [CrossRef]
5. Rittenhouse, C.D.; Rissman, A.R. Changes in winter conditions impact forest management in north temperate forests. *J. Environ. Manag.* **2015**, *149*, 157–167. [CrossRef] [PubMed]
6. Labelle, E.R.; Poltorak, B.J.; Jaeger, D. The role of brush mats in mitigating machine-induced soil disturbances: An assessment using absolute and relative soil bulk density and penetration resistance. *Can. J. For. Res.* **2019**, *49*, 164–178. [CrossRef]
7. Kremer, J.; Wolf, B.; Matthies, D.; Borchert, H. Bodenschutz Beim Forstmaschineneinsatz [Soil Protection when Using Forest Machinery]. *LWF Merkblatt* **2012**, *22*. Available online: http://www.lwf.bayern.de/mam/cms04/service/dateien/mb-22-bodenschutz.pdf (accessed on 6 August 2020). (In Germany).
8. Marra, E.; Cambi, M.; Fernandez-Lacruz, R.; Giannetti, F.; Marchi, E.; Nordfjell, T. Photogrammetric estimation of wheel rut dimensions and soil compaction after increasing numbers of forwarder passes. *Scand. J. For. Res.* **2018**, *33*, 613–620. [CrossRef]
9. Ziesak, M. Entwicklung eines Informationssystems zum bodenschonenden Forstmaschineneinsatz. [Development of an information system for soil-conserving forestry machine use]. Ph.D. Thesis, Technische Universität München, Freising, Germany, 13 October 2004. Available online: https://mediatum.ub.tum.de/doc/603531/603531.pdf (accessed on 6 August 2020). (In Germany).
10. Sakai, H.; Nordfjell, T.; Suadicani, K.; Talbot, B.; Bøllehuus, E. Soil Compaction on Forest Soils from Different Kinds of Tires and Tracks and Possibility of Accurate Estimate. *Croat. J. For. Eng.* **2008**, *29*, 15–27.
11. Bygdén, G.; Eliasson, L.; Wästerlund, I. Rut depth, soil compaction and rolling resistance when using bogie tracks. *J. Terramech.* **2003**, *40*, 179–190. [CrossRef]
12. John Deere. 1010G Forwarder. Available online: https://www.deere.co.uk/en/forwarders/1010g/ (accessed on 12 December 2019).
13. PONSSE. PONSSE Forwarder Systems. Available online: https://www.ponsse.com/products/information-systems/product#/forwarder_systems (accessed on 15 August 2019).
14. EMB Baumaschinen GmbH. Elliator. Available online: https://www.elliator.com/ (accessed on 12 December 2019).
15. Talbot, B.; Björheden, R.; Poikela, A.; Lazdins, A.; Kopetzky, M. Forest Operations in Response to Environmental Challenges: The OnTrack forwarder—Results and experiences from constructing and testing a rubber-tracked forwarder with low ground pressure. In Proceedings of the Nordic-Baltic Conference on Operational Research (NB-NORD), Honne, Norway, 3–5 June 2019; Volume 5, pp. 32–34. Available online: https://nordicforestresearch.org/wp-content/uploads/2019/06/Proceedings-Lowres-Honne-2019.pdf (accessed on 6 August 2020).
16. Wahlers Forsttechnik GmbH. PONSSE 10w. Available online: https://wahlersforsttechnik.de/Produkte/ponsse-10w/ (accessed on 28 January 2020).
17. Edlund, J.; Keramati, E.; Servin, M. A long-tracked bogie design for forestry machines on soft and rough terrain. *J. Terramech.* **2013**, *50*, 73–83. [CrossRef]
18. Edlund, J.; Bergsten, U.; Arvidsson, H. A forest machine bogie with a bearing capacity dependent contact area: Acceleration and angular orientation when passing obstacles and drawbar pull force and free rolling resistance on firm ground. *Silva Fenn.* **2013**, *47*. [CrossRef]
19. HSM Hohenloher Spezial-Maschinenbau GmbH und Co. KG. HSM 208F 10t 2020—Datasheet, 2018.
20. Cambi, M.; Certini, G.; Neri, F.; Marchi, E. The impact of heavy traffic on forest soils: A review. *For. Ecol. Manag.* **2015**, *338*, 124–138. [CrossRef]

21. Porinsky, T.; Pentek, T.; Bosner, A.; Stanki, I. Ecoefficient Timber Forwarding on Lowland Soft Soils. In *Global Perspectives on Sustainable Forest Management: Ecoefficient Timber Forwarding on Lowland Soft Soils*; Okia, C.A., Ed.; Intec: Rijeka, Croatia, 2012; ISBN 978-953-51-0569-5.
22. Labelle, E.R. Effects of brush mats and machine characteristics on dynamic peak loads and surface contact pressures exerted by forest machinery. Ph.D. Thesis, The University of New Brunswick, Fredericton, NB, Canada, 2012. [CrossRef]
23. Venanzi, R.; Picchio, R.; Spinelli, R.; Grigolato, S. Soil Disturbance and Recovery after Coppicing a Mediterranean Oak Stand: The Effects of Silviculture and Technology. *Sustainability* **2020**, *12*, 4074. [CrossRef]
24. Labelle, E.R.; Jaeger, D. Soil Compaction Caused by Cut-to-Length Forest Operations and Possible Short-Term Natural Rehabilitation of Soil Density. *Soil Sci. Soc. Am. J.* **2011**, *75*, 2314–2329. [CrossRef]
25. Cullen, S.J.; Montagne, C.; Ferguson, H. Timber Harvest Trafficking and Soil Compaction in Western Montana. *Soil Sci. Soc. Am. J.* **1991**, *55*, 1416–1421. [CrossRef]
26. Frutig, F.; Lüscher, P. Was bringt das Rücken mit halber Ladung? Bodenverdichtung und Spurbildung auf Rückegassen. [What's the point of working with half a load? Soil compaction and rut development on skid trails]. *Wald Holz* **2015**, *8*, 49–52. (In Germany)
27. Lüscher, P.; Frutig, F.; Thees, O. *Physikalischer Bodenschutz im Wald: Waldbewirtschaftung im Spannungsfeld zwischen Wirtschaftlichkeit und Erhaltung der physikalischen Bodeneigenschaften*; [Physical soil protection in forests: Forest management in the area of conflict between economic efficiency and preservation of physical soil properties]; Bundesamt für Umwelt: Bern, Switzerland, 2016; Umwelt-Wissen Nr. 1607; p. 159. Available online: https://www.bafu.admin.ch/dam/bafu/de/dokumente/wald-holz/uw-umwelt-wissen/physikalischer_bodenschutzimwald.pdf.download.pdf/physikalischer_bodenschutzimwald.pdf (accessed on 6 August 2020). (In Germany)
28. Aust, W.M.; Burger, J.A.; Carter, E.A.; Preston, D.P.; Patterson, S.C. Visually Determined Soil Disturbance Classes Used as Indices of Forest Harvesting Disturbance. *South. J. Appl. For.* **1998**, *22*, 245–250. [CrossRef]
29. Spinelli, R.; Magagnotti, N.; Nati, C. Benchmarking the impact of traditional small-scale logging systems used in Mediterranean forestry. *For. Ecol. Manag.* **2010**, *260*, 1997–2001. [CrossRef]
30. Talbot, B.; Rahlf, J.; Astrup, R. An operational UAV-based approach for stand-level assessment of soil disturbance after forest harvesting. *Scand. J. For. Res.* **2018**, *33*, 387–396. [CrossRef]
31. Giannetti, F.; Chirici, G.; Travaglini, D.; Bottalico, F.; Marchi, E.; Cambi, M. Assessment of Soil Disturbance Caused by Forest Operations by Means of Portable Laser Scanner and Soil Physical Parameters. *Soil Sci. Soc. Am. J.* **2017**, *81*, 1577–1585. [CrossRef]
32. Koren, M.; Slančík, M.; Suchomel, J.; Dubina, J. Use of terrestrial laser scanning to evaluate the spatial distribution of soil disturbance by skidding operations. *iForest* **2015**, *8*, 386–393. [CrossRef]
33. Naghdi, R.; Solgi, A.; Zenner, E.K.; Tsioras, P.A.; Nikooy, M. Soil disturbance caused by ground-based skidding at different soil moisture conditions in Northern Iran. *Int. J. For. Eng.* **2016**, *27*, 169–178. [CrossRef]
34. Sirén, M.; Salmivaara, A.; Ala-Ilomäki, J.; Launiainen, S.; Lindeman, H.; Uusitalo, J.; Sutinen, R.; Hänninen, P. Predicting forwarder rut formation on fine-grained mineral soils. *Scand. J. For. Res.* **2019**, *34*, 145–154. [CrossRef]
35. Eliasson, L. Effects on Forwarder Tyre Pressure on Rut Formation and Soil Compaction. *Silva Fenn.* **2005**, *39*, 549–557. [CrossRef]
36. Heubaum, F.; Padberg, A. Praxisobjekt zur bodenschonenden Holzernte mit Bogie-Bändern. [Field trial for soil-conserving timber harvesting with bogie tracks]. *AFZ/Wald* **2014**, *17*, 16–20. (In Germany)
37. Pierzchała, M.; Talbot, B.; Astrup, R. Measuring wheel ruts with close-range photogrammetry. *Forestry* **2016**, *89*, 383–391. [CrossRef]
38. Cambi, M.; Giannetti, F.; Bottalico, F.; Travaglini, D.; Nordfjell, T.; Chirici, G.; Marchi, E. Estimating machine impact on strip roads via close-range photogrammetry and soil parameters: A case study in central Italy. *iForest* **2018**, *11*, 148–154. [CrossRef]
39. Haas, J.; Hagge Elhöft, K.; Schack-Kirchner, H. *Spurgleisbildung bei der Befahrung mit Forwardern. Vergleich Unterschiedlicher Bereifungsvarianten Hinsichtlich Erhaltung der Technischen Befahrbarkeit von Rückegassen in der Mechanisierten Holzernte*; [Rut development in forwarder use. Comparison of different tyre variants with regard to maintaining the technical usability of skid trails in mechanised timber harvesting]; University of Freiburg: Freiburg, Germany, 2013. Unpublished work. (In Germany)

40. Cambi, M.; Grigolato, S.; Neri, F.; Picchio, R.; Marchi, E. Effects of Forwarder Operation on Soil Physical Characteristics: A Case Study in the Italian Alps. *Croat. J. For. Eng.* **2016**, *37*, 233–239.
41. Poršinsky, T.; Sraka, M.; Stankić, I. Comparison of two approaches to soil strength classifications. *Croat. J. For. Eng.* **2006**, *27*, 17–26.
42. Jacke, H.; Hittenbeck, J.; Stiehm, C. Spuren im Wald: Zur Spurrillen-Diskussion um und auf Rückegassen. [Forest tracks: Contributing to the wheel rut discussion on skid trails]. In *Ifa Mitteilungen, Reihe A Heft 9*; Abteilung Arbeitswissenschaften und Verfahrenstechnologie der Universität Göttingen: Göttingen, Germany, 2011. (In Germany)
43. Heubaum, F. Bodenscherfestigkeit auf Rückegassen. [Soil shear strength on skid trails]. *Forst Tech.* 18–22. Available online: https://www.waldwissen.net/technik/holzernte/boden/sbs_scherfestigkeit/sbs_scherfestigkeit_originalpdf.pdf (accessed on 6 August 2020). (In Germany).
44. Jansson, K.-J.; Johansson, J. Soil changes after traffic with a tracked and a wheeled forest machine: A case study on a silt loam in Sweden. *Forestry* **1998**, *71*, 57–66. [CrossRef]
45. Wüthrich, M. Vorteile des Triplebogies im Forwardereinsatz. [Advantages of the triple-bogie in forwarder use]. Bachelor's Thesis, Bern University of Applied Sciences, Zollikofen, Switzerland, 10 August 2018.
46. Holzerntetechnologien. Richtlinie zur Anwendung im Staatswald des Freistaates Sachsen. [Timber harvesting technologies. Guideline for application in the state forest of the Free State of Saxony]. Guideline. 2006. Available online: https://docplayer.org/58632673-Holzerntetechnologien-richtlinie-zur-anwendung-im-staatswald-des-freistaates-sachsen-staatsbetrieb-sachsenforst.html (accessed on 7 August 2020). (In Germany).
47. Najafi, A.; Solgi, A.; Sadeghi, S.H. Soil disturbance following four wheel rubber skidder logging on the steep trail in the north mountainous forest of Iran. *Soil Tillage Res.* **2009**, *103*, 165–169. [CrossRef]
48. World reference base for soil resources. *International Soil Classification System for Naming Soils and Creating Legends for Soil Maps*; FAO: Rome, Italy, 2014; ISBN 978-92-5-108369-7.
49. Benning, R.; Petzold, R.; Danigel, J. Umweltdatenbank Bodenprofile. [Environmental data repository soil profiles]. BWI 2012, 2019.
50. HSM Hohenloher Spezial-Maschinenbau GmbH und Co. KG. HSM 208F BigFoot. Available online: https://www.hsm-forest.net/hsm-208f-bigfoot-en.html (accessed on 17 January 2020).
51. Geiger, C.; Geimer, M. Efficiency Optimisation of a Forestry Crane by Implement Hydraulics with Energy Recovery. In Proceedings of the VDI-MEG Conference: Agricultural Engineering—Land. Technik AgEng, Hannover, Germany, 10–11 November 2017; VDI Verlag: Düsseldorf, Germany, 2017.
52. Gabriel, O. Triple Bogie. *Forst und Technik*; 2019, *Sonderheft 2019*, 54–57. (In Germany).
53. HSM Hohenloher Spezial-Maschinenbau GmbH und Co. KG. HSM 208F 10t Kombi short/long. Available online: https://www.hsm-forest.net/hsm-208f-10-t-kombi-short-long.html (accessed on 20 January 2019).
54. The CRAN R-Project. The CRAN R-Project. Available online: https://cran.r-project.org/ (accessed on 23 April 2020).
55. Heubaum, F. Bodenschutz im Staatsbetrieb Sachsenforst. Projekte zur Technologiererprobung. [Soil protection in the state enterprise Sachsenforst. Projects for testing technology]. 2014. Available online: https://www.sbs.sachsen.de/download/Bodenschutz_Projekte_2015_09_30.pdf (accessed on 30 August 2018). (In Germany).
56. Björheden, R. Arbetsrapport 989–2018. Rutting and Vibration Levels of the on Track Concept Forwarder on Standardised Test Tracks. 2018. Available online: https://www.skogforsk.se/contentassets/54df1878341a454ea81955692d7dd4f2/arbetsrapport-989-2018.pdf (accessed on 5 February 2020).
57. McDonald, T.; Way, T.; Löfgren, B.; Seixas, F.; Landström, M. Load and inflation pressure effects on soil compaction of forwarder tires. In Proceedings of the Joint Conference of the Canadian Woodlands Forum, Canadian Pulp and Paper Association and the International Union of Forest Research Organisations, Quebec City, QC, Canada, 9–11 September 1996; pp. 67–70.

sustainability

MDPI

Article

Applications of GIS-Based Software to Improve the Sustainability of a Forwarding Operation in Central Italy

Rodolfo Picchio [1], Francesco Latterini [2,*], Piotr S. Mederski [3], Damiano Tocci [1], Rachele Venanzi [1], Walter Stefanoni [2] and Luigi Pari [2]

1 Department of Agriculture, Forests, Nature and Energy, Tuscia University, DAFNE, Viterbo-Italy; Via San Camillo de Lellis, 01100 Viterbo, Italy; r.picchio@unitus.it (R.P.); toccidamiano91@gmail.com (D.T.); venanzi@unitus.it (R.V.)

2 Consiglio per la Ricerca in Agricoltura e l'Analisi dell'Economia Agraria (CREA), Centro di Ricerca Ingegneria e Trasformazioni Agroalimentari, Via della Pascolare 16, 00015 Monterotondo (RM), Italy; walter.stefanoni@crea.gov.it (W.S.); luigi.pari@crea.gov.it (L.P.)

3 Department of Forest Utilisation, Faculty of Forestry, Poznań University of Life Sciences (PULS), ul. Wojska Polskiego 71A, 60-625 Poznań, Poland; piotr.mederski@up.poznan.pl

* Correspondence: francesco.latterini@crea.gov.it; Tel.: +39-0690675205

Received: 10 June 2020; Accepted: 14 July 2020; Published: 16 July 2020

Abstract: Reducing potential soil damage due to the passing of forest machinery is a key issue in sustainable forest management. Limiting soil compaction has a significant positive impact on forest soil. With this in mind, the aim of this work was the application of precision forestry tools, namely the Global Navigation Satellite System (GNSS) and Geographic Information System (GIS), to improve forwarding operations in hilly areas, thereby reducing the soil surface impacted. Three different forest study areas located on the slopes of Mount Amiata (Tuscany, Italy) were analyzed. Extraction operations were carried out using a John Deere 1410D forwarder. The study was conducted in chestnut (*Castanea sativa* Mill.) coppice, and two coniferous stands: black pine (*Pinus nigra* Arn.) and Monterey pine (*Pinus radiata* D. Don). The first stage of this work consisted of field surveys collecting data concerning new strip roads prepared by the forwarder operator to extract all the wood material from the forest areas. These new strip roads were detected using a GNSS system: specifically, a Trimble Juno Sb handheld data collector. The accumulated field data were recorded in GIS Software Quantum GIS 2.18, allowing the creation of strip road shapefiles followed by a calculation of the soil surface impacted during the extraction operation. In the second phase, various GIS tools were used to define a preliminary strip road network, developed to minimize impact on the surface, and, therefore, environmental disturbance. The results obtained showed the efficiency of precision forestry tools to improve forwarding operations. This electronic component, integrated with the on-board GNSS and GIS systems of the forwarder, could assure that the machine only followed the previously-planned strip roads, leading to a considerable reduction of the soil compaction and topsoil disturbances. The use of such tool can also minimize the risks of accidents in hilly areas operations, thus allowing more sustainable forest operations under all the three pillars of sustainability (economy, environment and society).

Keywords: GIS; GNSS; forwarder; precision forestry; sustainable forest operations

---

1. Introduction

To fulfil the relevant ecological, economic and social functions, sustainable forest management (SFM) [1] should include effective [2,3] and environmentally-acceptable forest operations [4]. Considering the above mentioned functions, SFM should minimize the negative impact of harvesting on the environment without limiting the productivity while assuring forest workers' safety [4–7]. Modern machines, such as harvesters and forwarders (with wide, rubber tyres), have become common in forest utilization [8], also because they reduce the environmental impact in comparison with others utilization systems characterized by lower mechanization level [9]. The application of precision forest harvesting (PFH) may contribute significantly to the enhancement of efficient cut-to-length technology, i.e., a harvesting system in which trees are delimbed and bucked into assortments prior to subsequent transport to the landing site [10], and optimize SFM. PFH may be implemented by using interdisciplinary concepts, integrating the use of new technologies to create innovative solutions for efficient forest operations [11]. With particular reference to forwarding, the integration of Global Navigation Satellite System (GNSS) technology, Geographic Information System (GIS) and the on-board computing (OBC) hardware and software of modern forwarders, as well as advanced Information and Communications Technology (ICT), could enhance the future development of forest utilization [12–20].

Electronic devices integrated into modern forest machines used for forest operations do not only guarantee higher work productivity, but they could also reduce the environmental impact and enhance the safety of the workers [21–25].

Nowadays, the integration of electronic solutions with forestry practice (which can be seen as part of precision forestry) can contribute significantly to SFM and this creates a new best practice. It is possible for electronic devices to be implemented in all phases of the forest value chain, from the intervention planning to the product traceability. GIS technology could be used to analyze the topographic, ecological and morphological characteristics of the study area. GIS can help to design strip road network for timber harvesting and alternative extraction systems, with particular attention to economic aspects, minimizing negative impact on environment and providing a guarantee safety for operators [26–32]. GIS developed files can be implemented on the modern forwarders' information and communication technology (ICT) system; therefore, the designed strip road pattern can be displayed on the on-board screen and, thanks to the GNSS device, the operator can follow this strip road network, thereby limiting soil compaction [33]. Moreover, geo-data from the GNSS transformed in GIS could be integrated with work productivity and recorded using the standard for forest machine data and communication (StanForD) to carry out an economic evaluation of the entire study area [33,34]. In addition, a radio frequency identification (RFID) system allows for the identification of trees and marking them individually [35]. This technology showed good performances and moreover there are many possibilities of implementation [36].

Hence, aims of the present study were: (1) to apply GNSS and GIS technologies for the design of strip roads for forwarding operations in central Italy, and (2) to compare the net of the electronically-designed strip roads with those established in the forest by a forwarder operator according to his experience. Thus, it aimed to evaluate the effectiveness of precision forestry technology in the improvement of forwarding.

2. Materials and Methods

2.1. Study Areas

Three study areas were located on the slopes of Mount Amiata (Tuscany, Italy) in the Piancastagnaio district, in the Province of Siena.

Average annual rainfall in the study areas is approximately 1400 mm/yr^{-1} with an average annual temperature of 11.5 °C. The stands were dominated by *Pinus radiata* D. Don, *Pinus nigra* Arn. and *Castanea sativa* Mill. in which three different forest treatments were conducted: clear-cutting (CC), thinning (TH) and coppicing with standards (CS), respectively (Table 1). In all three study areas, felling and processing were performed with a John Deere 1070G harvester, whilst extraction was carried out using a John Deere 1410D forwarder.

Table 1. Main characteristics of three analyzed study areas.

Study Area	Dominant Species	Intervention	Study Area Surface [ha]	Standing Timber Mass [$Mg \cdot ha^{-1}$]	Harvested Total Timber Mass [$Mg \cdot ha^{-1}$]
CC	*Pinus radiata* D.Don.	Clear-cutting	0.31	400	400
TH	*Pinus nigra* Arn.	Thinning	1.10	365	148
CS	*Castanea sativa* Mill.	Coppicing with standards	1.93	220	200

2.1.1. Clear Cutting (CC)

The total study area amounted to 0.31 ha dominated by *Pinus radiata* D.Don. (Figure 1), with a standing volume of 400 $Mg \cdot ha^{-1}$; all the timber was harvested as clear-cut (Table 1). The stand was located in a hilly area, the prevalent elevation was 618 m a.s.l. with a maximum elevation of 622 m a.s.l. and a minimum of 611 m a.s.l., with a predominately north-westerly aspect. The degree of the slope was between 5% (I Class) and 22% (II Class) (Figure 1).

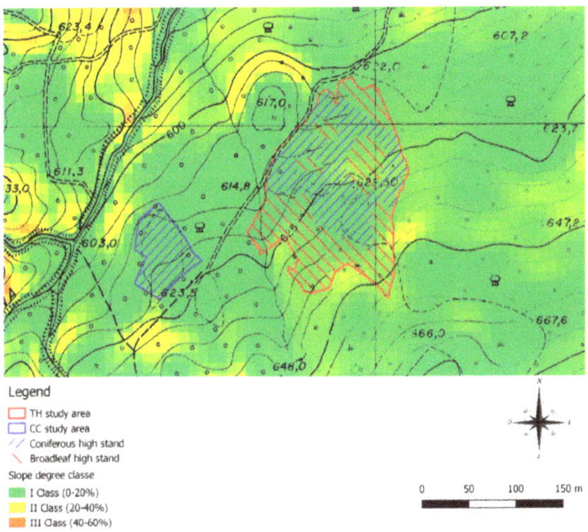

Figure 1. Land cover and slope map of clear cutting (CC) and thinning (TH) study areas. CRS: WGS84-UTM32T. EPSG 32632.

2.1.2. Thinning (TH)

The total area of the forest sub-compartment equaled 2.30 ha, from which 1.10 ha was dominated by *Pinus nigra* Arn. and 1.20 ha of high stands of *Castanea sativa* Mill., *Quercus cerris* L. and *Fraxinus ornus* L. derived from a natural regeneration after an artificial pine stand (Figure 1). The total standing mass was 365 $Mg \cdot ha^{-1}$, from which 148 $Mg \cdot ha^{-1}$ were harvested as thinning (Table 1). Only pine trees were cut, while all the broadleaved individuals were left upstanding. The area was also hilly, with a prevalent elevation of 626 m a.s.l. (maximum 653 m a.s.l. and minimum 613 m a.s.l.), mainly with a northwest aspect. The slope degree was between 5% (I Class) and 33% (II Class) (Figure 1).

2.1.3. Coppicing with Standards (CS)

The total study area was 1.93 ha, fully covered with *Castanea sativa* Mill coppice with standards (Figure 2) with a standing mass of 220 $Mg \cdot ha^{-1}$, from which coppicing of 200 $Mg \cdot ha^{-1}$ had been conducted with the release of 50 standard trees per hectare. This area had the steepest slopes, with a prevalent elevation of 1,019 m a.s.l. (from 999 m a.s.l. to 1077 m a.s.l.), and a predominant northeasterly aspect. The degree of the slope ranged between 5% (I Class) and 45% (III Class) (Figure 2).

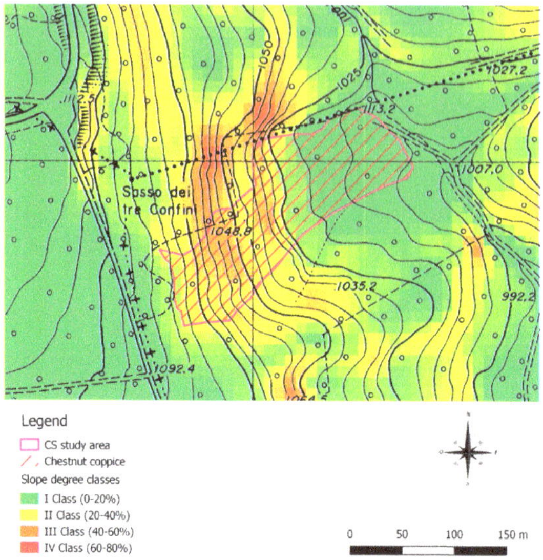

Figure 2. Land cover and slope map of coppice with standards (CS) study area. CRS: WGS84-UTM32T. EPSG 32632.

2.2. Field Reliefs

A preliminary survey was conducted by recording the existing road network with a GNSS device. Then, in each of the three forest areas, six sample plots were randomly identified. Each sample plot had a circular shape and a radius of 10 m (surface of ca. 314 m^2). Within the plots, all the strip roads open for harvesting (and forwarding) were detected using a GNSS device. Additionally, the length and average width of the strip roads were measured by using a measuring tape. Between the two systems used for distance measurement, there were no statistical differences. Finally, the coordinates of the center point of each sample area were recorded using the GNSS device to make it possible to transfer the sample area surfaces and locations on GIS. For data collection, a Trimble Juno Sb handheld was used. The Trimble Juno Sb was powered by the Windows Mobile 6.1 operating system and a 533 MHz Samsung S3C3443 processor. According to device's specifications, a real-time accuracy of 2 to 5 m was possible thanks to the integrated SBAS receiver. Subsequent post processing of the data using Trimble Delta Phase technology made it possible to reach a positional accuracy of 1–3 m [37]. However, for the aims of this study, post processing the data was unnecessary; therefore, the field data showed a positional accuracy of 2–5 m.

The data characterizing the relief were collected and recorded by GIS software: in particular, the open-source software Quantum GIS 2.18 Las Palmas, that allowed the creation of a line shape file of the strip road pattern, within the six sample areas for the three study areas. All the GIS files were geo-referenced in WGS84-UTM32T CRS (EPSG 32632). The data collected from the field surveys and elaborated using GIS technology were used to calculate three crucial parameters for the designed experiment: the surface impacted by forwarder passes, the length of the strip roads and strip road density.

2.3. GIS Implementation

2.3.1. Preliminary GIS Steps

The GIS procedure developed for and applied in the design of a new improved network of strip roads needed two basic elements, i.e., a line shape file of the existing forest road network and a digital terrain model (DTM) of the area. The line shape file of the existing forest road network was derived from the GNSS survey. The DTM was built based on a topographic vector map of Tuscany, with a scale of 1:5,000. More precisely, a 2 m resolution DTM was derived with a QGIS plugin, using triangulated irregular network (TIN) interpolation. It should be noted here that the best DTM resolution freely available for the whole of Italy is currently 10 m [38,39]. Considering the size of the three study areas, it was decided in this case that it was inappropriate to use a 10 m DTM resolution; therefore, a DTM with a 2 m resolution was built using local geo-data.

2.3.2. New Strip Road Pattern Development and Determination of the Forwarder Passes Needed for Extraction

For the creation of the GIS-planned strip road network, the QGIS tool, Forest Road Designer (FRD), [40] was used. This is a GIS plugin that relies on a DTM and on points or a line shapefile reporting the zones. It generates another polyline meeting a series of design requirements established by the user (longitudinal slope and curvature radius among others) [40].

One of the most important parameters to be set using the FRD was the maximum slope gradient characterizing the new strip roads. This parameter had to be defined taking into account the characteristics of the forest machine that is supposed to be used in the strip roads. Considering the 1410D forwarder, a maximum slope of 45% was defined for stretches perpendicular to the contour lines and 25% for stretches parallel to them. In this way, machine-tipping risk was minimized. Moreover, with the FRD plugin, it was also possible to indicate some areas over which the newly designed strip roads should not pass (for example, high-value conservation areas). This was done by simply indicating such areas with a polygon shapefile. It was necessary in this case to indicate certain areas over which driving was forbidden as the study areas were surrounded by the properties of other owners. Another GIS procedure was developed and implemented to define the number of forwarder's passes needed for the extraction of all the timber from the three study areas, according to the GIS-planned strip road pattern. The first step was a calculation of the forwarding areas (i.e., forwarder accessible areas) where timber was within reach of the forwarder's boom. This was 12 m from the middle of each strip road, taking into account the fact that the working distance of the forwarder boom was 12 m. Once the forwarding areas were identified using the QGIS plugin fixed-distance buffer, it was possible to divide them using another QGIS tool: the polygon divider, which differentiated an input polygon layer (forwarding area) into a number of squarish polygons of a defined size. Knowing the forwarder loading capacity, which was 13 tons for the 1410D, and the harvested mass for each study area, it was possible to define the dimension of the sub-polygons or forwarding pixels (FPx), into which each forwarding area was divided. Each FPx surface corresponded to one forwarder load. The number of forwarder passes (NPs) corresponded to:

$$\text{NPs: } 2 \cdot \text{FPx} - 2 \quad (1)$$

Finally, using the dedicated QGIS tool, the GIS-based strip road network was converted from an ESRI shape file format to .kml and .gpx ones, in order for them to be compatible with the forwarder's computer and to be visible on the screen.

2.3.3. GIS Data Elaboration and Statistical Analysis

Having defined the GIS-based new strip road network, it was possible to calculate the soil disturbance parameters: the impacted surface, strip road length and strip road density for the GIS-planned study areas. These were calculated using the geo-fences of six sample plots from each study area. Following this, the data obtained were analyzed with Statistica 7.0 software: after checking for normality and homoscedasticity, both a one-way ANOVA and HSD Tukey test were used to find out if there were statistically significant differences among data collected manually in the forest and data obtained from the GIS.

3. Results

3.1. Clear Cutting (CC)

In the clear-cutting study area, the mean strip road length in reality was 31.93 m, corresponding to 1017 m· ha^{-1} of strip road density and 35% of impacted surface. With GIS planning, the mean strip road length obtained from the sample plots was only 15.50 m, with a strip road density of 494 m·ha^{-1} and 17% of surface impacted.

With regard to the harvested volume in the CC study area, the surface of each FPx was 325 m^2, GIS analysis returned 9 FPx corresponding to 16 NPs. The forest road (fuchsia line, 3D model developed using QGIS tool QGIS2threejs, Figure 3b) is located along the southeast side of the study area and connected to a secondary truck road (light blue line) located to the northeast (Figure 3a).

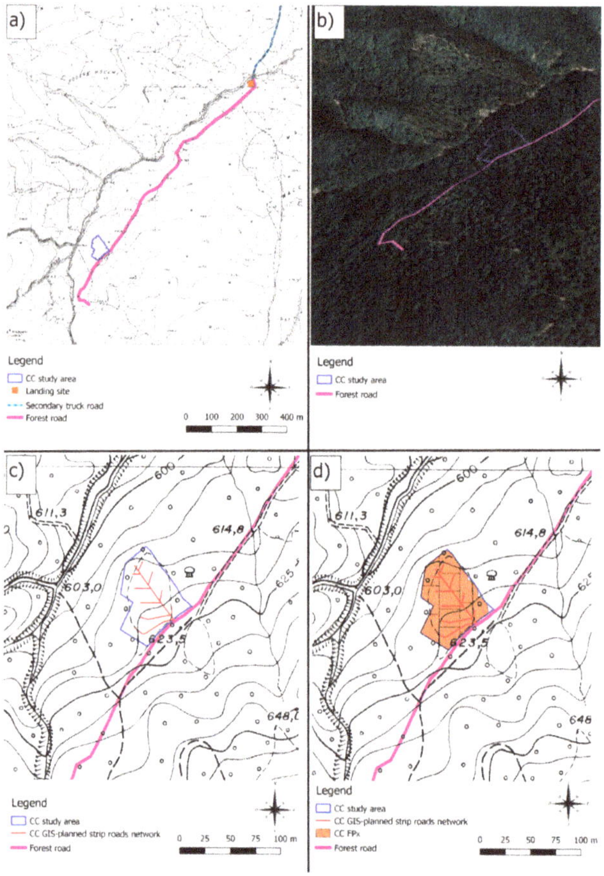

Figure 3. (**a**) Actual forest road network in clear-cutting (CC) study area; (**b**) CC on 3D-model built using QGIS plugin QGIS2threejsbased on digital terrain model (DTM) and orthophoto map from 2013; (**c**) CC GIS (Geographic Information System)-planned strip road network; (**d**) CC study area divided into small areas contributing to one load, forwarder pixel (FPx). CRS: WGS84-UTM32T. EPSG 32632.

The GIS-planned strip road network (red lines) was created in a fir-shape (Figure 3c) reaching basically all the FPx (orange rectangles) of the CC study area (Figure 3d). The GIS-planned strip road network started from the existing forest road with a central axis, from which various branches departed to extract all the timber from the whole sub-compartment.

3.2. Thinning (TH)

In the thinning study area, the manually measured mean strip road length was 25.00 m, which corresponded to a strip road density of 796 m·ha^{-1} and 28% of surface impacted. With GIS planning (Figure 4), lower values were obtained likewise: mean strip road length was 12.50 m, strip road density equaled 398 m·ha^{-1} and the surface impacted was 14%.

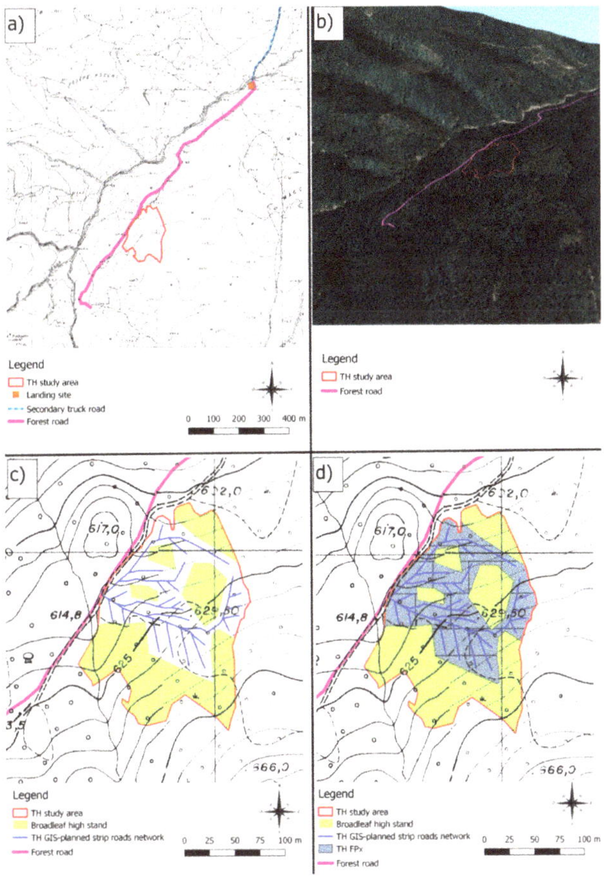

Figure 4. (**a**) Actual forest road network in thinning (TH) study area; (**b**) TH on 3D-model built using QGIS plugin QGIS2threejs based on DTM and orthophoto map from 2013; (**c**) TH GIS-planned strip road network; (**d**) TH study area divided into small areas contributing to one load, forwarder pixel (FPx). CRS: WGS84-UTM32T. EPSG 32632.

Considering the harvested volume in the TH study area, the surface of each FP was 866 m^2, the GIS analysis returned 14 FPx corresponding to 26 NPs. The forest road was attached to the northwest side of the study area (fuchsia line, Figure 4a) and it was linked to a secondary truck road (light blue line), located to the northeast (3D model developed using QGIS tool QGIS2threejs, Figure 4b). The GIS-planned strip road network (blue lines) also had a fir-shape (Figure 4c), which gave access to all the FPx (blue rectangles, Figure 4d). The GIS-planned strip road network started from the existing forest road, but in contrast to the CC, it was not possible to only develop a single central axis because of the presence of broadleaf groups (green shading), which had to remain upstanding. As a consequence, the GIS-planned strip road network had a more developed dendritic pattern.

3.3. Coppice with Standards (CS)

In the coppice study area (six sample plots), the manually measured mean strip road length was 36.50 m, the strip road density was 1,162 m·ha^{-1} and 41% of the area was impacted by the forwarder's driving. It was possible to reach the CS study area from driving from two sides: from the top of the slope and from the valley. Therefore, two GIS-planned strip road networks were designed: herringbone (CS_Herr) and high-low (CS_HL). The CS_Herr strip roads basically ran parallel to the contour lines, while the CS_HL ran in a perpendicular direction towards the contour lines. The GIS-planned strip road mean length (from the sample plots) came to only 13.00 m and 10.83 m, for CS_Herr and CS_HL, respectively. The strip road density was 414 and 345 m·ha^{-1} for each design, while the impacted surface amounted to only 14.49% and 12% of the area for Herr and HL, respectively. The harvested timber mass of 200 Mg·ha^{-1} required 650 m^2 of FPx, which in the GIS analysis amounted to 30 FPx, corresponding to 58 NPs for CS_Herr, and 27 FPx corresponding to 52 NPs for CS_HL. The study area visible on the topographic map was surrounded by the existing forest road network: from the west and east sides (Figure 5a). To the west, there was the main truck road (green line), from which a forest road (fuchsia line) departed and ran near the western side of the sub-compartment. To the east, there was the presence of a secondary truck road (light blue line), from which another forest road (red dotted line) started running along the southern side of the study area. A 3D model of road locations was developed using the QGIS tool QGIS2threejs (Figure 5b). The CS_Herr GIS-planned strip road network (blue lines) had a perpendicular layout in comparison with the CS_HL one (red lines, Figure 5c,d). The FPx of CS_Herr (purple rectangles) were slightly different to the CS_HL FPx (orange rectangles, Figure 5e,f). The GIS-planned strip road network departed from the existing forest road, developing in a parallel manner to the contour lines in CS_Herr and perpendicularly to those in CS_HL.

3.4. Statistical Analysis

There were statistically significant differences between the real and the GIS-planned strip road networks in the CC and CS study areas, for both the CS_Herr and CS_HL models (Table 2). At the same time, there were no statistically significant differences in the TH area (the ANOVA p-value was 0.07, close to the level of significance).

However, for all three study areas, GIS-planning brought a considerable reduction in the area of impacted soil, with a percentage reduction between 50% and 71%. In the CS study area, there were no significant differences between the two GIS-based strip road patterns (CS_Herr and CS_HL). Therefore, GIS-planning considerably decreased the area of impacted soil.

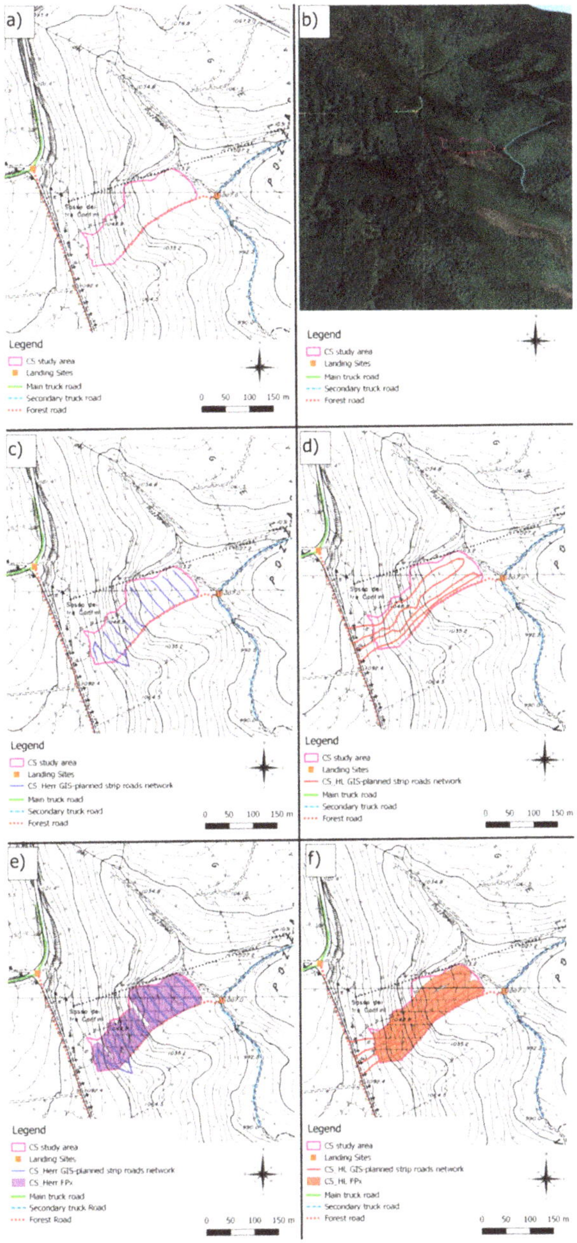

Figure 5. (**a**) Actual forest road network in coppice with standards (CS)study area; (**b**) CS on 3D-model built using QGIS plugin QGIS2threejs based on DTM and Orthophoto from 2013; (**c**) herringbone (CS_Herr) GIS-planned strip road network. (**d**) high–low (CS_HL) GIS-planned strip road network; (**e**) CS study area divided into small areas contributing to one load, forwarder pixel (FPx) according to CS_Herr strip roads network; (**f**) CS study area divided into small areas contributing to one load, forwarder pixel (FPx) according to CS_HL strip roads network CRS: WGS84-UTM32T. EPSG 32632.

Table 2. Overall results of one-way ANOVA and HSD Tukey test conducted separately for each forest study area. "***" in the first column indicates p-value significance at 0.1%. Letter "a" or "b" within various cells indicate HSD Tukey test homogeneous groups.

Study Area	Impacted Surface [%]			Strip Road Length [m]			Strip Road Density [m·ha^{-1}]		
	Planned (GIS)		Real (Forest)	Planned (GIS)		Real (Forest)	Planned (GIS)		Real (Forest)
CC ***	17.3% ± 8.9 a		34.9% ± 6.7 b	153 ± 79 a		292 ± 68 b	494 ± 254 a		1017 ± 191 b
CS ***	CS_Herr	CS_HL	40.7% ± 12.5 b	CS_Herr	CS_HL	2243 ± 686b	CS_Herr	CS_HL	1162 ± 356 b
	14.5% ± 5.9 a	12.1% ± 10.3 a		799 ± 323 a	665 ± 567a		414 ± 167 a	345 ± 294 a	
TH	13.9% ± 16.1a		27.9% ± 3.7a	437 ± 525a		876 ± 117a	398 ± 477a		796 ± 107a

4. Discussion

As found and demonstrated in several other studies, reducing the area of impacted soil during forest utilization is a good indication of SFM standards [41–43]. In this study, the effectiveness of the advanced electronic systems in reducing soil impact has been demonstrated. Thanks to the application of GNSS and GIS precision forestry tools for the planning of strip road networks, there was a reduction of 50%–70% in the area impacted in comparison with the plots on which the strip roads were created during the harvesting operation.

A GIS planned strip road pattern can also be beneficial from a social point of view. For instance, it was helpful to plot strip roads on slopes with a limited gradient, improving safety and maneuverability. Thanks to technological progress, which in the last years has led to an efficient integration of electronic devices in modern forest machines, such as harvesters and forwarders, it is possible to take one step further and transfer GIS files onto these machines.

Integrated GNSS technologies and modern ICT systems can visualize an optimal strip road pattern on the machine's display and help the operator drive in a comfortable, safe and efficient way in the forest (Figure 6).

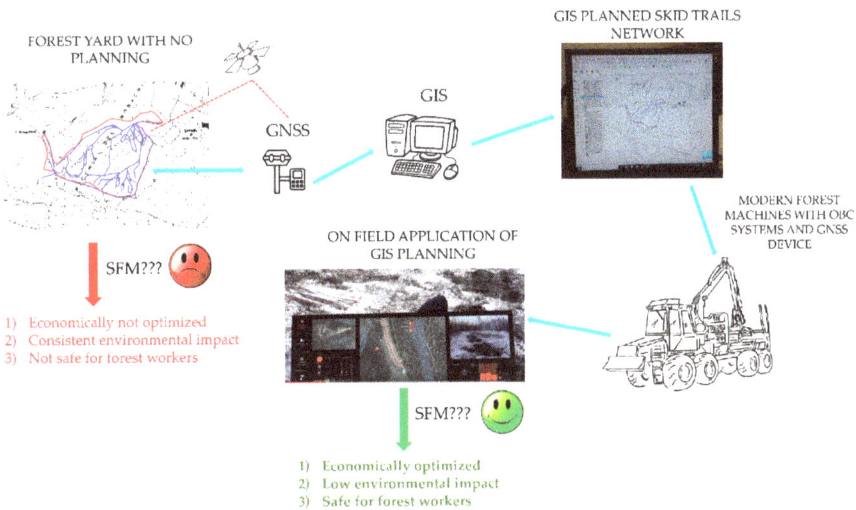

Figure 6. Integrated GNSS, GIS, OBC and ICT—precision forestry optimizing workflow and contributing to good SFM standards.

Such a possibility is still available in modern harvester and forwarder models, which have OBC with dedicated software, such as John Deere TimberMatic Maps or TimberOffice [44,45], Ponsse Opti2 [46] or Komatsu MaxiXT [47]. These OBC systems can record the data from the harvested or processed timber through the StanForD standard, thus also providing the operator information about the work productivity and quality [34]. Furthermore, integrating the positioning data from GNSS and GIS, with productivity data from the StanForD data, acquired by the harvester or forwarder OBC system, can be helpful for the forest inventory [48] or for building decision support systems [49]. Moreover, vibration and ultrasonic sensors applied to a forwarder's OBC can record data on vehicle stability [50] and rut depth [51]. Although the suitability of modern technologies to improve the sustainability of forest operations has already been highlighted by scientific research, very little has been conducted in the Mediterranean region. This study therefore aimed to be a starting point in central Italy, demonstrating the effectiveness of a GIS-GNSS approach in decreasing the negative impact of forwarding.

Considering the above, another important aspect to be underlined is the possibility of using these technologies in small-scale forestry, though with rather lower level of accuracy. A feasible example of this could be smartphone use for improving forest utilization [52]. Smartphones are able to act as low-cost GNSS receivers, also under forest canopy cover, with sufficient precision, i.e., about 9 m of accuracy, which should be sufficient for small-scale forestry use [52,53]. Many smartphone applications, developed both for Android and for iOS systems, are able to display geo-data, geopoints, geolines and geo-fences files in .kml or .gpx format, and locate the operator's position. However, even if ca. 9 m accuracy is not sufficient for a forestry-fitted farm tractor driving (following a GIS-based strip road pattern displayed on the smartphone's screen), there are other useful functions which may be available. It can be very helpful for forest workers, for example, to display the geo-fence of the treatment area on the smartphone screen, allowing them to remain within the land boundaries or to avoid restricted areas, such as biodiversity hotspots.

A further step ahead in the integration of navigation technologies on forest machines could be represented by the development of tele-operated or unmanned forest vehicles. To reach this goal, which has been achieved in agriculture [54], there is the need to integrate in forest machines differential GNSS (DGNSS) technology, such as radio-beacon differential GNSS (RBDGNSS, or real time kinematic (RTK) [55,56], inertial measurement unit (IMU) sensors [18,57] and simultaneous localization and mapping (SLAM) algorithms [19].

5. Conclusions

In recent years, several improvements have been observed in forestry, mainly the growing interest in sustainability, due to the importance of forests as environmental and social value [58,59].

Consequently, one of the most important purposes of the scientific research on forest utilization is to minimize the negative impact of forest operations, specifically on soil disturbances. Cutting edge technology and electronic devices could be powerful instruments used to reach this goal. In fact, using technological innovations, which are often mistakenly considered as something negative for the environment, may turn very helpful in forest operations.

In the presented research, the study confirmed that GNSS and GIS were useful technologies for forest operations and could improve SFM. GNSS and GIS resulted in being very helpful, both for real strip roads (established in the forest) detection and for electronically-designed strip road network.

Additionally, the use of the GIS-planned strip roads showed that soil impact due to forwarding may be decreased by 50%–70%. The herein presented precision forestry approach may be considered an efficient strategy for improving forest operations and SFM. Obviously, the practical implementation of such an approach in real forest yards in central Italy requires further steps in the training of operators, but could be very helpful in improving the sustainability of forest operations. Therefore, the presented findings can be used to improve forest utilization, also with the application of advanced technology, such as GIS and GNSS, in order to reach the effective sustainability of the whole value chain.

Author Contributions: Conceptualization, R.P. and F.L.; methodology, R.P. and F.L.; software, F.L.; validation, R.P., P.S.M., R.V.; formal analysis, F.L., D.T., W.S.; data curation, F.L., D.T.; writing—original draft preparation, F.L. and R.V.; writing—review and editing, R.P., F.L., P.S.M., R.V., D.T., W.S. and L.P.; supervision, R.P. and L.P.; funding acquisition, M.P. and L.P. All authors have read and agreed to the published version of the manuscript.

Funding: The study was performed within the framework of the Polish Ministry of Science and Higher Education's programme: "Regional Initiative Excellence" in the years 2019–2022, Project No. 005/RID/2018/19. The APC was funded by the project AGROENER ("Energia dall'agricoltura: innovazioni sostenibili per la bioeconomia"), founded by Italian Ministry of Agriculture and Forestry DD n. 26329/2016.

Acknowledgments: The authors gratefully acknowledge MIUR (the Italian Ministry for Education, University and Research) for support (Law 232/2016, Italian University Departments of Excellence).

Conflicts of Interest: The authors declare no conflict of interest.

References

1. European Community. General Guidelines for the Sustainable Management of Forests in Europe. In Proceedings of the 2nd Ministerial Conference on the Protection of Forests in Europe, Helsinki, Finland, 16–17 June 1993.
2. Mederski, P.S.; Bembenek, M.; Karaszewski, Z.; Łacka, A.; Szczepańska-Álvarez, A.; Rosińska, M. Estimating and modelling harvester productivity in pine stands of different ages, densities and thinning intensities. *Croat. J. For. Eng.* **2016**, *37*, 27–36.
3. Mederski, P.S.; Bembenek, M.; Karaszewski, Z.; Pilarek, Z.; Łacka, A. Investigation of log length accuracy and harvester efficiency in processing of oak trees. *Croat. J. For. Eng.* **2018**, *39*, 173–181.
4. Marchi, E.; Chung, W.; Visser, R.; Abbas, D.; Nordfjell, T.; Mederski, P.S.; McEwan, A.; Brink, M.; Laschi, A. Sustainable Forest Operations (SFO): A new paradigm in a changing world and climate. *Sci. Total Environ.* **2018**, *634*, 1385–1397. [CrossRef] [PubMed]
5. Picchio, R.; Magagnotti, N.; Sirna, A.; Spinelli, R. Improved winching technique to reduce logging damage. *Ecol. Eng.* **2012**, *47*, 83–86. [CrossRef]
6. Picchio, R.; Spina, R.; Calienno, L.; Venanzi, R.; Lo Monaco, A. Forest operations for implementing silvicultural treatments for multiple purposes. *Ital. J. Agron.* **2016**, *11*, 156–161.
7. Cambi, M.; Certini, G.; Fabiano, F.; Foderi, C.; Laschi, A.; Picchio, R. Impact of wheeled and tracked tractors on soil physical properties in a mixed conifer stand. *IForest* **2015**, *9*, 89–94. [CrossRef]
8. Mederski, P.S.; Karaszewski, Z.; Rosinska, M.; Bembenek, M. Dynamics of harvester fleet change in Poland and factors determining machine occurrence. *Sylwan* **2016**, *160*, 795–804.
9. Labelle, E.R.; Lemmer, K.J. Selected Environmental Impacts of Forest Harvesting Operations with Varying Degree of Mechanization. *Croat. J. For. Eng.* **2019**, *40*, 239–257. [CrossRef]
10. Hakkila, P. Logging in Finland. *Acta For. Fenn.* **1989**, *207*, 39. [CrossRef]
11. Ziesak, M. Precision Forestry-An overview on the current status of Precision Forestry. A literature review. In Proceedings of the Precision Forestry in Plantations, Semi-Natural and Natural Forests, Munich, Germany, 5–6 March 2006; pp. 5–10.
12. Picchio, R.; Latterini, F.; Mederski, P.S.; Venanzi, R.; Karaszewski, Z.; Bembenek, M.; Croce, M. Comparing accuracy of three methods based on the gis environment for determining winching areas. *Electronics* **2019**, *8*, 53. [CrossRef]
13. Picchio, R.; Pignatti, G.; Marchi, E.; Latterini, F.; Benanchi, M.; Foderi, C.; Venanzi, R.; Verani, S. The application of two approaches using GIS technology implementation in forest road network planning in an Italian mountain setting. *Forests* **2018**, *9*, 277. [CrossRef]
14. Olivera, A.; Visser, R. Using the harvester on-board computer capability to move towards precision forestry. *NZ J. For.* **2016**, *60*, 3.
15. Pellegrini, M.; Ackerman, P.; Cavalli, R. On-board computing in forest machinery as a tool to improve skidding operations in South African softwood sawtimber operations. *South. For.* **2013**, *75*, 89–96. [CrossRef]
16. Manner, J.; Palmroth, L.; Nordfjell, T.; Lindroos, O. Load level forwarding work element analysis based on automatic follow-up data. *Silva Fenn.* **2016**, *50*, 1–19. [CrossRef]
17. Ding, X.; Kong, J.; Yan, L.; Liu, J.; Yu, Z. A novel stumpage detection method for forest harvesting based on multi-sensor fusion. *Signal Image Video Process.* **2015**, *9*, 1843–1850. [CrossRef]
18. Lindroos, O.; La Hera, P.; Häggström, C. Drivers of advances in mechanized timber harvesting—A selective review of technological innovation. *Croat. J. For. Eng.* **2017**, *38*, 243–258.
19. Billingsley, J.; Visala, A.; Dunn, M. *Springer Handbook of Robotics*; Springer: Berlin/Heidelberg, Germany, 2008. [CrossRef]
20. Möller, J.J.; Arlinger, J.; Hannrup, B.; Larsson, W.; Barth, A. Harvester data as a base for management of forest operations and feedback to forest owners. In Proceedings of the 4th Forest Engineering Conference: Innovation in Forest Engineering—Adapting to Structural Change, White River, South Africa, 5–7 April 2011.
21. Wempe, A.M.; Keefe, R.F.; Newman, S.M.; Paveglio, T.B. Intent to adopt location sharing for logging safety applications. *Safety* **2019**, *5*, 7. [CrossRef]
22. Wempe, A.M.; Keefe, R.F. Characterizing rigging crew proximity to hazards on cable logging operations using GNSS-RF: Effect of GNSS positioning error on worker safety status. *Forests* **2017**, *8*, 357. [CrossRef]

23. Zimbelman, E.G.; Keefe, R.F.; Strand, E.K.; Kolden, C.A.; Wempe, A.M. Hazards in motion: Development of mobile geofences for use in logging safety. *Sensors* **2017**, *17*, 822. [CrossRef]
24. Zimbelman, E.G.; Keefe, R.F. Real-time positioning in logging: Effects of forest stand characteristics, topography, and line-of-sight obstructions on GNSS-RF transponder accuracy and radio signal propagation. *PLoS ONE* **2018**, *13*, e0191017. [CrossRef]
25. Newman, S.M.; Keefe, R.F.; Brooks, R.H.; Ahonen, E.Q.; Wempe, A.M. Human factors affecting logging injury incidents in Idaho and the potential for real-time location-sharing technology to improve safety. *Safety* **2018**, *4*, 43. [CrossRef] [PubMed]
26. Enache, A.; Pentek, T.; Ciobanu, V.D.; Stampfer, K. GIS based methods for computing the mean extraction distance and its correction factors in Romanian mountain forests. *Sumar. List* **2015**, *139*, 35–46.
27. Enache, A.; Kühmaier, M.; Stampfer, K.; Ciobanu, V.D. An integrative decision support tool for assessing forest road options in a mountainous region in Romania. *Croat. J. For. Eng.* **2013**, *34*, 43–60.
28. Picchio, R.; Proto, A.R.; Civitarese, V.; Di Marzio, N.; Latterini, F. Recent Contributions of Some Fields of the Electronics in Development of Forest Operations Technologies. *Electronics* **2019**, *8*, 1465. [CrossRef]
29. Contreras, M.A.; Parrott, D.L.; Chung, W. Designing skid-trail networks to reduce skidding cost and soil disturbance for ground-based timber harvesting operations. *For. Sci.* **2016**, *62*, 48–58. [CrossRef]
30. Gumus, S.; Turk, Y. A new skid trail pattern design for farm tractors using linear programing and geographical information systems. *Forests* **2016**, *7*, 306. [CrossRef]
31. Synek, M.; Klimánek, M. Proposal of using GIS for multi-criteria evaluation of environmentally friendly use of skidding technologies in forestry. *J. For. Sci.* **2014**, *60*, 51–60. [CrossRef]
32. Laschi, A.; Neri, F.; Montorselli, N.B.; Marchi, E. A methodological approach exploiting modern techniques for forest road network planning. *Croat. J. For. Eng.* **2016**, *37*, 319–331.
33. Olivera, A. Exploring Opportunities for the Integration of GNSS with Forest Harvester Data to Improve Forest Management. Ph.D. Thesis, University of Canterbury, Christchurch, New Zealand, 2016.
34. Olivera, A.; Visser, R.; Acuna, M.; Morgenroth, J. Automatic GNSS-enabled harvester data collection as a tool to evaluate factors affecting harvester productivity in a Eucalyptus spp. harvesting operation in Uruguay. *Int. J. For. Eng.* **2016**, *27*, 15–28. [CrossRef]
35. Sandak, J.; Sandak, A.; Marrazza, S.; Picchi, G. Development of a sensorized timber processor head prototype—Part 1: Sensors description and hardware integration. *Croat. J. For. Eng.* **2019**, *40*, 25–37. [CrossRef]
36. Alabrah, A.; Bassiouni, M. A tree-based authentication scheme for a cloud toll/traffic RFID system. In Proceedings of the 2015 IEEE Vehicular Networking Conference (VNC), Kyoto, Japan, 16–18 December 2015.
37. Trimble Juno T41™ / Juno 5 / Slate / Site Mobile, Series: Windows® Embedded Handheld 6.5 Operating System. 2018. Available online: https://geospatial.trimble.com/products-and-solutions/juno-5 (accessed on 20 January 2020).
38. Tarquini, S.; Isola, I.; Favalli, M.; Mazzarini, F.; Bisson, M.; Pareschi, M.T.; Boschi, E. TINITALY/01: A new triangular irregular network of Italy. *Ann. Geophys.* **2007**, *50*, 407–425.
39. Tarquini, S.; Vinci, S.; Favalli, M.; Doumaz, F.; Fornaciai, A.; Nannipieri, L. Release of a 10-m-resolution DEM for the Italian territory: Comparison with global-coverage DEMs and anaglyph-mode exploration via the web. *Comput. Geosci.* **2012**, *38*, 168–170. [CrossRef]
40. QGIS Python Plugins Repository. Available online: https://plugins.qgis.org/plugins/forestroaddesigner/ (accessed on 13 September 2019).
41. Venanzi, R.; Picchio, R.; Grigolato, S.; Latterini, F. Soil and forest regeneration after different extraction methods in coppice forests. *For. Ecol. Manag.* **2019**, *454*, 117666. [CrossRef]
42. Sohrabi, H.; Jourgholami, M.; Jafari, M.; Shabanian, N.; Venanzi, R.; Tavankar, F.; Picchio, R. Soil recovery assessment after timber harvesting based on the sustainable forest operation (SFO) perspective in iranian temperate forests. *Sustainability* **2020**, *12*, 2874. [CrossRef]
43. Venanzi, R.; Picchio, R.; Spinelli, R.; Grigolato, S. Soil Disturbance and Recovery after Coppicing a Mediterranean Oak Stand: The Effects of Silviculture and Technology. *Sustainability* **2020**, *12*, 4074. [CrossRef]
44. John Deere TimberMatic Maps. Available online: https://www.deere.co.uk/en/forestry/timbermatic-manager/ (accessed on 27 June 2020).
45. John Deere TimberOffice. Available online: https://www.timberoffice.com/ (accessed on 27 June 2020).

46. Ponsse Opti. Available online: https://www.ponsse.com/products/information-systems/product/-/p/forwarder_systems#/ (accessed on 27 June 2020).
47. Komatsu MaxiXT. Available online: https://www.komatsuforest.com/forest-machines/control-and-information-systems (accessed on 27 June 2020).
48. Melander, L.; Einola, K.; Ritala, R. Fusion of open forest data and machine fieldbus data for performance analysis of forest machines. *Eur. J. For. Res.* **2020**, *139*, 213–227. [CrossRef]
49. Fardusi, M.J.; Chianucci, F.; Barbati, A. Concept to practice of geospatial-information tools to assist forest management and planning under precision forestry framework: A review. *Ann. Silvic. Res.* **2017**, *41*, 3–14.
50. Marinello, F.; Proto, A.R.; Zimbalatti, G.; Pezzuolo, A.; Cavalli, R.; Grigolato, S. Determination of forest road surface roughness by Kinect depth imaging. *Ann. For. Res.* **2017**, *60*, 217–226. [CrossRef]
51. Pužuls, K.; Štāls, T.; Zimelis, A.; Lazdiņš, A. Preliminary conclusions on application of ultrasonic sensors in evaluation of distribution and depth of ruts in forest thinning. *Agron. Res.* **2018**, *16*, 1209–1217. [CrossRef]
52. Kennedy, R.; McLeman, R.; Sawada, M.; Smigielski, J. Use of Smartphone Technology for Small-Scale Silviculture: A Test of Low-Cost Technology in Eastern Ontario. *Small-Scale For.* **2014**, *13*, 101–115. [CrossRef]
53. Tomaštík, J.; Saloň, Š.; Piroh, R. Horizontal accuracy and applicability of smartphone GNSS positioning in forests. *Forestry* **2017**, *90*, 187–198. [CrossRef]
54. del Rey, J.C.; Vega, J.A.; Pérez-Ruiz, M.; Emmi, L. Comparison of Positional Accuracy betweenRTK and RTX GNSS Based on the Autonomous Agricultural Vehicles under Field Conditions. *Appl. Eng. Agric.* **2014**, *30*, 361–366.
55. Zhang, H.; Zheng, J.; Dorr, G.; Zhou, H.; Ge, Y. Testing of GPS Accuracy for Precision Forestry Applications. *Arab. J. Sci. Eng.* **2014**, *39*, 237–245. [CrossRef]
56. Soycan, M. A quality evaluation of precise point positioning within the Bernese GPS software version 5.0. *Arab. J. Sci. Eng.* **2012**, *37*, 147–162. [CrossRef]
57. Kaartinen, H.; Hyyppä, J.; Vastaranta, M.; Kukko, A.; Jaakkola, A.; Yu, X.; Pyörälä, J.; Liang, X.; Liu, J.; Wang, Y.; et al. Accuracy of kinematic positioning using global satellite navigation systems under forest canopies. *Forests* **2015**, *6*, 3218–3236. [CrossRef]
58. Schweier, J.; Blagojević, B.; Venanzi, R.; Latterini, F.; Picchio, R. Sustainability assessment of alternative strip clear cutting operations for wood chip production in renaturalization management of pine stands. *Energies* **2019**, *12*, 3306. [CrossRef]
59. Mederski, P.S.; Venanzi, R.; Bembenek, M.; Karaszewski, Z.; Rosińska, M.; Pilarek, Z.; Luchenti, I.; Surus, M. Designing Thinning Operations in 2nd Age Class Pine Stands—Economic and Environmental Implications. *Forests* **2018**, *9*, 335. [CrossRef]

sustainability

MDPI

Article

Prediction of Fuel Loading Following Mastication Treatments in Forest Stands in North Idaho, USA

Ryer M. Becker * and Robert F. Keefe

Forest Operations Research Lab, College of Natural Resources, University of Idaho, 875 Perimeter Drive, Moscow, ID 83844-4264, USA; robk@uidaho.edu

* Correspondence: rbecker@uidaho.edu; Tel.: +1-208-874-2698

Received: 7 August 2020; Accepted: 21 August 2020; Published: 28 August 2020

Abstract: Fuel reduction in forests is a high management priority in the western United States and mechanical mastication treatments are implemented common to achieve that goal. However, quantifying post-treatment fuel loading for use in fire behavior modeling to forecast treatment effectiveness is difficult due to the high cost and labor requirements of field sampling methods and high variability in resultant fuel loading within stands after treatment. We evaluated whether pre-treatment LiDAR-derived stand forest characteristics at 20 m × 20 m resolution could be used to predict post-treatment surface fuel loading following mastication. Plot-based destructive sampling was performed immediately following mastication at three stands in the Nez Perce Clearwater National Forest, Idaho, USA, to correlate post-treatment surface fuel loads and characteristics with pre-treatment LiDAR-derived forest metrics, specifically trees per hectare (TPH) and stand density index (SDI). Surface fuel loads measured in the stand post-treatment were consistent with those reported in previous studies. A significant relationship was found between the pre-treatment SDI and total resultant fuel loading ($p = 0.0477$), though not between TPH and fuel loading ($p = 0.0527$). SDI may more accurately predict post-treatment fuel loads by accounting for both tree number per unit area and stem size, while trees per hectare alone does not account for variations of tree size and subsequent volume within a stand. Relatively large root-mean-square errors associated with the random forest models for SDI (36%) and TPH (46%) suggest that increased sampling intensity and modified methods that better account for fine spatial variability in fuels resulting from within-stand conditions, treatment prescriptions and machine operators may be needed. Use of LiDAR to predict fuel loading after mastication is a useful approach for managers to understand the efficacy of fuel reduction treatments by providing information that may be helpful for determining areas where treatments can be most beneficial.

Keywords: LiDAR; mastication; SDI; stand density index; fuels; random forest

1. Introduction

Due to the variability of species, management objectives, spatial configuration of management areas, regulatory restrictions, landowner funding availability, fuels characteristics, and other geographic and vegetative factors, developing a one-size-fits-all approach for wide-scale fire management is challenging [1–4]. However, forests with high fire risk must be actively managed [5]. Over the previous century, forest management practices such as fire exclusion have resulted in historically uncharacteristic stand attributes in many forests in the western United States, including dense, small-diameter stands with increased surface fuel loads [5,6]. Fuel reduction in stands that have lacked prior density management is a high priority in many areas of the western United States, especially on federal lands. Understanding the unique challenges and selecting strategies to best suit the needs of each management area, typically applied through one or more treatments applied at the stand level, is vital to long-term management success [2]. Properly designed and implemented fuel treatments have been found to increase fire resilience and resistance while simultaneously changing the behavior of wildfires that impact treated areas [5,7–9].

Given the variability of management factors, including forest composition, topography, climatic conditions, and management history, not all silvicultural and fuel reduction treatments are feasibly implemented. In its most basic form, creating fire resilient stands generally involves three objectives: reducing surface fuels, reducing ladder fuels, and reducing crown density [5]. The complexity of planning fuel treatments for influencing the behavior of large fires must also account for spatial configuration and density of treatments when determining how to effectively and efficiently treat landscapes [4]. Fuel treatment programs have been implemented across the Western United States and include prescribed burning and mechanical treatments such as regeneration harvest, precommercial and commercial thinning, and mastication. The risk of fire escape, smoke restrictions, and poor public perception may limit the feasibility of large-scale prescribed burning efforts [10]. The need to meet particular site and climatic conditions to successfully implement prescribed burning often results in limited availability of windows in prescription, unlike mechanized treatments, that are less dependent on these factors [2]. To mimic the changes to forest structure created by fire, mechanical treatments are widely used to reduce crown fire risk, particularly in the Wildland Urban Interface (WUI) [10]. Mechanical treatments modify the vertical distribution of fuels and reduce overall canopy fuels to levels that are less susceptible to crown fires and rapid fire spread, without the risks associated with prescribed burning.

In many cases, stands that are at risk for severe fire are overstocked and may have high levels or mortality, which reduces the merchantable stand volume component and reduces the profitability of commercial timber harvest in the context of salvage harvesting. Alternative mechanical fuel treatment options, including mastication and chipping, are used in these instances when revenue from timber harvest may not be a core management objective, but fuel loads nevertheless need to be reduced. In these operations, which occur commonly on federal lands, fuels generated are left on site, unlike final harvest and commercial thinning treatments, where harvested materials are most commonly removed, treated at the landing, or treated in piles within the unit. The size of mechanical fuel treatments is dependent on the overall management objectives of the area. Mastication may target relatively small, high fire risk areas possessing dangerous fuel loads, where harvesting is not feasible. Alternatively, larger mastication operations may be used to reduce canopy density and ladder fuels while simultaneously reducing competition, removing undesirable trees, and preparing the stand for future harvest. Mastication entails the grinding, shredding, chunking, or by other means reducing the size of both standing and downed materials via boom-mounted mulching attachments of excavators, skid-steers, or other machines [11–13]. The risk of active crown fires is reduced by altering the vertical distribution and continuity of forest fuels and compacting them on the forest floor as irregularly shaped chips, though dead surface fuel loads are increased in the process [10,14]. Changes in physical properties of woody material resulting from mastication can influence fire behavior, including the rate of spread, flame length, and intensity [14]. According to Agee and Skinner [5], reducing surface fuels, increasing live crown height, retaining large, fire resistant trees, and decreasing crown bulk density are all important factors in producing fire-resistant stands. Altering the amount and condition of ladder fuels similarly influences fire intensity and burn severity [12]. When implemented correctly, mechanized fuel treatments address all points, excluding the reduction of surface fuel.

Despite the widespread use of mastication treatments [15] and past research, the spatial variability of masticated fuel beds has not been previously studied [10,11]. Studies have found surface fuel loadings in mulched treatments to range from 16 to 65 Mg ha^{-1}, with woody fuels concentrated in the 1-h and 10-h time-lag classes, which have average diameters < 2.54 cm [10,14,16–18]. Relative to untreated stands, mulched fuel beds with fuels concentrated in these classes have reduced rate of spread and flame lengths, but increased smoldering and flaming duration [10]. Quantifying masticated fuel loads is challenging, however, given the wide variability in masticated fuel physical structures and site, ecosystem, and regional fuel characteristics [15]. In masticated stands, fuel loads are also highly variable, leading to challenges when predicting and modeling loading and fire behavior. Fuel loads rely heavily on multiple factors, including vegetation type, pretreatment stand conditions, machinery and mastication attachment, operator and the treatment objectives, and desired conditions post-treatment [10,14–17]. As a result, the spatial heterogeneity of fuels within and across masticated stands remains unclear [15]. Several studies have successfully used fuel depth and/or fuel coverage in mulched areas to estimate surface fuel loads [10,14,17,18]. These techniques provide total surface fuel estimates more easily and accurately than planar transect sampling, but still require visiting the site following treatments. While less labor-intensive than past methods, the widespread assessment of surface fuel loading following mastication still requires in-person site visitation and assessment, which can be time-intensive and is only possible post-treatment. Efficient and effective methods to map post-treatment fuel loadings using pre-treatment conditions across stand, site, forest, and broader extents could provide valuable information to landowners when developing fuel management programs and evaluating their potential cost-effectiveness.

Fuel mapping is a difficult and often infeasible process across broad spatial and temporal scales due to high fuel variability, and costs and time constraints associated with field sampling [19–22]. Given the additional variability of masticated fuel loads, prediction models will likely need to focus on relatively small geographic extents and account for various forest characteristics. Micro-site predictions may better address site and forest variability and result in more accurate models. To better understand fuel loads resulting from treatments, it is important to first determine what forest conditions existed prior to mastication. On large landscape and regional scales, it is infeasible to perform a forest inventory to determine pre-treatment conditions due to labor and cost restrictions. Therefore, remotely sensed data, specifically LiDAR, has been repeatedly shown to provide large-scale forest metric predictions and enable the future extrapolation of models [23]. LiDAR has been used in part or entirely for assessing forest fuels characteristics including canopy [24–26], surface fuel loading [27–31], ladder fuels [32,33] and parameters including overall loading, spatial distribution, composition, vertical and horizontal arrangement, bulk density, and hazard ratings [34–38]. The authors were unable to find any relevant studies for LiDAR applications in masticated surface fuel loadings. Additionally, no models have currently been developed to predict post-mastication surface fuel loads from pre-treatment LiDAR-derived stand conditions.

In this study, we evaluate whether pre-treatment LiDAR-derived forest metrics can be used to predict surface fuel loads and associated fuel characteristics following mastication treatments. Masticated fuels were assessed across varying pre-treatment stand conditions to determine if relationships exist between the masticated fuel loads and LiDAR-derived forest metrics prior to treatment. Additionally, relationships between pre-treatment metrics and additional fuel bed characteristics including depth, size class distribution, and bulk density were assessed. Masticated fuels were sorted and quantified based on time-lag classes to determine if pre-treatment stand characteristics impact these distributions within sample plots. If remotely sensed forest metrics relate directly to mastication fuel characteristics, these models could be used to predict fuel loading and fuel bed characteristics for areas of similar forest composition, prior to mastication treatments. This information provides valuable insight to natural resources managers when selecting potential forest and fuel treatment options, ensuring both the economic and ecological sustainability of mechanical fuel treatments and other concurrent forest operations. Economically and ecologically unsustainable fuel treatments are financially and operationally infeasible, unsuccessful in achieving the desired operational, environmental and management results, and lead to the inefficient and ineffective use of limited financial resources. Understanding the potential impacts of mastication treatments based on existing forest conditions will help assess areas where this treatment option can be implemented cost-effectively when—coupled with existing fire behavior models.

2. Materials and Methods

2.1. Study Site

The study sites were located in three stands in the Nez Perce-Clearwater National Forest, in north central Idaho, following fuel treatments to gather data pertaining to the resultant surface fuel loads. These treatments were part of the larger Orogrande timber sale and consisted of approximately 38 hectares (95 acres) of mechanical fuel treatment. The management units were predominantly mixed conifer forest type with slopes averaging 35% throughout the units. Mastication was successfully implemented in the three stands as a management alternative to timber harvest. The three stands treated for this study were originally planned for timber harvest. Harvesting was found to be financially infeasible due to the low value of harvested products and long-haul distances to the mill. The prescription developed for the project was intended to release remaining trees to increase timber value for future harvest while simultaneously decreasing stand density and increasing canopy base height to reduce the risk of crown fire using mastication. According to forest personnel, the management approach used in this project was its first application on the Nez Perce-Clearwater National Forest. The machine used to perform the mechanical fuel treatment was a Takeuchi TB290 compact excavator with a Fecon Bull Hog mastication head (Figure 1). The machine weighs 8685 kg, is 2.2 m wide and 2.9 m long at the undercarriage, has a maximum reach of 7.4 m, and creates only 37.9 kPa of ground pressure when equipped with 450-mm-wide rubber tracks.

Figure 1. Mini-excavator with a horizontal shaft masticator head used in mastication treatments.

The treatment prescription for the stands included a target range for post-treatment stocking level. For the units in this study, the operator was instructed to leave 40 to 80 trees per hectare after treatment while removing only stems less than 18 cm (7 inches) in diameter. Further, all dead and down material up to 30 cm (12 inches) were masticated [39]. Post-treatment surface fuel sampling occurred in the masticated portions of three replicate stands: 117 (13 hectares), 120 (15 hectares), and 147 (10 hectares), within the management boundary. Due to many downed trees in stand 147, meeting the prescription specification for dead and downed material was not operationally feasible. Therefore, the mastication intensity for downed trees was reduced after stand 147 was partially treated. This prescription adjustment was used for treating the remainder of stand 147 and for the entirety of stands 117 and 120 [39] (Figure 2).

2.2. LiDAR Processing and Sample Plot Selection

The Orogrande timber sale and the three stands (117, 120, 147) were within the extent of the 18,450-hectare (45,600-acre) Crooked River LiDAR acquisition flown in 2012 with a pulse density return of ≥4 points per square meter (Figure 3). Field sampling inventory data from 91 20 × 20-m (1/10 acre) plots were run through the USFS Forest Vegetation Simulator [40] to summarize stand composition and structure. These forest inventory data were part of a previous sampling effort and were collected using field methods described in Falkowski et al. (2005) [22]. Random forest models [41,42] describing trees per hectare, total volume (m^3 ha^{-1}), basal area (m^2 ha^{-1}), and stand density index (SDI) were then developed using LiDAR metrics encompassing identical extents to the field sampling plots. These methods are consistent with those described in Becker et al. [43]. All random forest development and metric predictions were performed in the open source statistical analysis program, R, using the randomForest package [42,44].

Figure 2. Visual comparison of treated stands and a representation of surface fuels in stand 147 before (left pane) and after (right pane) adjustments to treatment intensity.

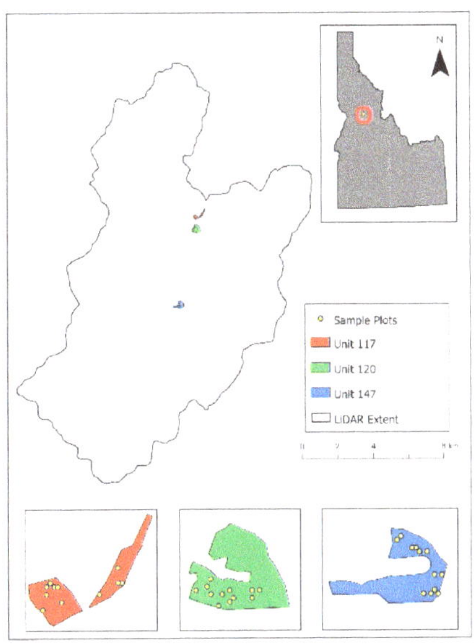

Figure 3. Crooked River LiDAR acquisition area and Orogrande T.S. mastication stands located in the Nez Perce-Clearwater National Forest, ID, USA.

Random forest ensemble learning algorithms were used because they provide excellent classification results, speed of processing, ability to reduce bias, and correlation and reduce overfitting compared to other classification and regression trees (CART) models, making them a widely used machine learning solution [45–48]. Random forest produces multiple decision trees using bagging and randomly selected subsets of training samples and variables to provide a majority vote from which a prediction is made. The default number of trees (n_{tree} = 500) and the default number of variables split at each node (m_{try} = square root of the total number of input variables) were used when building random forest models for each forest characteristic. 177 LiDAR metrics were available to develop each unique random forest algorithm, but only a subset of metrics were used in the final models, based on importance, defined using rfUtilities [49]. Random forest models were built using 2/3 of the data and validated using the remaining 1/3 of the data.

The entire Crooked River LiDAR acquisition was processed using the USDA LiDAR processing software FUSION version 3.60 to create an identical post-processed data structure as the initial 91 sampled plots [50]. This enabled the random forest models to be applied directly to the whole area to develop predicted metrics in 20 × 20-m pixels. Stand metrics derived from the LiDAR analysis included trees per hectare (TPH), total cubic foot volume (m^3 ha^{-1}), total basal area (m^2 ha^{-1}), and stand density index. SDI has been used in even-aged monocultures, and more recently in uneven-aged, mixed species stands to assess stand density as a function of quadratic mean diameter and stem density [51–55]. This metric was selected in addition to TPH to provide a more descriptive indication of stand density. The trees per hectare vector map was then stratified into four classes: 0–247; 248–494; 495–740; 741+. These classes were used to select sample plots within the study stands. Trees per hectare classes were used to stratify the selection of a broad distribution of relative stocking in sampled areas prior to mastication.

2.3. Field Sampling Procedures

Twelve plots were sampled within each of the three stands, with three representing each of the four levels of pre-treatment trees per hectare derived from the LiDAR data. The 20 × 20-m pixels chosen for sampling were randomly selected from all available pixels of the trees per hectare class within the stand boundaries. The resulting sampled pixels amounted to 36, with nine plots representing each of the four classes of pre-treatment trees per hectare. Trees per hectare was selected as the stand metric by which to select sample plots, due to the mastication treatment prescription being based on a goal trees per hectare post-treatment. All mastication treatments and sampling of fuel loading occurred during summer 2017.

Center points within the 20 × 20-m pixels were determined via ArcMap, and the resultant coordinates were used to locate the field plot centers. A simple method of plot center relocation was established to address situations where plot centers occur in areas that prohibited the sampling of fuel loading including tree stumps, roadways, rock outcroppings, and exposed mineral soil due to machinery movement. In these instances, plot centers were moved due north 3 m. If needed, plots were moved due west from original plot centers 3 m if the movement due north did not resolve the issue with the obstruction. A variation on destructive plot-based sampling was used to quantify fuel characteristics and fuel loading following mastication treatments within the three stands at each of the 36 plots [10,14,17].

Within each of the 36 sampled plots, fuel size classes were sampled in four quadrats. Once the plot center was located via GPS coordinates, 5-m vectors extending directly north, south, east, and west of the plot center were marked and established as the corner points for the quadrats. For instances in which uncharacteristic site conditions occurred within the quadrats, the frame was reflected over the transect. If this quadrat reflection did not resolve the issue and fuel collection within the quadrat was still not possible, the quadrat was excluded from sampling overall. Situations that would permit quadrat reflection over the transect or exclusion included buried logs, stumps, and rock outcroppings. In addition to the collection of fuels within the 25-cm squared quadrats, the fuel depth of masticated fuels was measured at two locations along the 5-m vectors (2.5 m and 5 m) and at the overall plot center (Figure 4). To measure fuel depths, a cross-section of the forest floor was cleared using a trowel, and the depths were manually recorded. For the depths of the woody/masticated material, any branch or piece of woody debris above the measurement point was included in the depth measurement. Where site conditions prevented the measuring of fuel depths, the depth was measured 0.5 m from the original measurement point moving away from the plot center along the transect. If the depth was still not measurable at the second location, the measurement was omitted. Each of the 36 plots contained four separate fuel collection quadrats and nine fuel depth measurements.

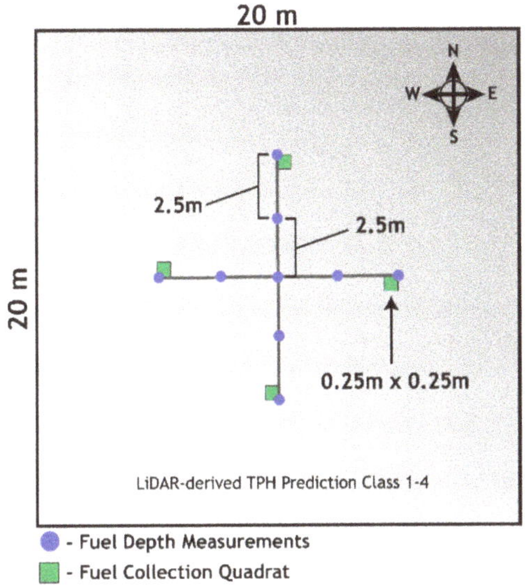

Figure 4. Destructive plot-based sampling design used for surface fuel collection and field measurements.

Frames made of PVC pipe measuring 25 cm by 25 cm were built to establish the sampling area extent. Within the 25 × 25-cm collection quadrats, all fuel down to bare mineral soil was gathered, and any pieces extending outside the collection frame were trimmed using hand shears, and thus only pieces completely within the frame were collected (Figure 5). All fuels collected were stored in paper bags labeled by their collection point and were brought back to the lab for detailed fuel composition analysis. Within the sampling quadrats, downed trees and logs were not sampled due to their irregular occurrence and collection difficulty for returning to the lab.

Figure 5. Fuel collection quadrat used in destructive fuel sampling.

It was assumed that when locating sample plots in the field, there may be instances where mastication, though planned, does not occur. This was a result of inaccessibility, due to the steep or very uneven terrain where the operator chose not to treat the area for safety reasons. In these instances, three supplemental sample points for each classification level of trees per hectare were randomly selected in each of the three stands. If an originally designated sample plot was found to be within an un-masticated area, a randomly selected supplemental sample plot of the same TPH class was selected for sampling instead. These supplemental plots were randomly generated from the remaining available pixels not included in the initial stratification prior to field sampling, using the same method used in the initial plot selection. Supplemental plots were used once in unit 117, four times in unit 120, and four times in unit 147. The final sampled plots across stands 117, 120, and 147 are shown in Figure 6.

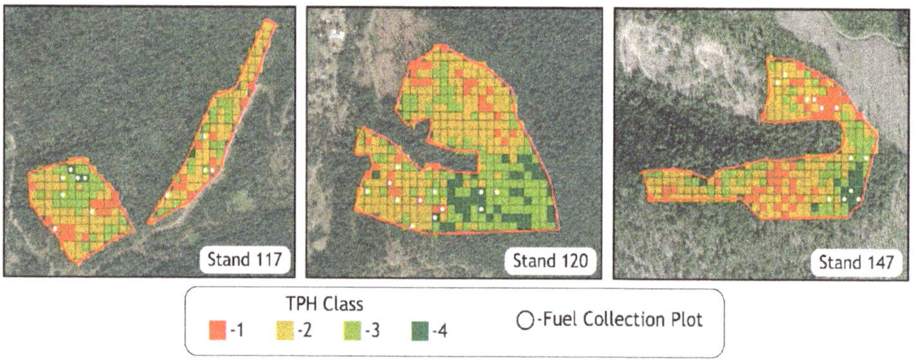

Figure 6. Field sampling plots within masticated stands 117, 120, and 147, selected based on stand density (TPH) prior to treatment.

2.4. Lab Measurements and Fuel Characterization

After field sampling was conducted, fuel collection bags were brought back to the lab to be processed by drying and sorting. A total of 133 quadrats of fuel was brought from the field for processing. If all plots had all quadrats collected, there would have been a total of 144. However, some quadrats were excluded from sampling because they were located on rock outcroppings, stumps, or other obstructions. One complete plot of quadrat fuel collections from stand 117, trees per hectare class 1, was misplaced during sampling, which constituted four of the eleven missing quadrat samples. Due to the omitted samples, stand 117 had two plots for trees per hectare class 1, resulting in 35 total plots rather than 36. Each collection sample was oven-dried at 105 degrees Celsius for 48 h, or until the sample weight stabilized, and was then weighted to the nearest gram. All fuels were then sorted, by quadrat, into five time-lag fuel classes: duff/litter and woody/masticated (1-h [<0.64 cm], 10-h [0.64–2.54 cm], 100-h [2.54–7.62 cm] and 1000-h [>7.62 cm]) [56] (Figure 7). Sorted fuels were then individually weighed to the nearest gram to determine the proportion of overall mass that each fuel class represented. These proportions for each quadrat were averaged with corresponding plot quadrats to determine the fuel composition proportions by mass for the entire plot. For each of the 35 sample plots, the fuel bed volume was calculated by multiplying the average of the fuel depths at the nine measured locations within each plot by the dimensions of the collection frame. We then determined the bulk density of the fuels in each plot by dividing the average oven-dried weight of the fuel classes in the four collection quadrats by the corresponding volume. Plot level values were calculated using the averages of each quadrat within the plot for fuel loading (Mg ha^{-1}) for the whole stand and by fuel class, fuel depth (cm), and bulk density (kg m^{-3}).

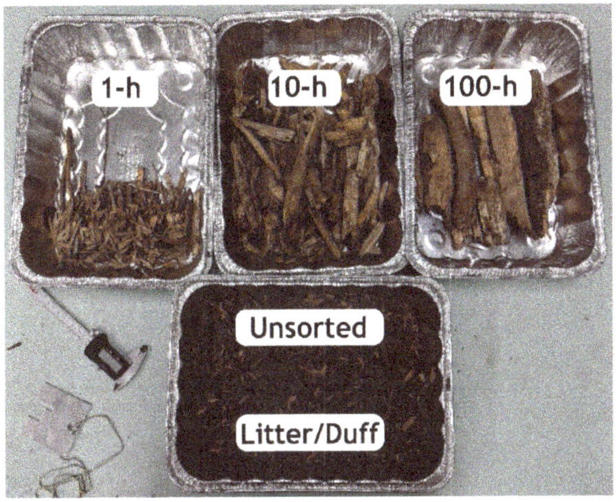

Figure 7. Sorting of dried surface fuels based on time-lag class (1-h, 10-h, 100-h, 1000-h, litter/duff).

To assess the correlation between pre-treatment LiDAR-derived forest metrics and post-treatment fuel conditions, all statistical analyses were performed using the R statistical programming environment. Pearson correlation coefficients were calculated to evaluate the strength of association between predictors. Additionally, linear mixed effects models were used to model the relationship between fuel loading following mastication treatments and the trees per hectare, stand density index, and basal area of plots prior to mastication, using the nlme R package [57]. The general equation for mixed effect models is described as:

$$\begin{aligned} y_i &= X_i\beta + Z_i u_i + \varepsilon_i \\ u_i &\sim N(0, D) \\ \varepsilon_i &\sim N(0, R_i) \end{aligned}$$

where β are fixed effects, u are random effects, X is the model matrix for fixed effects, Z is the model matrix for random effects, ε is the vector of errors, R is the variance-covariance matrix of within-individual measurements, and D is the variance-covariance matrix of random effects [58]. In linear mixed effects, models evaluating predictors, stand, and TPH class were treated as random effects, with the TPH class nested within the stand. Random intercepts were used when fitting models. Due to the inconsistency of the mastication treatment in stand 147, two different mixed effects models were fit for each of the LiDAR predictors. One model contained all three stands, while the second model contained only stand 117 and 120. This was in order to avoid potential influential data artifacts associated with the treatment change in stand 147.

3. Results

Parameter estimates for the random forest models used in the pretreatment derivation of forest characteristics from LiDAR metrics for 20 × 20-m pixels are shown in Table 1. In this study, an acceptable maximum root-mean-square error (RMSE) of 50% of the prediction means was used based on values derived in previous studies [59,60]. The RMSE was within the acceptable range for the density (TPH), basal area (m^2 ha^{-1}), and stand density index models. For the random forest model predicting stand volume (m^3 ha^{-1}), an RMSE of 166.93 was about 54% of the predicted mean and just outside of the desired range. Predicted volume was therefore excluded from use in subsequent analyses. Model accuracies for forest metrics were 71.5%, 77.4%, 74.3%, and 79% for trees per hectare, basal area, volume, and stand density index, respectively, which are comparable to those obtained by Falkowski et al. [61] and

Hudak et al. [60]. These pretreatment maps of predicted stand characteristics provided the basis for study plot selection and the subsequent regression modeling of post-treatment fuel loading.

Table 1. Random forest model quality assessment for pre-treatment forest metrics.

Random Forest	Prediction Mean	RMSE	R-Squared	Accuracy (%)
Stand Density (TPH)	468.00	217.36	0.55	71.5
Basal Area (m^2 ha^{-1})	30.56	12.95	0.63	77.4
Total Volume (m^3 ha^{-1})	307.99	165.35	0.57	76.3
Stand Density Index (SDI)	299.17	110.296	0.45	79.0

Table 2 shows the summary data of the stands for the fuel collection as averages of the sampled plots and quadrats within each stand. Surface fuel loadings range from 9.3–83.4 Mg ha^{-1}, 1.8–34.5 Mg ha^{-1}, 5.4–80.5 Mg ha^{-1}, 0–48.1 Mg ha^{-1}, and 0–8.2 Mg ha^{-1} for litter/duff, to 1-h, 10-h, 100-h and 1000-h fuel classes, respectively, across all plots and stands. Fuel depths ranged from 6.4–26.3 cm and bulk densities ranged from 22.2 to 154.2 kg m^{-3}. Across all plots, there was a significant ($p = 0.029$) moderate positive (0.369) correlation between trees per hectare and fuel loading (Mg m^{-1}), found by performing a Pearson's correlation test. Additionally, there was a positive (0.3577) and significant ($p = 0.0349$) correlation between SDI and fuel loading. No significant relationship was found between pre-treatment TPH and bulk density of resulting fuels (kg m^{-3}); SDI and bulk density of resulting fuels; basal area (m^2 h^{-1}) and loading or bulk density; nor between pre-treatment total volume (m^3 ha^{-1}) and fuel loading or bulk density (Table 3). Based on the results of the correlation tests, the linear mixed effects model was fitted to evaluate the relationship of pre-treatment TPH and the resulting fuel loads as well as SDI and the resulting fuel loads.

Table 2. Stand-level summary data representing stand averages and standard errors for pre-treatment trees per hectare and post-treatment destructive plot-based surface fuel characteristics for stands 117, 120, and 147.

Stand	Pre-TPH Avg. (SE)	Fuel Loading (Mg ha^{-1}) Avg. (SE)						Fuel Depth (cm) Avg. (SE)	Bulk Density (kg m^{-3}) Avg. (SE)
		Litter/Duff	1-h	10-h	100-h	1000-h	Total		
117	530 (77)	43.4 (6.7)	6.7 (1.2)	30.3 (4.3)	13.2 (3.7)	0.0 (0.0)	93.7 (13.1)	15.6 (1.5)	59.0 (5.8)
120	515 (77)	31.9 (3.3)	5.5 (1.0)	25.3 (3.2)	13.4 (2.6)	0.7 (0.7)	76.8 (8.8)	16.1 (1.2)	48.2 (4.4)
147	516 (93)	32.9 (3.6)	9.1 (2.5)	34.3 (5.9)	22.8 (3.6)	0.0 (0.0)	99.2 (10.4)	18.4 (1.5)	59.2 (10.6)

Table 3. Pearson's correlation assessments for pre-treatment forest characteristics and surface fuel characteristics following mastication treatments, where T is the t-test statistic and DF is the degrees of freedom.

Correlation	T	DF	*p*-Value	Coefficient
Density (TPH)/Loading (Mg ha^{-1})	2.2812	33	0.0291	0.3691
Density (TPH)/Bulk Density (kg m^{-3})	1.566	33	0.1269	0.2630
Volume (m^3 ha^{-1})/Loading (Mg ha^{-1})	1.8018	33	0.0807	0.2993
Volume (m^3 ha^{-1})/Bulk Density (kg m^{-3})	1.1251	33	0.2687	0.1922
Basal Area (m^2 h^{-1})/Loading (Mg ha^{-1})	1.9676	33	0.0576	0.3240
Basal Area (m^2 h^{-1})/Bulk Density (kg m^{-3})	1.3492	33	0.1865	0.2286
Stand Density Index/Loading (Mg ha^{-1})	2.2004	33	0.0349	0.3577
Stand Density Index/Bulk Density (kg m^{-3})	1.8702	33	0.0704	0.3096

The linear mixed effects models predicting the total fuel loading of all time-lag classes from pre-treatment TPH showed no significant relationship between the two factors (test statistic = 0.05318, df = 22, *p*-value = 0.0527) when accounting for all three stands. In the reduced model, pre-treatment trees per hectare and fuel loading were significant (*p*-value = 0.0066) (Table 4). The linear mixed effects models predicting the total fuel loading of all time-lag classes from pre-treatment SDI showed significant relationships between the two factors for all three units ($p = 0.0477$) and when assessing stand 117 and 120 alone ($p = 0.0337$) (Table 4). Figure 8 shows the associated relationships between SDI and the resulting fuel load for all three units. The black line represents the regression line of the complete data set and the individual regression line for each stand.

Table 4. Mixed effects model summary assessing pre-treatment trees per hectare (TPH) and stand density index (SDI) impact on fuel loading (Mg ha^{-1}) post-mastication. The influence of stand 147 on the overall significance of the factors is shown.

		Stand 117, 120, 147		
Predictor	**Estimate**	**Std. Error**	**DF**	***p*-Value**
TPH	0.05318	0.025974	22	0.0527
SDI	0.16736	0.079809	22	0.0477
		Stand 117, 120		
Predictor	**Estimate**	**Std. Error**	**DF**	***p*-Value**
TPH	0.09524	0.029924	14	0.0066
SDI	0.227069	0.09647	14	0.0337

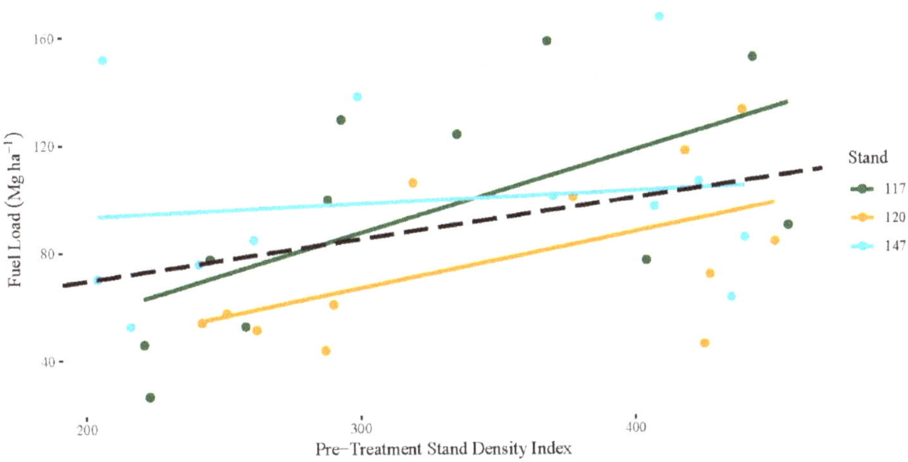

Figure 8. Surface fuel loading (Mg ha^{-1}) of all sampled plots by pre-treatment stand density index (SDI). The black dotted line represents the regression for all data, and the colored solid lines indicate the regression lines for each stand individually.

Additional linear mixed effects models were developed to further assess the impact of pre-treatment TPH and SDI on resulting fuels loads for each of the 5 time-lag fuel classes (litter/duff, 1-h, 10-h, 100-h, 1000-h). Only the litter/duff fuel class loading was found to be significantly correlated to pre-treatment trees per hectare for all three stands (Table 5). The litter/duff fuel class is generally independent of the mastication process, as most fuels in this class were present before treatment. However, when assessing stands 117 and 120, litter/duff ($p = 0.0242$), 1-h ($p = 0.0232$), and 100-h ($p = 0.0059$) were found to be significant. When assessing the relationship between SDI and the resulting fuel loading across all five time-lag cases, the model containing all three units showed that both litter/duff ($p = 0.0042$) and 100-h ($p = 0.0293$) were significant, while the model describing units 117 and 120 showed that only 100-h ($p = 0.0476$) was significant. The data for the SDI model are shown in Figure 9.

Table 5. Mixed effects model summary assessing pre-treatment trees per hectare (TPH) and stand density index (SDI) impact on fuel loading (Mg ha^{-1}) sorted by time-lag fuel class post-mastication. The influence of stand 147 on the overall significance of the predictors for each of the time-lag classes is shown.

Stand 117, 120, 147					
Fuel Class	**Predictor**	**Estimate**	**Std. Error**	**DF**	***p*-Value**
Litter/Duff	TPH	0.0033284	0.0010120	22	0.0034
1-h	TPH	0.0000789	0.0003873	22	0.8405
10-h	TPH	0.0000719	0.0009999	22	0.9433
100-h	TPH	0.0015591	0.0008138	22	0.0685
1000-h	TPH	−0.0000106	0.0000901	22	0.9072
Litter/Duff	SDI	0.096484	0.030246	22	0.0042
1-h	SDI	0.001137	0.012228	22	0.9267
10-h	SDI	0.004637	0.031801	22	0.8854
100-h	SDI	0.0562922	0.024141	22	0.0293
1000-h	SDI	−0.0007695	0.002888	22	0.7924
Stand 117, 120					
Fuel Class	**Predictor**	**Estimate**	**Std. Error**	**DF**	***p*-Value**
Litter/Duff	TPH	0.0038925	0.0015411	14	0.0242
1-h	TPH	0.0008194	0.0003215	14	0.0232
10-h	TPH	0.0021593	0.0010591	14	0.0608
100-h	TPH	0.0025606	0.0007890	14	0.0059
1000-h	TPH	−0.0000245	0.0001527	14	0.8747
Litter/Duff	SDI	0.095805	0.046646	14	0.0592
1-h	SDI	0.019364	0.010368	14	0.0829
10-h	SDI	0.058114	0.034173	14	0.1111
100-h	SDI	0.057603	0.026532	14	0.0476
1000-h	SDI	−0.002020	0.004787	14	0.6794

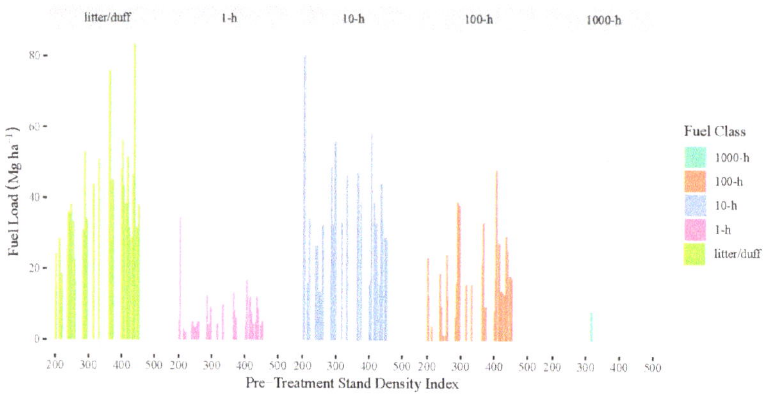

Figure 9. Surface fuel loading (Mg ha^{-1}) distribution for all sample plots by pre-treatment stand density index (SDI) arranged by time-lag fuel class.

Our second research objective explored whether the in-plot distribution of masticated fuels among the five time-lag fuel classes was impacted by the stand density of the plot prior to treatment. Mixed effects models were developed to assess these questions, with results being found in Table 6. 10-h fuels in the sampled plots were the only fuel class found to change significantly as the pre-treatment TPH changed (test statistic = −0.01581, p = 0.0178, df = 22). SDI was shown to have a significant relationship with post-mastication fuel loading for both the 10-h (p = 0.0302) and 100-h (p = 0.0406) fuel classes (Table 6).

Table 6. Mixed effects model summary assessing pre-treatment stand density (TPH) and stand density index (SDI) impact on the percentage of total fuel load (Mg ha^{-1}) by time-lag fuel class.

Stand Density (TPH)				
Fuel Class	**Estimate**	**Std. Error**	**DF**	***p*-Value**
Litter/Duff	0.00560	0.009796	22	0.5736
1-h	−0.00190	0.0021351	22	0.3825
10-h	−0.01581	0.006169	22	0.0178
100-h	0.01161	0.006380	22	0.0823
1000-h	−0.0001	0.000850	22	0.9072
Stand Density Index (SDI)				
Fuel Class	**Estimate**	**Std. Error**	**DF**	***p*-Value**
Litter/Duff	0.00766	0.03098	22	0.8069
1-h	−0.007688	0.006679	22	0.2620
10-h	−0.04619	0.019931	22	0.0302
100-h	0.042089	0.019346	22	0.0406
1000-h	−0.0007254	0.0027224	22	0.7924

The data distribution for the five time-lag classes as a percentage across the range of stand density indices and fuel loading are shown in Figure 10. Woody and mulched fuels (1-h, 10-h, 100-h, and 1000-h) were found to be between 30–85% of the overall fuel loads across all plots, which shows the variability found within the stands. 1-h, 10-h, 100-h, and 1000-h fuels contained 6–27%, 36–94%, 0–57%, and 0–13% of the woody and mulched fuels, respectively. On average, woody fuels made up about 58.2% of the overall surface fuel loading across all sites, with 1-h, 10-h, 100-h, and 1000-h fuel classes averaging 7.5%, 33.2%, 17.3%, and 0.2% of the loading, respectively.

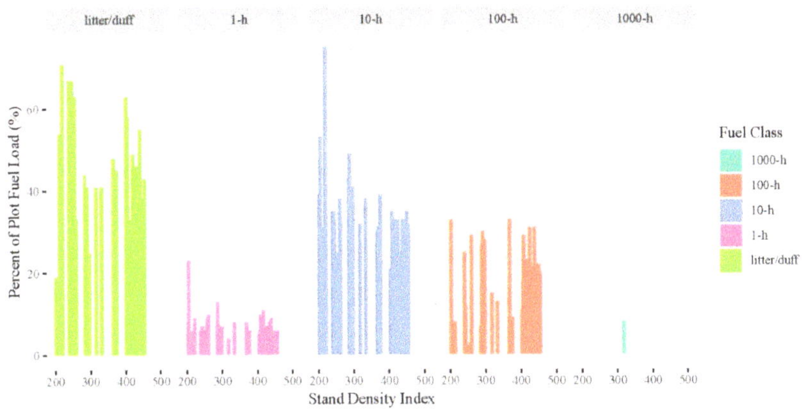

Figure 10. Percentage of total surface fuel load for all sample plots by pre-treatment stand density index (SDI) arranged by time-lag fuel class.

4. Discussion

Total surface fuel loadings varied widely across our study plots (26.4 to 158.2 Mg ha^{-1}), with woody/masticated surface fuels representing a similarly wide range (7.7 to 127.5 Mg ha^{-1}). Across all stands and plots, total surface fuel loading averaged 89.8 Mg ha^{-1}, and woody surface fuels averaged 53.9 Mg ha^{-1}. These total surface fuel loads were similar to those reported by Stephens and Moghaddas [16] in the Sierra Nevada Mountains (87.1 and 93.8 Mg ha^{-1}), Reiner et al. [18] in the Sierra Nevada (81.6 Mg ha^{-1}), Kane et al. [14] at study sites in Northern California and southwestern Oregon (83.6, 83.8 and 71.1 Mg ha^{-1}), and Hood and Wu [17] in the Northern Rockies (82.0–95.9 Mg ha^{-1}), but were higher than those reported by Brewer et al. [62] in mixed conifer Idaho stands (58.4 Mg ha^{-1}) and lower than those reported by Battaglia et al. [10] in mixed conifer stands in Colorado (110.4 Mg ha^{-1}). This finding was not surprising, as masticated fuel beds and characteristics have a wide variability across sites and regions. The attempt to develop the surface fuel prediction model at a 20 × 20-m resolution in this study was meant to help address this site variability.

Based on results reported by previous mastication studies, the average total fuel depth of 16.7 cm recorded in the mixed conifer stands we studied was significantly higher than those in Battaglia et al. [10], but only slightly higher than those shown by Stephens and Moghaddas [16]: 14.6 and 14.7 cm. All fuel sampling in this study occurred within a month of mastication treatment, so fuels were still green at the time of collection and had not settled to the forest floor, whereas sampling of masticated fuels for other studies occurred 2–6 years post-mastication [10,14]. These temporal changes in masticated fuel beds make the generalization of loadings difficult, especially across broad geographic extents. For example, a recently masticated stand may indicate greater fuel depths than a stand masticated several years ago, due to the decomposition and deterioration of fuel structural integrity. The additional compaction of fuel beds over time as they settle may affect subsequent fire behavior. Therefore, the development of the model estimating surface fuel characteristics directly after masticating would provide a consistent expectation of fuel loads, as was done in this study.

4.1. Relationship between Pre-Treatment Stand Characteristics and Fuel Loading

Through the analysis performed across the three stands we studied—117, 120, and 147—no significant relationship was found between overall fuel loading following mastication and the pre-treatment tree per hectare we derived from LiDAR. However, SDI was found to be a significant predictor variable for post-mastication fuel loading when accounting for all management units and time-lag classes jointly. Initially, we expected to see an increase in the fuel loading as the pre-treatment TPH increased. It was believed that, given a consistent prescription implementation, greater TPH would result in more fuel, as a greater number of standing trees were mulched to meet treatment objectives. The contrary findings for absolute stand density may have resulted from the inconsistency in the initial treatment of stand 147, which was then corrected. When excluding stand 147, increasing pre-treatment TPH resulted in greater loading, as expected. However, even when including data from stand 147, the p-value of 0.0526 was just outside the level of significance needed to reject the null. The significance of both mixed effects models for SDI ($p = 0.0477$ and $p = 0.0337$) indicates fuel loading may be more accurately predicted using a metric that accounts for both tree size and number, as opposed to simply using trees per hectare where only the number of stems is accounted for.

Stand 147 was the first stand treated and was initially treated to prescription specifications. It was found that once treatment began, the original degree to which large downed woody debris was to be treated was operationally infeasible due to the increased treatment time. Further, stand 147 contained a small pocket of lodgepole pine killed by beetle, with a significant portion of downed trees which were, under the original treatment specs, to be masticated heavily. This resulted in a larger amount of masticated fuels in plots with relatively low stand densities.

The decision to retain stand 147 in the analysis was made to maximize the data available for assessment and provide a realistic portrayal of the large variability of mastication treatments. As a relatively new treatment option being deployed over large areas, it is likely that similar inconsistency in operational treatments may occur during implementation and administration of mechanical fuel treatments, particularly as operators familiarize themselves with prescription requirements in new treatment areas. However, when the treatment prescription and execution was consistent for the entire stand, as seen in stands 117 and 120, a clear relationship between pre-treatment trees per hectare and fuel loading was seen. Given the potential for variability of mastication treatments and the heterogeneity of stand conditions in practice, the predictive success of SDI in a "real-world" management scenario is valuable for future modeling efforts.

Given the extent of our results, it remains unclear if a relationship exists between the fuel loading following mastication treatments and the pre-treatment stand density based solely on TPH or stand basal area, but SDI is a useful predictor. In mastication, the conservation of mass must be considered, as materials are not removed from the stand after treatment but rearranged in different physical forms. A larger masticated tree will understandably produce a larger amount of masticated material than a tree of smaller size. For example, two stands may both have similar numbers of trees per unit area, but one stand may have a larger average tree diameter than the other. If both stands are treated to the same prescription and reduced to a defined tree per unit area, it would be expected that the stand with the larger average stand diameter would produce heavier masticated fuel loadings. Accounting for both stem number per unit area and tree size in a single pre-treatment stand metric, SDI addresses this issue. Therefore, alternative approaches to modeling landscape scale fuel loading following mastication based on pre-treatment stand conditions that incorporate both stem numbers and size may offer improved prediction in future research and should be the focus of future study design and implementation.

When assessing the fuel loading for each time-lag class (litter/duff, 1-h, 10-h, 100-h, 1000-h), the litter duff class showed a significant relationship for the TPH and SDI models (Table 5). This may be a result of greater stand density, leading to higher amounts of organic material and litter on the forest floor. In all, minimal 1000-h fuels were collected at the plots, limiting the available data for the particular classes and making predictions difficult. This finding corroborates Kane et al.'s [11], who found that the plot-based method of surface fuel sampling does not assess a large enough area to effectively capture the presence of 1000-h fuels as well as planar intersect methods. This is a result of 1000-h fuels generally occurring less frequently than other fuel classes in fuel beds.

4.2. Relationship between Pre-Treatment Stand Density and Fuel Class Distribution

As shown above (Table 6), only the 10-h fuels expressed as a percentage of the overall surface fuel loading were found to change as the TPH increased. It is unclear why the percentage of 10-h fuels would decrease with increasing TPH, but this may be a result of changes in treatment implementation. For example, the operator may spend less time masticating trees to maintain production in a denser stand, resulting in an increase in the proportion of larger fuel classes. It would be expected that, with one fuel class decreasing over increasing TPH, another fuel class would increase. This was seen in the SDI model, where the significant decrease in 10-h fuels (coefficient = −0.04616, p = 0.0302) was matched by a significant increase in 100-h fuels (coefficient = 0.042089, p = 0.0406). With increasing stand density, it is possible the operator attempted to maintain the desired operational production by decreasing the time spent masticating each tree. As a result, trees would be masticated less thoroughly, and there would be a larger percentage of larger fuel particles. The 10-h fuel class accounted for the highest fuel loads across all classes by a considerable amount in our study, which is consistent with other studies [10,14]. Given the variability of the mastication as a whole, and the wide range of fuel loadings across 1-h, 10-h, 100-h, and 1000-h fuel classes, our results show that the distribution of surface fuels among time-lag fuel classes was not clearly modeled as a function of changes in TPH and SDI alone, apart from the 10-h and 100-h classes, in the case of this study.

4.3. Study Limitations and Future Work

Mastication is a highly variable operation impacted by many factors, and it is understood that there are some limitations to the scope of our research that should be addressed in future studies. One factor to consider in future applications of this methodology is the pixel size at which the LiDAR metrics were predicted. In the study development, it was believed that maximizing prediction resolution was the best option. Mastication, however, is a variable process, resulting in a scattered distribution of fuels on the forest floor. The directionality, travel distance, and particle size of comminuted materials may be affected by the type of mastication head (disk vs. drum), equipment type (all surface vehicle vs. excavator carrier), equipment horsepower, boom or attachment height, local topography within the stand, or other factors. While the sampling method developed by Hood and Wu [17] attempts to address the variability in stand and site conditions by sampling across multiple quadrats within the same plot, what was not accounted for in our study design was the possibility of fuels from adjoining pixels being distributed inter-pixel. During the observation of the treatment, fuels were clearly distributed more irregularly and further than anticipated. The 20 m × 20-m pixels used in the plot selection may have been too small to limit the influence of surrounding pixels in the resulting fuels found during sampling. For instance, a stand with a high stand density may have resulted in fuels initially in the stand as standing trees being distributed to an adjacent stand of a lower stand density, or to areas within the same stand that were not accounted for in our sampling design. During sampling, it would then appear that the pixel with the lower stand density was responsible for creating greater fuel loads than was possible. By decreasing the resolution and increasing the pixel size, this may be avoided.

In the fuel collection process, future studies should incorporate a hybrid, plot-based, and planar intersect method, as suggested by Kane et al. [14]. Doing so may help to ensure a more accurate representation of the fuel classes, as 1-h and 10-h fuels are more accurately represented in plot-based sampling [14], while planar intersect methods cover greater proportions of the overall masticated area, properly representing the 100-h and 1000-h fuels that may be missed in plot-based approaches [11]. Supplemental planar intersect sampling was not performed in this study due to the small 20 m × 20-m pixel size and the concern that sufficiently long intersect paths would extend too far to sample plot edges and be impacted by the distribution of fuels from adjacent pixels. Increasing the pixel size used in predicting the stand characteristics from the LiDAR, as described above, would enable an easier implementation of supplemental planar intersect sampling. Additionally, due to the variability of fuel distribution across the forest floor, plot-based sampling in future studies should use larger sampling quadrats than the 25 × 25-cm ones used in this study. Alternatively, a larger number of 25 × 25-cm sampling quadrats may also provide a greater representation of overall fuel variability within the sampling plot. The goal in using a smaller sampling quadrat in this study than those described in previous studies [17,63] was to create an efficient and effective sampling procedure. However, larger quadrats will provide a greater representation of overall surface fuel loadings and should be studied.

5. Conclusions

The ability to quickly, efficiently, and effectively predict surface fuel loads resulting from mastication treatments is a valuable tool, as increased implementation of this management technique occurs. A variety of research and management questions regarding the longevity, fuel bed characteristics, and fire behavior within masticated fuels exist and will increase in relevance as LiDAR data become more widespread, along with the use of mastication to reduce fuels in stands where commercial thinning may be infeasible or more difficult to implement administratively. Existing methods for predicting surface fuel loads rely on intensive, time-consuming sampling following treatment. While existing methods are effective for estimating fuel loading, methods based on remote sensing may help managers to proactively plan and predict post-treatment fire behavior over large areas to optimize treatments in ways that incorporate topography and stand adjacency.

The results from this study showed that pre-treatment stand density metrics that account only for tree number per unit area, such as TPH, were not good predictors of resulting surface fuel loads following mechanical fuel treatments with the sampling design and sample size we evaluated. TPH prior to treatment was not directly related to the distribution of fuel time-lag classes within the fuel bed, although the percentage of 10-h fuels could be predicted from pre-treatment conditions. However, stand density index, which accounts for both the relative stem number and DBH of the stand, effectively predicts post-treatment fuel loading across the whole study area. Further, SDI predicted that as the density of a stand increases, a greater percentage of the overall fuel load consisted of 100-h fuels, while 10-h fuels decreased in percentage, likely a result of operational adjustments. Future modeling efforts to predict post-mastication fuel loading should account for both the stem number and stem size, as stand density alone may not provide the necessary predictive ability. Attempting to predict resulting fuels from the number of trees per unit area alone does not account for variable volumes of materials in trees of different diameters. Stand density measures, such as SDI, provide greater insight into stand composition and overall stand biomass, which is significant when predicting fuel load volumes resulting from the physical conversion of standing biomass to surface-based mulched materials. Two stands with identical TPH may contain varying amounts of biomass as standing trees, whereas it is expected that two stands with identical SDI would have the same amount of overall biomass given similar forest types and species.

We believe that revisiting these methods, while taking into account the sampling considerations mentioned in the discussion, is an important undertaking and could lead to the increased implementation and effectiveness of mastication treatments. The rapid onset of LiDAR-derived models to map individual-tree locations and stem characteristics, coupled with onboard GNSS mapping of spatially, explicit, real-time equipment activities, offer the promise of improved high-resolution fuel bed prediction in the immediate future. Further expanding the scope of the field sampling to multiple, unique forest types, operators, and prescriptions would better capture the variability associated with the masticated surface fuel loads. Future work should address these factors more closely, though the determination of their impacts will likely require sampling at a higher intensity than that performed in this study, or with a sampling design that directly accounts for the spatial resolution at which comminuted material is scattered as a function of localized stand density, treatment prescription, topography, equipment type and size, and the pattern of equipment movements.

Author Contributions: Conceptualization, R.M.B. and R.F.K.; methodology, R.M.B. and R.F.K.; formal analysis, R.M.B. and R.F.K.; writing—original draft preparation, R.M.B.; writing—review and editing, R.M.B., R.F.K.; visualization, R.M.B.; supervision, R.F.K.; funding acquisition, R.F.K. All authors have read and agreed to the published version of the manuscript.

Funding: This research was funded by Joint Fire Sciences, JFSP PROJECT ID: 13-1-05-7.

Acknowledgments: The authors would like to thank Tom McLeod and Bob Vermey from the Nez Perce-Clearwater National Forest for help in facilitating and locating field sampling locales. Further, we would like to thank Mackenzie Smith for the time given during the field sampling. Publication of this article was funded by the University of Idaho's Open Access Publishing Fund. All authors have read and agreed to the published version of the manuscript.

Conflicts of Interest: The authors declare no conflict of interest.

References

1. Reinhardt, E.D.; Keane, R.E.; Calkin, D.E.; Cohen, J.D. Objectives and considerations for wildland fuel treatment in forested ecosystems of the interior western United States. *For. Ecol. Manag.* **2008**, *256*, 1997–2006. [CrossRef]
2. Hartsough, B.R.; Abrams, S.; Barbour, R.J.; Drews, E.S.; McIver, J.D.; Moghaddas, J.J.; Schwilk, D.W.; Stephens, S.L. The economics of alternative fuel reduction treatments in western United States dry forests: Financial and policy implications from the National Fire and Fire Surrogate Study. *For. Policy Econ.* **2008**, *10*, 344–354. [CrossRef]
3. Oliveira, T.M.; Barros, A.M.G.; Ager, A.A.; Fernandes, P.M. Assessing the effect of a fuel break network to reduce brunt area and wildfire risk transmission. *Int. J. Wildland Fire* **2016**, *25*, 619–632. [CrossRef]
4. Alcasena, F.J.; Ager, A.A.; Salis, M.; Day, M.A.; Vega-Garcia, C. Optimizing prescribed fire allocation for managing fire risk in central Catalonia. *Sci. Total. Environ.* **2018**, *621*, 872–885. [CrossRef] [PubMed]
5. Agee, J.K.; Skinner, C.N. Basic principles of forest fuel reduction treatments. *For. Ecol. Manag.* **2005**, *211*, 83–96. [CrossRef]
6. Stephens, S.L.; Moghaddas, J.J.; Edminster, C.; Fiedler, C.E.; Haase, S.; Harrington, M.; Keeley, J.E.; Knapp, E.E.; McIver, J.D.; Metlen, K.; et al. Fire treatment effects on vegetation structure, fuels, and potential fire severity in western U.S. forests. *Ecol. Appl.* **2009**, *19*, 305–320. [CrossRef]
7. Agee, J.K. The influence of forest structure on fire behavior. In Proceedings of the 17th Annual Forest Vegetation Management Conference, Redding, CA, USA, 16–18 January 1996; pp. 52–68.
8. Stephens, S.L. Effects of fuels and silvicultural treatments on potential fire behavior in mixed conifer forests of the Sierra Nevada, CA. *For. Ecol. Manag.* **1998**, *105*, 21–34. [CrossRef]

9. Huggett, R.J., Jr.; Abt, K.L.; Shepperd, W. Efficacy of mechanical fuel treatments for reducing wildfire hazard. *For. Policy Econ.* **2008**, *10*, 408–414. [CrossRef]
10. Battaglia, M.A.; Rocca, M.E.; Rhoades, C.C.; Ryan, M.G. Surface fuel loadings within mulching treatments in Colorado conifer forests. *For. Ecol. Manag.* **2010**, *260*, 1557–1566. [CrossRef]
11. Kane, J.M.; Knapp, E.E.; Varner, J.M. Variability in loadings of mechanically masticated fuel beds in northern California and southwestern Oregon. In *Fuels Management-How to Measure Success: Conference Proceedings, Proceedings RMRS-P-41, Portland, OR, USA, 28–30 March 2006*; Andrews, P.L., Butler, B.W., Eds.; U.S. Department of Agriculture, Forest Service, Rocky Mountain Research Station: Fort Collins, CO, USA, 2006; pp. 341–350.
12. Jain, T.; Graham, R.T.; Sandquist, J.; Butler, M.; Brockus, K.; Frigard, D.; Cobb, D.; Sup-Han, H.; Halbrook, J.; Denner, R.; et al. Restoration of northern Rocky Mountain moist forests: Integrating fuel treatments from the site to the landscape. In *Integrated Restoration of Forested Ecosystems to Achieve Multiresource Benefits, Proceedings of the 2007 National Silviculture Workshop, Gen. Tech. Rep. PNW-GTR-733, Ketchikan, AK, USA, 7–10 May 2007*; Deal, R.L., Ed.; U.S. Department of Agriculture, Forest Service, Pacific Northwest Research Station: Portland, OR, USA, 2008.
13. Jain, T.; Sikkink, P.; Keefe, R.; Byrne, J. *To Masticate or Not: Useful Tips for Treating Forest, Woodland, and Shrubland Vegetation*; Gen. Tech. Rep. RMRS-GTR-381; U.S. Department of Agriculture, Forest Service, Rocky Mountain Research Station: Fort Collins, CO, USA, 2018.
14. Kane, J.M.; Varner, J.M.; Knapp, E.E. Novel fuelbed characteristics associated with mechanical mastication treatments in northern California and south-western Oregon, USA. *Int. J. Wildland Fire* **2009**, *18*, 688–697. [CrossRef]
15. Kreye, J.K.; Brewer, N.W.; Morgan, P.; Varner, J.M.; Smith, A.M.S.; Hoffman, C.M.; Ottmar, R.D. Fire behavior in masticated fuels: A review. *For. Ecol. Manag.* **2014**, *314*, 193–207. [CrossRef]
16. Stephens, S.L.; Moghaddas, J.J. Fuel treatment effects on snags and course woody debris in a Sierra Nevada mixed conifer forest. *For. Ecol. Manag.* **2005**, *214*, 53–64. [CrossRef]
17. Hood, S.; Wu, R. Estimating fuelbed loadings in masticated areas. In *Fuels Management-How to Measure Success: Conference Proceedings, Proceedings RMRS-P-41, Portland, OR, USA, 28–30 March 2006*; Andrews, P.L., Butler, B.W., Eds.; U.S. Department of Agriculture, Forest Service, Rocky Mountain Research Station: Fort Collins, CO, USA, 2006; pp. 333–340.
18. Reiner, A.L.; Vaillant, N.M.; Fites-Kaufman, J.; Dailey, S.N. Mastication and prescribed fire impacts on fuels in a 25-year old ponderosa pine plantation, southern Sierra Nevada. *For. Ecol. Manag.* **2009**, *258*, 2365–2372. [CrossRef]
19. Arroyo, L.A.; Pascual, C.; Manzanera, J.A. Fire models and methods to map fuel types: The role of remote sensing. *For. Ecol. Manag.* **2008**, *256*, 1239–1252. [CrossRef]
20. Keane, R.E.; Burgan, R.; van Wagtendonk, J. Mapping wildland fuels for fire management across multiple scales: Integrating remote sensing, GIS, and biophysical modelling. *Int. J. Wildland Fire* **2001**, *10*, 301–319. [CrossRef]
21. Rollins, M.G.; Keane, R.E.; Parsons, R.A. Mapping fuel and fire regimes using remote sensing, ecosystem simulation, and gradient modeling. *Ecol. Appl.* **2004**, *14*, 75–95. [CrossRef]
22. Falkowski, M.J.; Gessler, P.E.; Morgan, P.; Hudak, A.T.; Smith, A.M.S. Characterizing and mapping forest fire fuels using ASTER imagery and gradient modeling. *For. Ecol. Manag.* **2005**, *217*, 129–146. [CrossRef]
23. Maltamo, M.; Eerikainen, K.; Pitkanen, J.; Hyyppa, J.; Vehmas, M. Estimation of timber volume and stem density based on scanning laser altimetry and expected tree size distribution functions. *Remote Sens. Environ.* **2004**, *90*, 319–330. [CrossRef]
24. Andersen, H.; McGaughey, R.J.; Reutebuch, S.E. Estimating forest canopy fuel parameters using LIDAR data. *Remote Sens. Environ.* **2005**, *94*, 441–449. [CrossRef]
25. Skowronski, N.S.; Clark, K.L.; Duveneck, M.; Hom, J. Three-dimensional canopy fuel loading predicted using upward and downward sensing LiDAR systems. *Remote Sens. Environ.* **2011**, *115*, 703–714. [CrossRef]
26. Erdody, T.L.; Moskal, L.M. Fusion of LiDAR and imagery for estimating forest canopy fuels. *Remote Sens. Environ.* **2010**, *114*, 725–737. [CrossRef]
27. Chen, Y.; Zhu, X.; Yebra, M.; Harris, S.; Tapper, N. Development of a predictive model for estimating forest surface fuel load in Australian eucalypt forests with LiDAR data. *Environ. Model. Softw.* **2017**, *97*, 61–71. [CrossRef]

28. Jakubowski, M.K.; Guo, Q.; Collins, B.; Stephens, S.; Kelly, M. Predicting surface fuel models and fuel metrics using lidar and CIR imagery in a dense, mountainous forest. *Photogramm. Eng. Remote Sens.* **2013**, *79*, 37–49. [CrossRef]
29. Mutlu, M.; Popescu, S.C.; Stripling, C.; Spencer, T. Mapping surface fuel models using lidar and multispectral data fusion for fire behavior. *Remote Sens. Environ.* **2008**, *112*, 274–285. [CrossRef]
30. Van Aardt, J.A.N.; Arthur, M.; Sovkoplas, G.; Swetnam, T.L. LiDAR-based estimation of forest floor fuel loads using a novel distributional approach. In Proceedings of the SilviLaser 2011, Hobart, Australia, 16–20 October 2011.
31. Mutlu, M.; Popescu, S.C.; Zhao, K. Sensitivity analysis of fire behavior modeling with LIDAR-derived surface fuel maps. *For. Ecol. Manag.* **2008**, *256*, 289–294. [CrossRef]
32. Kramer, H.A.; Collins, B.M.; Kelly, M.; Stephens, S.L. Quantifying ladder fuels: A new approach to using LiDAR. *Forests* **2014**, *5*, 1432–1453. [CrossRef]
33. Skowronski, N.; Clark, K.; Nelson, R.; Hom, J.; Patterson, M. Remotely sensed measurements of forest structure and fuel loads in the Pinelands of New Jersey. *Remote Sens. Environ.* **2007**, *108*, 123–129. [CrossRef]
34. Garcia, M.; Riano, D.; Chuvieco, E.; Salas, J.; Danson, F.M. Multispectral and lidar data fusion for fuel type mapping using Support Vector Machine and decision rules. *Remote Sens. Environ.* **2011**, *115*, 1369–1379. [CrossRef]
35. Koetz, B.; Morsdorf, F.; van der Linden, S.; Curt, T.; Allgower, B. Multi-source land cover classification for forest fire management based on imaging spectrometry and lidar data. *For. Ecol. Manag.* **2008**, *256*, 263–271. [CrossRef]
36. Price, O.F.; Gordon, C.E. The potential for LiDAR technology to map fire fuel hazard over large areas of Australian forests. *J. Environ. Manag.* **2016**, *181*, 663–673. [CrossRef]
37. Garcia, M.; Saatchi, S.; Casas, A.; Koltunov, A.; Ustin, S.L.; Ramirez, C.; Balzter, H. Extrapolating forest canopy fuel properties in the California Rim Fire by combining airborne LiDAR and Landsat OLI data. *Remote Sens.* **2017**, *9*, 394. [CrossRef]
38. Chen, Y.; Zhu, X.; Yebra, M.; Harris, S.; Tapper, N. Strata-based forest fuel classification for wild fire hazard assessment using terrestrial LiDAR. *J. Appl. Remote Sens.* **2016**, *10*, 046025. [CrossRef]
39. Vermey, B.; (Nez Perce-Clearwater National Forest Red River Ranger District, Elk City, ID, USA). Personal communication, 2017.
40. Dixon, G.E. *Essential FVS: A User's Guide to the Forest Vegetation Simulator. Internal Report*; U.S. Department of Agriculture, Forest Service, Forest Management Service Center: Fort Collins, CO, USA, 2002.
41. Breiman, L. Random Forests. *Mach. Learn.* **2001**, *45*, 5–32. [CrossRef]
42. Liaw, A.; Wiener, M. Classification and Regression by randomForest. *R News* **2002**, *2*, 18–22.
43. Becker, R.M.; Keefe, R.F.; Anderson, N.M.; Eitel, J.U.H. Use of LiDAR-derived landscape parameters to characterize alternative harvest system options in the Inland Northwest. *Int. J. For. Eng.* **2018**, *29*, 179–191. [CrossRef]
44. R Core Team. *R: A Language and Environment for Statistical Computing. R Foundation for Statistical Computing*; R Core Team: Vienna, Austria, 2020. Available online: https://www.R-project.org/ (accessed on 12 February 2020).
45. Breiman, L. *Manual on Setting Up, Using, and Understanding Random Forests v3.1*; Statistics Department University of California: Berkeley, CA, USA, 2002. Available online: https://www.stat.berkeley.edu/~{}breiman/RandomForests/cc_manual.htm (accessed on 13 February 2020).
46. Brosofske, K.D.; Froese, R.E.; Falkowski, M.J.; Banskota, A. A review of methods for mapping and prediction of inventory attributes for operational forest management. *For. Sci.* **2014**, *60*, 733–756. [CrossRef]
47. Cinaroglu, S. Comparison of performance of decision tree algorithms and random forest: An application of OECD countries health expenditures. *Int. J. Comput. Appl.* **2016**, *138*, 37–41. [CrossRef]
48. Belgiu, M.; Dragut, L. Random forest in remote sensing: A review of applications and future directions. *ISPRS J. Photogramm.* **2016**, *114*, 24–31. [CrossRef]
49. Evans, J.S.; Murphy, M.A. rfUtilities: Random Forests Model Selection and Performance Evaluation. R Package Version 2.0-1. Available online: http://rdocumentation.org/packages/rfUtilities/versions/2.0-1.html. (accessed on 15 May 2017).

50. McGaughey, R.J. *FUSION/LDV: Software for LiDAR Data Analysis and Visualization. February 2013–FUSION Version 3.60*; USDA Forest Service, Pacific Northwest Research Station, University of Washington: Seattle, WA, USA, 2017; Available online: http://forsys.sefs.uw.edu/fusion/fusionlatest.html (accessed on 15 June 2017).
51. Reineke, L.H. Perfecting stand-density index for even-aged forests. *J. Agric. Res.* **1933**, *46*, 627–638.
52. Woodall, C.W.; Miles, P.D.; Vissage, J.S. Determining maximum stand density index in mixed species stands for strategic-scale stocking assessments. *For. Ecol. Manag.* **2005**, *216*, 367–377. [CrossRef]
53. Shaw, J.D. Reineke's stand density index: Where are we and where do we go from here? In Proceedings of the Society of American Foresters National Convention, Fort Worth, TX, USA, 19–23 October 2005; Society of American Foresters: Bethesda, MD, USA, 2006.
54. Ducey, M.J.; Knapp, R.A. A stand density index for complex mixed species forests in the northeastern United States. *For. Ecol. Manag.* **2010**, *260*, 1613–1622. [CrossRef]
55. VanderSchaaf, C.L. Reineke's stand density index: A quantitative and nonunitless measure of stand density. In Proceedings of the 15th Biennial Southern Silvicultural Research Conference, Hot Springs, AK, USA, 17–20 November 2008; Southern Research Station: Asheville, NC, USA, 2013; pp. 577–579.
56. Brown, J.K. *Handbook for Inventorying Downed Woody Material*; Intermountain Forest and Range Experiment Station: Ogden, UT, USA, 1974.
57. Pinheiro, J. Fit and Compare Gaussian Linear and Nonlinear Mixed-Effects Models. 2020. Available online: https://svn.r-project.org/R-packages/trunk/nlme/ (accessed on 15 August 2019).
58. Laird, N.M.; Ware, J.H. Random-effects models for longitudinal data. *Biometrics* **1982**, *38*, 963–974. [CrossRef] [PubMed]
59. Fekety, P.A.; Falkowski, M.J.; Hudak, A.T.; Jain, T.B.; Evans, J.S. Transferability of lidar-derived basal area and stem density models within a Northern Idaho ecoregion. *Can. J. Remote Sens.* **2018**, *44*, 131–143. [CrossRef]
60. Hudak, A.T.; Bright, B.C.; Pokswinski, S.M.; Loudermilk, E.L.; O'Brien, J.J.; Hornsby, B.S.; Klauberg, C.; Silva, C.A. Mapping forest structure and composition from low-density LiDAR for informed forest, fuel, and fire management at Englin Air Force Base, Florida, USA. *Can. J. Remote Sens.* **2016**, *42*, 411–427. [CrossRef]
61. Falkowski, M.J.; Hudak, A.T.; Crookston, N.L.; Gessler, P.E.; Uebler, E.H.; Smith, A.M.S. Landscape-scale parameterization of a tree-level forest growth model: A k-nearest neighbor imputation approach using LiDAR data. *Can. J. For. Res.* **2010**, *40*, 184–199. [CrossRef]
62. Brewer, N.W.; Smith, A.M.S.; Hatten, J.A.; Higuera, P.H.; Hudak, A.T.; Ottmar, R.D.; Tinkman, W.T. Fuel moisture influences on fire-altered carbon in masticated fuels: An experimental study. *J. Geophys. Res. Biogeosci.* **2013**, *118*, 30–40. [CrossRef]
63. Lyon, Z.D.; Morgan, P.; Stevens-Rumann, C.S.; Sparks, A.M.; Keefe, R.F.; Smith, A.M.S. Fire behavior in masticated forest fuels: Lab and prescribed fire experiments. *Int. J. Wildland Fire* **2018**, *27*, 280–292. [CrossRef]

sustainability MDPI

Article

Performance Comparison for Two Cable Extraction Machines in a *Larix kaempferi* (Lamb.) Carr. Plantation

Seung-An Baek, Koo-Hyun Cho and Eunjai Lee *

Forest Technology and Management Research Center, National Institute of Forest Science, Seoul 11186, Korea; bsa1201@korea.kr (S.-A.B.); khcho@korea.kr (K.-H.C.)
* Correspondence: ejaylee@korea.kr; Tel.: +82-31-540-1132

Received: 28 September 2020; Accepted: 19 October 2020; Published: 27 October 2020

Abstract: Forests in Korea are mainly located in steep mountainous areas, where small-shovel-based extraction technology is widely used, with the level of mechanization undoubtedly low due to financial limitations. On this steep terrain, a better approach may be to use cable yarders, which can offer high revenues through cable-based extraction. Therefore, improving the efficiency of cable yarding activities in good-quality timber forests is necessary. The main objectives of this study were to (1) evaluate the productivity and cost of a cable yarder operation for tree-length clearcut treatment of a *Larix kaempferi* (Lamb.) Carr. stand and (2) compare the productivity efficiency of two yarder (K301-4 and HAM300) types. The productivity rates of the K301-4 ranged from 10.2 to 12.5 m^3/productive machine hours, with corresponding costs of US $12.6–15.4 /$m^3$. The productivity of the HAM300 was 26% lower than that of the K301-4 for a 30% lower cycle log volume while yarding and a comparable lateral distance. This study provides insights to support production and management decisions in the forest supply chain associated with planning cable-yarding operations.

Keywords: clearcut harvesting; tree-length logging; cable yarder; time study technique; efficiency

1. Introduction

Commercial planting of deciduous needle conifer species, primarily the *Larix kaempferi* (Lamb.) Carr., also known as the Japanese larch (about 14,000 hectares (ha) in 2018, representing 30% of the total coniferous plantation area), is important due to its sustainable economic value in South Korea (hereafter Korea) [1]. These Japanese larch forests cover approximately 0.3 million ha, representing 18% of the total coniferous forests (1.7 million ha), with an average stand volume of 172 m^3/ha in 2018. This stand volume exceeds that of other conifer species (160 m^3/ha in 2018). Previous studies including Cáceres et al. [2], Nagamitsu et al. [3] and Marmet et al. [4] reported that the genus Larix is resistant to climatic changes, exhibits rapid juvenile growth, and provides wood for many products (e.g., lumber, pulp and paper). Therefore, the Japanese larch contributes high-quality timber with a high market value in Korea.

The mechanization of timber harvesting commonly involves several challenges: the required equipment is costly [5], and the harvesting activities considerably disturb the forest soil through compaction, rutting and displacement [6]. These activities are even more difficult to conduct on steep terrains due to their inaccessibility to forest vehicles [7]. An alternative is to pursue the cable-based yarding operation. With this technology, negative environmental impacts would be reduced. For example, this concept is more appealing than ground-based harvesting since it eliminates the costs of constructing extensive road networks and environmental compliance [8]. In fact, cable yarding is commonly utilized on high-value-yield stands to enhance productivity and cost effectiveness [6,9]. Thus, extending its usage to steep terrains can potentially ease operations, improve safety and lower cost. Further, by expanding cable-based yarding operations, environmental impacts may be considerably reduced.

In Korea, although approximately 80% of all forested areas are on steep terrains (>40% of the land surface), the timber harvesting methods involve a combination of motor-manual felling, limbing, bucking and small-shovel-based extraction techniques [10]. The harvesting technology choice depends primarily on the financial resources, forest road network, forest machinery and available labor. Although changing from small-shovel-based techniques to cable yarding operations in steep terrains may enhance the productivity and forest management sustainability, their technical feasibility and economic efficiency are debatable [8,11]. In cable yarding activities, the HAM300 and K301-4 (Koller Forsttechnik GmbH Austria) yarders (Figure 1) are the most frequently used. The HAM300 is a cable yarder powered by a 60 kW farm tractor manufactured by the National Forestry Cooperatives Federation of Korea [12], whereas the 84 kW diesel engine two-axle truck-mounted K301-4 system is produced by Koller Forsttechnik GmbH.

(a1) (b1)

(a2) (b2)

Figure 1. Exemplary cable extraction machines: (**a1**,**a2**) HAM300 yarder equipped with a HAM-C 1.0 remote controlled slack pulling carriage with a maximum load capacity of 1.0 tons, and (**b1**,**b2**) K301-4 yarder equipped with a Koller USKA 1.5 slack pulling carriage with a maximum load capacity of 1.5 tons.

Several studies investigating the efficiency improvement of timber production and associated supply chain optimization using cable yarding systems exist. These studies involve understanding the efficiency of cable-based extraction technologies [13,14], decision-making [8,15] and compiling production models [16]. In most of these studies, however, the simulations are generally based on empirical data. Thus, utilizing decision techniques from these studies remains challenging since the machinery, stand features, corridor characteristics, and yarding direction comprising the cable-based extraction activities vary widely.

In Korea, no previous study has explicitly compared the HAM300 and K301-4 to highlight their differences and advantages. Many studies have rather focused on the productivity and operation of the HAM300 [12,17] and K301-4 [18]. In addition, the need for cost-effective technologies for steep terrain harvesting is increasing as cable-based extraction methods expand and improve. Consequently, the objectives of this study are to (1) evaluate the efficiencies of the HAM300 and K301-4 cable-based extraction machines in a Japanese larch plantation, (2) develop a general productivity model for the cable yarding method from many observations and (3) compare the performances of the cable extraction machines based on different site conditions. Our study improves the understanding of the effects of the stand and machinery on the performance of each extraction method. This knowledge can help forest managers properly evaluate harvesting costs and make informed decisions concerning Japanese larch stand management to maximize economic benefits.

2. Materials and Methods

The tests were conducted in six harvest units in the Gangwon region (37°02′–38°37′ N and 127°05′–129°22′ E; Figure 2) in the center of the eastern part of the Korean Peninsula from 2014 to 2016. For all the units, a time and motion study (TMS) was employed on the cable-based, tree-length (TL) clearcutting of the *Larix kaempferi* (Lamb.) Carr. harvest units. Although the TMS is generally conducted for just a short-term period, it is a common and vital tool for understanding the time consumption and productivity of individual harvesting machines [5]. For all stand inventories, the mean diameter at breast height (DBH) of the trees exceeded 32 cm, while the mean basal area ranged from 4.1 to 5.1 m^2/ha (Table 1).

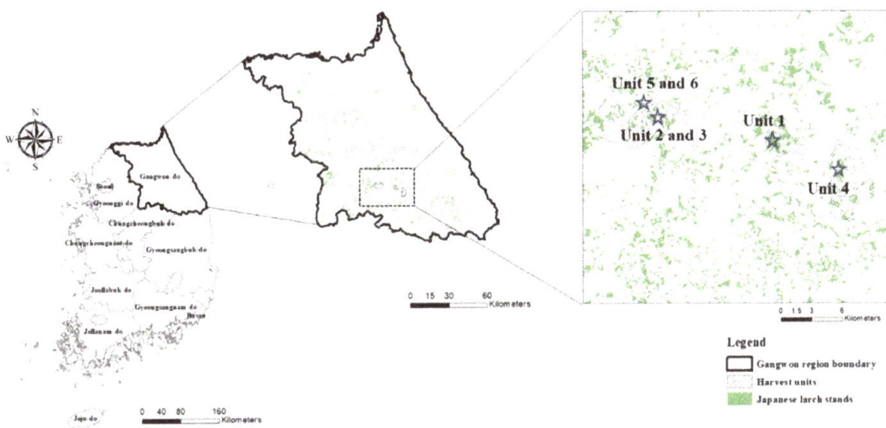

Figure 2. Site map of the study harvest unit located in the Gangwon region, Korea.

Table 1. Stand characteristics of the study harvest unit.

	K301-4			HAM300		
	Unit 1	Unit 2	Unit 3	Unit 4	Unit 5	Unit 6
Area (ha)	1.0	1.0	3.0	1.1	3.0	3.0
Mean DBH (cm)	37	37	34	32	34	34
Mean height (m)	26	23	23	22	23	23
Basal area (m^2/ha)	5.0	5.0	4.7	4.5	4.5	4.6
Trees per ha	175	179	177	187	172	175

The TL harvesting operation comprised motor-manual felling, limbing and topping at the stump and cable yarding from the stump to the landing using two yarders (K301-4 and HAM300). The yarders used for extraction, according to the studied stands, are presented in Table 2. The TL logs were extracted uphill to the landing or roadside, with the yarders transporting the profitable logs to the landing, while the toppings and residues were left on the stumps. Most of the harvested trees yielded one or two logs averaging 20 m in length. During the operation, the yarder operator was replaced by a chaser at the landing. In addition, an experienced crew of three choker setters hooked an optimum load.

Table 2. Specifications of the cable-based extraction machines.

	K301-4	HAM300
Manufacturer	Koller	National Forestry Cooperatives Federation in Korea
Skyline drum capacity (m)	400	350
Skyline diameter (mm)	18.0	16.0
Mainline drum capacity (m)	450	350
Mainline diameter (m)	9.0	9.5
Haul-back line drum capacity (m)	800	500
Haul-back line diameter (mm)	9.0	9.0
Tower height (m)	8.8	7.3
Carriage engine power (kW)	99	48 (tractor PTO)
Maximum pulling capacity (kN)	26	24
Maximum pulling speed (m/sec)	7.5	4.2

During the field tests, we recorded the total extraction cycle times for both machines using stopwatches. Independent variables, including the yarding distance, log diameter and length and number of logs, were also measured for each cycle. The yarding procedure was categorized into six main tasks, including the outhaul, lateral out, hook-up, lateral in, in-haul and unhook [19], with the installation and takeoff times excluded. In addition, for all logs of each study site, the small and large end diameter and length were measured for calculating the individual log volume.

The hourly cost (US $/scheduled machine hours, SMH) of each machine was estimated using standard machine rate measurement practices [20]. This cost is commonly divided into ownership and operating costs. In cost analysis, the ownership costs, also termed fixed costs, traditionally involve the depreciation, interest, insurance and taxes. Particularly, for comparing the operating costs, the assumed expected economic life was 1400 SMH/year [1]. An interest rate of 10% and a tax rate of 4% were used to evaluate the ownership costs (Table 3). Conversely, the operating costs, also known as variable costs, comprised fuel, lubrication, repair and maintenance costs, as well as wage and benefits. The overhead, indirect, profit allowance and shipping expenses were excluded from the hourly cost.

Table 3. Extraction equipment purchase price, annual depreciation, utilization rate and ownership cost for total cost calculation.

	Purchase Price ($)	Annual Depreciation ($/SMH [a])	Utilization Rate (%)	Ownership Cost ($/SMH [a])
K301-4	300,000	27,551	70	51.25
HAM300	134,000	12,306	70	22.89

[a] Scheduled machine hours.

Delay-free cycle times (DCT) data were used to construct productivity simulation models for evaluating the cable yarding activities of TL harvesting. We adopted the least squares linear regression technique, also known as linear regression for data analysis. This method has been employed for producing empirical models from large datasets involving independent variables [21]. The prediction equation obtained with the least squares method is preferable, although the regression results can be adversely affected because of outliers and multipolar data [22]. For each machine, we created two linear regression models to estimate the DCT and then compared the predicted and observed values using a paired *t*-test. During the model construction, two-thirds of the training data were randomly selected, while the remaining (one-third) data were used for model validation. All statistical analyses were performed using the *R* statistical software version 4.0.2.

After the regression analysis, we conducted a sensitivity analysis for the cable yarders to determine the responses of the yarding operations to different variables. We also tested the effects of these variables on productivity. To compare the performances of the yarders, the DCT changes were converted to productivity and cost patterns for different yarder activities. Thus, this test aided in understanding the impacts of these independent variables on the maximum productivity and least cost.

3. Results

3.1. Productivity and Costs of Cable-Based Extraction Operations

The DCT data varied widely for both cable-based machines in each unit (Table 4). Our results show that the K301-4 required substantially more time for a cycle than the HAM300 because it involved a higher yarding distance and cycle log volume. In addition, the average K301-4 yarding productivity values from harvest units 1, 2 and 3 were 12.3, 12.5, and 10.2 m^3/productive machine-hours (PMH), respectively (yarding distance ranged from 12 to 215 m; Table 4). The productivity values of individual unit operations differed by up to 20% for similar stands. The estimated production cost, including the sum of ownership, operation and labor costs, ranged between US $12.6 and 15.4/ m^3. On average, the HAM300 machine produced 8.9 to 9.8 m^3/PMH at conditions of US $12.8 and 14.1/$m^3$, respectively (yarding distance ranged from 9 to 137 m). The productivity and costs slightly differed between the harvest units, ranging from 2 to 9% (Table 4), but these values were lower than those for the K301-4 tests. Further, the operating conditions such as the cycle log volume, yarding distance and lateral distance were statistically the same for each test (K301-4 test: unit 1 vs. 2 vs. 3; HAM300 test: unit 4 vs. 5 vs. 6).

Table 4. Mean delay-free cycle times observed by cable extraction machines.

	K301-4			HAM300		
	Unit 1	**Unit 2**	**Unit 3**	**Unit 4**	**Unit 5**	**Unit 6**
Number of cycle times	86	98	162	64	72	108
	Machine cycle time					
Mean (sec/PMH [a])	317	280	342	282	319	288
Standard deviation	68.82	66.88	118.73	78.87	63.92	71.92
Productivity (m^3/PMH)	12.3	12.5	10.2	9.1	9.8	8.9
Cycle log volume (m^3)	1.1	1.0	1.0	0.7	0.9	0.7
Yarding distance (m)	94	61	118	63	75	89
Cost (US$/$m^3$)	12.8	12.6	15.4	13.8	12.8	14.1

[a] Productive machine-hour.

The performances of the cable yarders also differed considerably, with the average productivity of the K301-4 operation about 26% higher than that for the HAM300. The 1.01 m^3 of logs per cycle transported on average by the K301-4 was approximately 32% higher than the 0.76 m^3 of logs hauled by the HAM300, creating a statistically significant difference between the operations (p-value = 0.0155). The average yarding and lateral distances of 91 and 11 m, respectively, for the K301-4 operation tests, surpassed the corresponding 75 and 9 m for the HAM300 by 20%, without any statistically significant difference between the machines (p-value > 0.05).

3.2. Delay-Free Cycle Time Regression Models

Prior to the least squares regression analysis, we blended the DCT data from the harvest units into a predictive equation for each yarding technology. Evidently, no statistically significant differences were found to exist among the sites (p-value > 0.05) for individual variables and the DCT regression equations and ranges of independent variables for each machine are presented in Table 5. Clearly, all models were significant, with no serious violations at the 1% significance level (p-value < 0.01), and most of the independent variables were significant (p-value < 0.05). However, the trees per cycle was not a significant variable in the K301-4 yarding operation (p-value > 0.05), whereas, in the HAM300 model, the cycle log volume was the only non-significant variable (p-value = 0.403). For all DCT equations, after a paired t-test, the predicted DCT values did not have any statistically significant difference with the observed values (p-value > 0.05).

Table 5. Delay-free cycle time regression equations for K301-4 and HAM300. Moreover, a paired t-test was used for equation validation against observed data.

Parameter	Range Variable	Estimate	SE	t	Pr	Model adj. R^2	Model p-Value	t-Test
				K301-4				
Intercept		95.694	19.806	4.832	<0.01			
Cycle log volume (m^3)	0.2–2.4	34.489	10.184	3.386	<0.01			
Yarding distance (m)	12–215	1.206	0.076	15.702	<0.01	0.5934	<0.01	0.4079
Lateral distance (m)	0–46	4.889	0.487	10.030	<0.01			
No. of trees per cycle	1–2	14.901	15.110	0.986	0.325			
				HAM300				
Intercept		128.420	18.535	6.928	<0.01			
Cycle log volume (m^3)	0.2–2.0	11.167	13.321	0.838	0.403			
Yarding distance (m)	9–137	1.041	0.111	9.353	<0.01	0.4813	<0.01	0.2042
Lateral distance (m)	0–40	4.726	0.606	7.794	<0.01			
No. of trees per cycle	1–2	28.897	13.922	2.076	<0.05			

3.3. *Sensitivity Analysis*

We performed a sensitivity analysis to evaluate the impact of the yarding distance on the K301-4 and HAM300 models. The productivity of each machine changed with the yarding distance from 10 to 200 m in 5 m increment distance (Figure 3). The calculation was conducted under the following conditions: lateral distance of 10 m, cycle log volume of 0.9 m^3 and a trees-per-cycle value of 1. Overall, the estimated productivity decreased, and the cost increase as the yarding distance increased in both models.

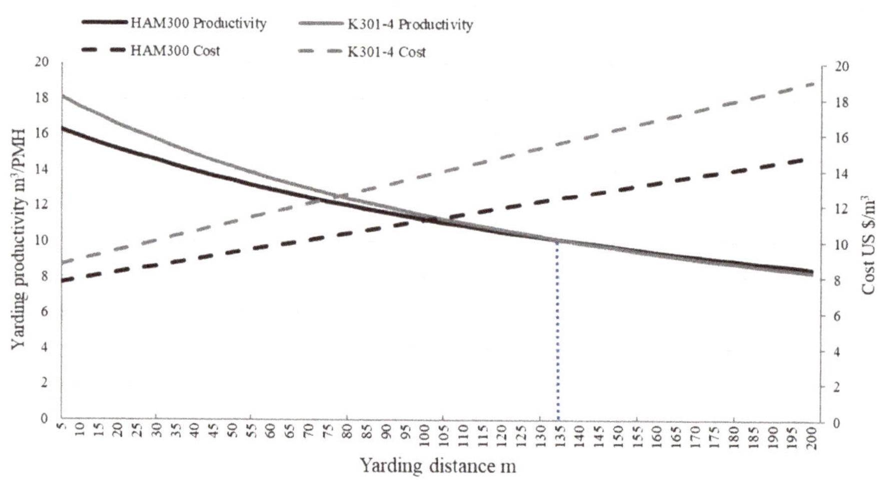

Figure 3. Changes in extraction productivity (left y-axis) and production cost (right y-axis) over different yarding distances by the K301-4 and HAM300 machines.

The sensitivity analysis enabled the evaluation of the break-even yarding distances of the two machines. The productivity values for the K301-4 were higher for yarding distances less than 135 m, while the costs were higher for all yarding distances compared with the HAM300 (Figure 3). This is because the purchase price of the K301-4 was more than two times that of the HAM300 yarder.

Further, we evaluated the impact of the cycle log volume on the performances of both machines (Figure 4). The productivity of each machine changed with yarding distances from 10 to 200 m under the following conditions: lateral distance of 10 m, cycle log volume between 0.9 and 1.5 m^3 and one tree per cycle. Overall, the estimated productivity increased as the cycle log volume increased for both models. Even if the DCT was longer for yarding a large cycle log volume, the productivity increased for both machines because of the payload per turn. In addition, the HAM300 exhibited higher productivity values for yarding distances higher than 55 m.

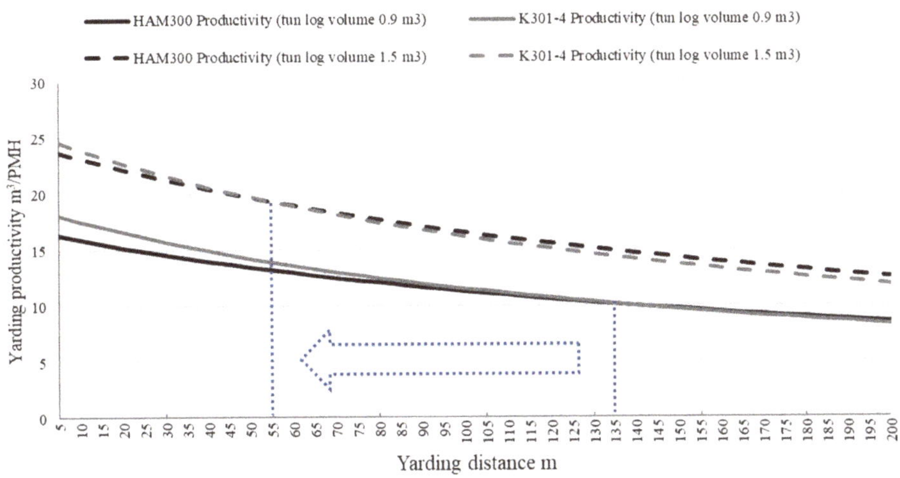

Figure 4. Changes in extraction productivity as a function of cycle log volume over different yarding distances by the K301-4 and HAM300 machines.

4. Discussion and Conclusions

An efficient solution for timber harvesting in a steep terrain should be primarily connected to cable operations. During the last two decades, these technologies have been widely employed in Europe and North America, while extraction in Korea has commonly involved a small-shovel harvesting system due to financial limitations. Although cable yarders are used for harvesting timber under variable conditions, the planning and design remain poor. Therefore, this study involved the following: (1) evaluating the productivity and costs associated with the K301-4 and HAM300, (2) developing productivity models using the least squares linear regression technique and (3) comparing the performances of the cable-yarding machines based on different site conditions. The production rates (10.2–12.5 m^3/PMH) for the K301-4 was approximately 25% higher than those for the HAM300 (8.9–9.8 m^3/PMH), with the operation cost correspondingly higher. In addition, for yarding distances less than 135 m at 0.9 m^3 cycle log volume, the K301-4 model yielded more productivity. Furthermore, the performance of the cable yarders changed in response to the cycle log volume and K301-4 potentially performed better for yarding distances less than 55 m when the cycle log volume was 1.5 m^3.

We showed that the productivity of each cable-based extraction operation (K301-4 and HAM300) varied by up to 20% between the three harvest units, with the performance differences linked to many reasons. For example, Lindroos and Cavalli [16], Erber et al. [23] and Schweier et al. [24] reported that the productivity variation in cable yarding operations was due to several reasons such as working conditions and load sizes. In our study, however, during the individual machine tests, the work conditions were similar for the three harvest units (units 1–3 and units 4–6). Several studies also indicate that the operator affects the harvesting equipment productivity more than the stand characteristics [25,26]. This is consistent with our results from tests with separate machine operators and chock setters during harvesting. Therefore, the performance of the yarders varied from 2% to 20%, consistent with the results in Kärhä et al. [27]. Thus, results can vary because of a cable yarding operation crew difference.

Our results imply that the heavier payload capacity of the K301-4 yarder offers a better performance as the yarding distance decreases compared to the HAM300, although its price is a limiting factor. The higher productivity efficiency of the K301-4 used in this study demonstrates its higher payload and faster line speed. According to Schweier et al. [24] and Engelbrecht et al. [28], the operation productivity was affected by the yarder type (heavy or medium) and piece size per cycle. The performance of the heavy yarder ranged from 20% to 30%, depending on the inclusion or exclusion of delay times, since the performance continuously improved with the cycle speed and payload. The differences in rates between the K301-4 and HAM300 in this study coincide with results from previous studies. For the K301-4 yarder, the higher productivity is attributed to its ability to haul large pieces of logs in a single cycle. Thus, the performance of the cable yarding operation is sensitive to the machine type.

We also evaluated the impact of the cycle log volume on the productivity for each yarding operation. The log production advanced with increasing cycle log volume for both yarding operations. This finding is consistent with previous studies such as Engelbrecht et al. [28] and Hiesl and Benjamin [29] reporting that the extraction (including cable yarding and forwarding) productivity significantly increases with timber size (0.2–1.2 m^3). The piece-volume has been reported earlier in many studies, such as Berendt et al. [6] and Ghaffariyan [30]. Their findings were that larger log sizes could increase the harvesting productivity, even though the time consumed per cycle in mechanized harvesting also increases. Consequently, the cycle log volume can considerably affect the productivity in cable extraction activities.

Transportation between a stump and a landing or forest roadside in steep terrain is technically difficult and expensive in terms of wood production. These may impact directly and indirectly change forest ecosystems and environments, such as soil, air, water, biodiversity and regeneration capacity. Therefore, adequate planning, including decision support technology for yarding/skidding, may be necessary to provide precise information for identifying the preferred conditions. Application to widely different conditions using empirical productivity models remains challenging [5,16,28]. Even though the validity of the empirical models may be limited to the K301-4 and HAM300 yarder configurations in this study, the results can highlight preferential conditions such as the yarding distance and cycle log volume for the cable extraction method selection. Further studies are needed to investigate improving the accuracy of the models based on spatial considerations. In addition, studies comparing timber supply chains (cable-based harvesting system vs. ground-based harvesting system) are also necessary to improve knowledge on the sustainability of the wood supply chain.

In conclusion, in this study, the performances of two yarder types (yarder mounted on a truck: K301-4 and yarder mounted on a tractor: HAM300) deployed in different harvest units were investigated. The K301-4 yarder operation involved a larger cycle log volume than that for the HAM300. Our results demonstrated that machine type considerably affected cable extraction performance. In addition, for a cycle log volume of 0.9 m^3 in our harvest units, the productivity of the K301-4 is higher up to 135 m, whereas it changes into a good performance option until 55 m when the cycle log volume is 1.5 m^3. Replications of the productivity models may be robust because the DCT data collected under different conditions involved many observations and long-term studies. Therefore, the models are potentially beneficial for evaluating the productivity of each machine and improving cable yarding planning. Future research should tackle the financial implications of the machine operating conditions (stand density, silvicultural prescriptions: thinning and partial harvest).

Author Contributions: Conceptualization, S.-A.B., K.-H.C. and E.L.; methodology, S.-A.B. and E.L.; data curation, S.-A.B. and K.-H.C.; formal analysis, S.-A.B. and E.L.; investigation, S.-A.B., K.-H.C. and E.L.; writing—original draft preparation, S.-A.B. and E.L.; writing—review and editing, K.-H.C. and E.L.; project administration, K.-H.C.; funding acquisition, K.-H.C. All authors have read and agreed to the published version of the manuscript.

Funding: This research received no external funding.

Conflicts of Interest: The authors declare no conflict of interest.

References

1. Korea Forest Service. Available online: http://www.forest.go.kr/kfsweb/kfs/idx/Index.do (accessed on 8 August 2020).
2. Cáceres, C.B.; Hernández, R.E.; Fortin, Y.; Beaudoin, M. Wood density and extractive content variation among Japanese larch (*Larix kaempferi* (Lamb.) Carr.) profenies/provenances trails in Eastern Canada. *Wood Fiber Sci.* **2017**, *49*, 363–372.
3. Nagamitsu, T.; Matsuzaki, T.; Nagasaka, K. Provenance variations in stem productivity of 30-year-old Japanese larch trees planted in Northern and Central Japan are associated with climatic conditions in the provenanves. *J. For. Res.* **2018**, *23*, 270–278. [CrossRef]
4. Marmet, S.D.; Brown, C.D.; Trant, A.J.; Laroque, C.P. Shifting global Larix distributions: Northern expansion and southern retraction as species respond to changing climate. *J. Biogeogr.* **2019**, *46*, 30–44. [CrossRef]
5. She, J.; Chung, W.; Kim, D. Discrete-event simulation of ground-based timber harvesting operations. *Forests* **2018**, *9*, 683. [CrossRef]
6. Berendt, F.; Fortin, M.; Suchomel, C.; Schweier, J. Productivity, costs, and selected environmental impacts of remote-controlled mini forestry crawlers. *Forests* **2018**, *9*, 591. [CrossRef]
7. Visser, R.; Berkett, H. Effect of terrain steepness on machine slope when harvesting. *Int. J. For. Eng.* **2015**, *26*, 1–9. [CrossRef]
8. Spinelli, R.; Maganotti, N.; Nati, C. Benchmarking the impact of traditional small-scale logging system used in Mediterranean forestry. *For. Ecol. Manag.* **2010**, *260*, 1977–2001. [CrossRef]
9. Ghaffariyan, M.R. A short review of cable yarding applications in Australia. *Eur. J. For. Eng.* **2019**, *5*, 47–51. [CrossRef]
10. Lee, E.; Han, S.-K.; Im, S. Performance analysis of log extraction by a small shovel operation in steep forests of South Korea. *Forests* **2019**, *10*, 585. [CrossRef]
11. Marchi, E.; Chung, W.; Visser, R.; Abbas, D.; Nordfjell, T.; Mederski, P.S.; McEwan, A.; Brink, M.; Laschi, A. Sustainable forest operations (SFO): A new paradigm in a changing world and climate. *Sci. Total Environ.* **2018**, *634*, 1385–1397. [CrossRef]
12. Lee, E.; Im, S.; Han, S.-K. Productivity and cost of a small-scale cable yarder in an uphill and downhill area: A case study in South Korea. *For. Sci. Technol.* **2018**, *14*, 16–22. [CrossRef]
13. Huber, C.; Stampfer, K. Efficiency of topping trees in cable yarding operations. *Croat. J. For. Eng.* **2015**, *36*, 185–194.
14. Enache, A.; Kühmaier, M.; Visser, R.; Stampfer, K. Forestry operations in the European mountains: A study of current practices and efficiency gaps. *Scand. J. For. Res.* **2015**, *31*, 412–427. [CrossRef]
15. Spinelli, R.; Maganotti, N.; Lambardini, C. Performance, capability, and costs of small-scale cable yarding technology. *Small Scale For.* **2010**, *9*, 123–135. [CrossRef]
16. Lindroos, O.; Cavalli, R. Cable yarding productivity models: A systematic review over the period 2000–2011. *Int. J. For. Eng.* **2016**, *27*, 79–94. [CrossRef]
17. Cho, M.; Cho, K.; Choi, B.; Cha, D. Yarding productivity of tree-length harvesting using a small cable-yarder in steep slope, South Korea. *For. Sci. Technol.* **2018**, *14*, 132–137. [CrossRef]
18. Kim, M.-K.; Baek, S.-A.; Cho, K.-H.; Jung, D.-H. Productivity, cost, and optimal forest road network density of tree-length yarding operations with tower yarder. *J. Korean For. Soc.* **2017**, *106*, 300–309.
19. Proto, A.R.; Skoupy, A.; Macri, G.; Zimbalatti, G. Time consumption and productivity of a medium size mobile tower yarder in downhill and uphill configurations: A case study in Czech Republic. *J. Agric. Eng.* **2016**, *47*, 216–221. [CrossRef]
20. Brinker, R.; Kinard, J.; Rummer, B.; Lanford, B. Machine rates for selected forest harvesting machines. In *Machine Rates for Selected Forest Harvesting Maines*; Auburn University: Auburn, AL, USA, 2002; pp. 1–32.
21. Costa, C.; Menesatti, P.; Spinelli, R. Performance modelling in forest operations through partial least square regression. *Silva. Fennica* **2012**, *46*, 241–252. [CrossRef]

22. Ünver-Okan, E. Modelling of work efficiency in cable traction with tractor implementing the least-squares methods and robust regression. *Croat. J. For. Eng.* **2020**, *41*, 109–117. [CrossRef]
23. Erber, G.; Haberl, A.; Pentek, T.; Stampfer, K. Impact of operational parameters on the productivity of whole tree cable yarding: A statistical analysis based on operation data. *Austrian J. For. Sci.* **2017**, *134*, 1–18.
24. Schweier, J.; Klein, M.-L.; Kristen, H.; Jaeger, D.; Brieger, F.; Sauter, U.H. Productivity and cost analysis of tower yarder systems using the Koller 507 and the Valentini 400 in Southwest Germany. *Int. J. For. Eng.* **2020**, 1–12. [CrossRef]
25. Ovaskainen, H.; Uusitalo, J.; Väätäinen, K. Characteristics and significance of a harvester operators' working technique in thinnings. *Int. J. For. Eng.* **2004**, *15*, 67–77. [CrossRef]
26. Lindroos, O. Scrutinizing the theory of comparative time studies with operator as a block effect. *Int. J. For. Eng.* **2010**, *21*, 20–30. [CrossRef]
27. Kärhä, K.; Rökkö, E.; Gumse, S.-I. Productivity and cutting costs of thinning harvesters. *Int. J. For. Eng.* **2004**, *15*, 43–56. [CrossRef]
28. Engelbrecht, R.; McEwan, A.; Spinelli, R. A robust productivity model for grapple yarding in fast-growing tree plantations. *Forests* **2017**, *8*, 396. [CrossRef]
29. Hiesl, P.; Benjamin, J.G. Applicability of international harvesting equipment productivity studies in Main, USA: A literature review. *Forests* **2013**, *4*, 898–921. [CrossRef]
30. Ghaffariyan, M.R. Review of European biomass harvesting technologies. *Silva. Balcanica* **2010**, *11*, 5–20.

sustainability

MDPI

Article

Harvester Productivity in Inclined Terrain with Extended Machine Operating Trail Intervals: A German Case Study Comparison of Standing and Bunched Trees

Ferréol Berendt [1,2,*], Eduardo Tolosana [3], Stephan Hoffmann [1], Paula Alonso [1,3] and Janine Schweier [1,4]

1 Chair of Forest Operations, Faculty of Environment and Natural Resources, University of Freiburg, 79085 Freiburg, Germany; stephan.hoffmann@canterbury.ac.nz (S.H.); paula.alonso.rod@gmail.com (P.A.); janine.schweier@wsl.ch (J.S.)

2 Department of Forest Utilization and Wood Markets, Faculty of Forest and Environment, Eberswalde University for Sustainable Development, 16230 Eberswalde, Germany

3 E.T.S.I. de Montes, Forestal y del Medio Natural, Universidad Politécnica de Madrid, 28040 Madrid, Spain; eduardo.tolosana@upm.es

4 Research Group Sustainable Forestry, Swiss Federal Institute for Forest, Snow and Landscape Research WSL, 8903 Birmensdorf, Switzerland

* Correspondence: ferreol.berendt@hnee.de

Received: 21 October 2020; Accepted: 2 November 2020; Published: 4 November 2020

Abstract: The complexity of highly structured forests with multiple tree species, especially when coniferous and broadleaved tree species are mixed, as well as stands with extended machine operating trail spacing and inclined terrain, create challenging operational conditions for mechanized timber harvesting and extraction. Motor-manually felling trees within the midfield and bunching them at the machine operating trails, prior to the arrival of a harvester-forwarder system, is a complex operation. The aim of this study was to assess and compare tethered harvester productivities of a thinning operation, for felling and processing standing trees and for processing bunched trees, through a time study in forest stands with 40-m distances between machine operating trails. Total operational costs of the analyzed thinning operation were 69 €/$m^3_{o.b.}$, including extraction using a multiple forwarder approach. Tree species, merchantable timber volume, and whether the trees were standing or presented as bunched logs all had a significant effect on the harvester time consumption. Moreover, harvester positioning time was significantly shorter when trees were already bunched at the machine operating trail. While the productivity of standing or bunched spruce trees did not differ significantly between the cases (approximately 18 $m^3_{o.b.}$/productive machine hours excluding all delays (PMH_0)), the productivity of standing broadleaved tree species (8.3 $m^3_{o.b.}$/PMH_0) was much lower than that of bunched trees (15.5 $m^3_{o.b.}$/PMH_0). Thus, the described timber harvesting and extraction system may be a valuable option for forest stands with high proportion of broadleaved trees.

Keywords: forest operations; machine operating trail; midfield; single-grip harvester; soil protection; tethering winch

1. Introduction

Through high structural and species diversity within managed forests, the resistance and resilience of the forests towards future climatic conditions can be improved [1]. However, such management approaches will lead to more complex forest stands, and thus the complexity and potential impacts caused by timber harvesting operations might increase as well [2]. Tree species and diameter, terrain factors such as slope, soil strength, and soil roughness, and operational factors, such as the forest road system, restrict in particular fully mechanized wood harvesting and extraction operations [3–5]. Timber extraction is usually conducted by using ground-based systems in flat and intermediate terrain and cable-based systems in steep terrain. However, depending on the wood species, amount, and assortments, the latter system might be a challenge in terms of economic viability [6,7]. At the same time, modern systems using tethering winches are becoming increasingly common and are recommended to be used in flat and intermediate terrain, but few studies were conducted on productivity, cost-effectiveness, environmental or work safety impacts, in order to fully rate these systems [7].

Forest operations are subject to technical limitations and personal constraints of the forest owners, associated with the individual management objective, but they also have to fulfil legal regulations and, if accredited by a forest certification scheme, the corresponding standard. Examples of international forest certification schemes are the Programme for the Endorsement of Forest Certification (PEFC) and the Forest Stewardship Council (FSC). The latter determines, for the German standard, that less than 13.5% of the forest area should be used as machine operating trails or forest roads, due to soil protection reasons [8]. The resulting 40-m machine operating trail spacing is also encouraged by many state forest enterprises, such as ForstBW of Baden-Württemberg [9]. Inevitably, such regulations lead to a machine operating trail system with midfields that are outside the boom reach of the harvester and where trees need to be felled motor-manually. Subsequently, full trees are bucked at the machine operating trail through a harvester, or—if too short—full trees or assortments are pre-winched to the machine operating trails using, for example, mini forestry crawlers (MFC) or tractors with a winch, and bunched for further processing [10,11].

The aim of this study is to gain knowledge on the application of timber harvesting and extraction systems with midfields, which are used for pre-winching and bunching trees. We hypothesized that the harvester productivity of bunched trees is higher than that of standing trees in the boom reach of the harvester. Thus, the specific goals of this study were:

1. to analyze the time consumption of ground-based wood harvesting and extraction on inclined terrain with midfields;
2. to compare the harvester productivity of standing and bunched trees;
3. to deduce further research goals from the study outcomes.

2. Materials and Methods

2.1. Study Approach and Site

We collected data as part of a case study in the German federal state of Baden-Württemberg (BW), more precisely in the municipality of Kleines Wiesental in the Black Forest mountains. The observed operation was a commercial thinning in terrain with slopes from 30% up to approximately 60% and was conducted by a private forestry contractor in October and November 2018. Forest operations in this kind of terrain are conventionally classified for cable yarder systems in Germany [12]. However, the forest manager did not consider using a cable yarder, due to (i) technical infeasibility and (ii) expected higher costs. Further, (iii) work safety was an important aspect because many trees were broken by snow.

To avoid long skidding distances and the resulting considerable stand damage, a harvester-forwarder system with a tethering winch was applied. Despite being at its upper slope limit of operation, the engaged machines were fully stable at the site while in a stationary position, as well as when the winch-cable was not tensioned, thus meeting requirements for safe operation of traction-assisted equipment by international standards [13]. Nevertheless, for both soil protection and societal acceptance reasons, the machine operating trail intervals were 40 m. Besides the harvesting and forwarding, motor-manual felling and pre-winching to the machine operating trail in the midfield were carried out by the forest company.

The study site, with the coordinates 47°47′19.2″ N and 7°46′58.4″ E, was located between 1100 and 1200 m above sea level. The slope inclination of the machine operating trails ranged between 60% and 70% and was estimated using a clinometer (PM-5, Suunto, Vantaa, Finland). The stand size was 9.9 ha and the dominant tree species was Norway spruce (*Picea abies* H. Karst), with a share of 65%. In addition, the mixed stand consisted of other conifers (16% Douglas fir (*Pseudotsuga menziesii* Franco) and 5% Scots pine (*Pinus sylvestris* L.)), as well as 14% broadleaved species (7% Sycamore maple (*Acer Pseudoplatanus* L.) and 7% European beech (*Fagus sylvatica* L.)). The mean diameter at breast height for these groups was measured prior to the harvesting and reached on average 27.2 ± 5.3 cm for Norway spruce, 33.4 ± 8.4 cm for Douglas fir, and 18.5 ± 10.6 cm for broadleaved tree species. The age of the two main tree species was on average 56 years for Norway spruce and 35 years for Douglas fir. The scheduled harvesting volumes of merchantable timber were 60 to 70 m^3 per ha. According to the billings of the forestry company, 670 $m^3_{o.b.}$ were harvested and processed.

2.2. Applied System

The distance between machine operating trails was 40 m and the overall operation was conducted as follows, with all working steps carried out one after the other: (1) Trees were felled motor-manually in the midfield using a chainsaw. Moreover, prior to mechanized felling and processing, some trees were felled motor-manually by a second chainsaw operator in order to facilitate the harvester's pass on the machine operating trail. (2) Full trees were partially bunched and pre-winched to the machine operating trails using a remote controlled MFC (Raup-Trac from Martin Alther, Switzerland) that was operated from the machine operating trails. This operation was carried out by a professional two-person team: one felled the trees using the chainsaw while the second operated the MFC. (3) A harvester (405FH4 8WD from HSM, Germany with H415 harvester head from Waratah, New Zealand and Finland) with a tethering winch first felled and processed the standing trees within the boom reach and then processed the bunched trees from the midfields. (4) A forwarder (208F from HSM, Germany) with a tethering winch extracted the processed logs from the machine operating trails to the next forest road, where a second forwarder (HSM 208F Big Foot from HSM, Germany) brought the logs to a roadside landing. The two units were necessary, as the first forwarder was supported by a tethering winch and thus could not leave the straight machine operating trails. Besides extracting timber to the landing site, the second forwarder acted when needed as anchorage for the tethering winch of the first forwarder. This was the case when no tree could be used as an anchor. The representation of the wood harvesting and extraction system according to the German Center for Forest Work and Technology (KWF) [14] is shown in Figure 1.

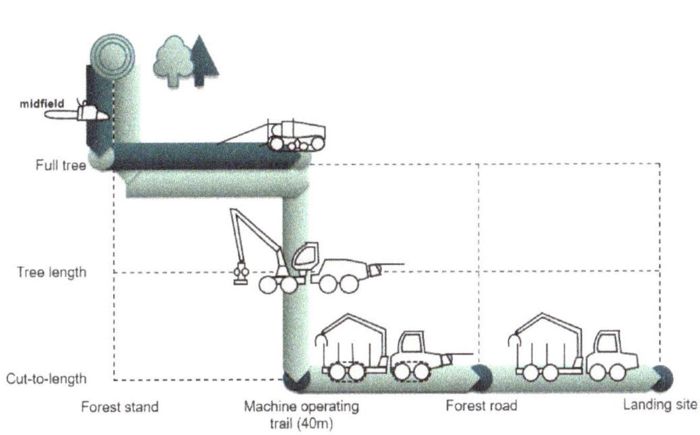

Figure 1. Representation of the wood harvesting and extraction operation.

2.3. Data Collection

The data collection concentrated on the pre-winching and harvesting operation (steps 2 and 3). An elemental time and motion study, applying the snapback timing approach [15], was conducted with an accuracy of 1/100 min through a digital stop watch in order to monitor the operations of both the MFC and harvester. All delays caused by the study were removed; delays for the lunch break (if any) and relocation to and from site were also not included in the data sets.

The work cycles of the MFC were divided into the following work elements: positioning on the machine operating trails, pulling, chocking, winching, unchocking, and waiting for the feller (including communication between the MFC operator and the feller). Moreover, mechanical delays as well as rest and personal times were recorded.

During data collection concerning the harvester, the cycles were divided into two mandatory work elements that occurred in each cycle: (a) positioning the harvester on the machine operating trail, moving the boom into the direction of the tree, and grabbing the tree ('positioning'); and (b) felling, delimbing and bucking, and stacking the crown ('felling & processing'). These elements correspond to the main work time and complementary work time according to the IUFRO (International Union of Forest Research Organizations) Forest Work Study Nomenclature [15]. Moreover, rest and personal times, as well as supportive work times, were recorded. Within the supportive work times, ancillary work times and time needed for rigging/unrigging the tethering winch cable were recorded. Along with time consumption, the merchantable wood volume in m^3 over bark ($m^3{}_{o.b.}$) was documented through the harvester measurements and on-board computers' records. The harvester operator separated the trees into Norway spruce, Douglas fir, and broadleaved tree species, with the corresponding assortments. Even though Scots pine was observed during the inventory prior to the harvesting operation, Scots pine was not harvested during the thinning. Thus, Scots pine was not considered further in the analysis.

Data concerning the other working steps (1 and 4) were not considered within the time study, since these processes are well studied and overall time consumption and corresponding costs were obtained from the forestry company.

2.4. Working Productivity and Costs

A regression analysis was applied in order to determine the effects of merchantable volume, tree species, and the presentation of the logs (i.e., whether the harvester operated with standing or bunched trees on the time consumption). The significance level (α) was set to 5%, with $\alpha \leq 1\%$ considered highly significant.

Overall productivity was calculated as the total volume of merchantable timber in m^3 over bark ($m^3{}_{o.b.}$), divided by the total time consumption, whereas the cycle productivity was calculated as the processed timber during one cycle divided by the cycle time consumption. The relationship between harvester productivity and wood volume was fitted through non-linear regression methods in which broadleaved species were excluded, as n number of broadleaved tree species was too low. Moreover, data with studentized residues greater than 3.0 were removed as outliers. These were mostly positive residues from cycles longer than normal.

Accurate accounting, with the number of working hours and costs for the whole area (9.9 ha), were obtained from billings of the forestry company and included both, fixed and variable costs. Thus, both machine and operator costs of all processes could be included. The costs were compartmentalized into chainsaw work, MFC, harvester, forwarder 1, forwarder 2, and machine transport from forestry company office to forest area.

3. Results

3.1. MFC and Harvester Time Consumption

The time consumption for the thinning operations was determined for each working step individually.

Considering the MFC, 178 winching cycles were observed with a mean load volume of 0.49 m^3. The recorded scheduled machine hours (SMHs) were 362 min, thereof 286 min referred to productive machine hours (PMH_0), resulting in a machine utilization rate (ratio of PMH_0 to SMH) of 79%. The overall gross productivity, based on SMHs, was 11.3 m^3/SMH, while net productivity was 14.3 m^3/PMH_0. Within the net cycle time, the longest working elements were winching (25.4%), positioning (24.9%), waiting for the feller (19.7%), and pulling (10.5%). Chocking time corresponded to 9.3% and included rechocking the cable if it was not properly fixed the first time.

For the harvester, total SMHs recorded were 1217 min. Within the 821 cycles, a total of 278.3 $m^3_{o.b.}$ was felled, processed, and bunched, corresponding to an overall gross productivity of 13.7 $m^3_{o.b.}$/SMH. When looking at the net cycle time, 912.4 min were used as productive time (Table 1). Thus, the machine utilization rate was 75%. The overall time consumption was divided into felling and processing, with 0.65 min per cycle (43.7%), followed by positioning (32.2%), supportive work time (14.8%), and rest and personal time (9.3%). Looking more specifically into the supportive work time, 49.5% was service time (mostly maintenance), 34.5% was ancillary work time (machine operating trails), and 15.9% was preparatory time (setting up and taking down the tethering winch). The mean time consumption for one whole harvester cycle was 1.12 ± 0.78 min. Considering the two work elements separately, mean time consumption was 0.47 ± 0.53 for 'positioning' and 0.65 ± 0.41 min for 'felling & processing'.

Table 1. Characteristics of the harvesting operations.

Characteristic	Bunched Trees	Standing Trees	Sum
Total productive machine hours (PMH_0, min)	496.7	415.6	912.4
Work cycles	469	352	821
Total volume ($m^3_{o.b.}$)	153.0	125.1	278.1
Mean volume per cycle ($m^3_{o.b.}$ ± SD)	0.33 ± 0.23	0.36 ± 0.21	0.34 ± 0.23
Number of spruces (pieces)	273	275	548
Number of Douglas firs (pieces)	108	50	158
Number of broadleaved trees (pieces)	88	27	115
Productivity ($m^3_{o.b.}/PMH_0$)	18.5	18.1	18.3

3.2. Influence of Log Presentation

The merchantable volume differed between species. The average volume of bunched broadleaved trees (0.24 ± 0.29 $m^3_{o.b.}$) was larger than the average volume of standing broadleaved trees (0.14 ± 0.09 $m^3_{o.b.}$; Figure 2). On the contrary, the average volume of conifers bunched next to the machine operating trails (0.45 ± 0.27 $m^3_{o.b.}$ for Douglas fir and 0.30 ± 0.18 $m^3_{o.b.}$ for spruce) was smaller than the average volume of standing coniferous trees (0.54 ± 0.33 $m^3_{o.b.}$ and 0.34 ± 0.16 $m^3_{o.b.}$ for Douglas fir and spruce, respectively). The lowest mean time consumption of one harvest cycle was 0.95 ± 0.63 min for bunched broadleaved trees and the highest mean time consumption was 1.24 ± 0.84 min for standing spruce trees (Figure 3).

Tree volume, tree species, and whether the trees were standing or bunched significantly affected the time consumption of a harvester cycle. Moreover, the tree species had a significant effect on both work elements ('positioning' and 'felling & processing'). When considering the work element 'positioning', the independent variable whether trees were standing or bunched had a highly significant effect; whereas it did not significantly affect the time consumption of the work element 'felling & processing'. On the opposite, the tree species did not affect the time consumption of the work element 'positioning' but did significantly influence the 'felling and processing' time.

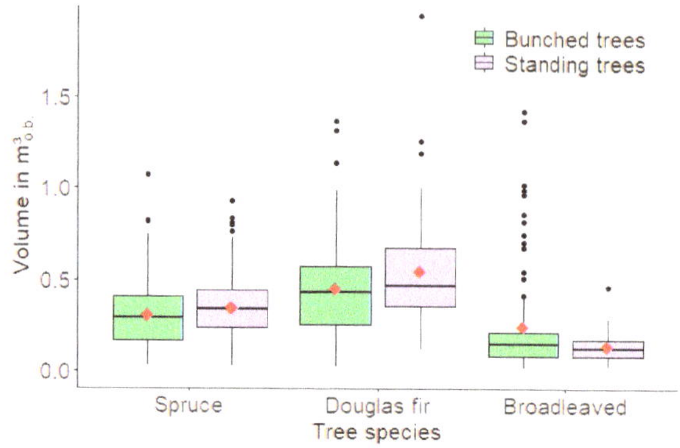

Figure 2. Boxplot of the merchantable volume in m$^3_{o.b.}$ for bunched (green) and standing (purple) trees of different species (spruce, Douglas fir and broadleaved tree species) during harvesting operation.

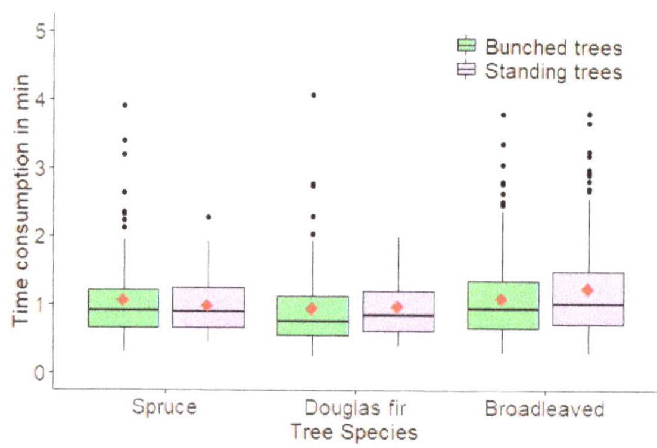

Figure 3. Boxplot of the time consumption (in min) for processing standing (purple) and bunched (green) trees of different species (spruce, Douglas fir and broadleaved tree species).

3.3. *Harvester Productivity*

Considering the productive working time, the overall productivity of the harvester was 18.3 m$^3_{o.b.}$/PMH$_0$ (Table 1). The overall net productivity of bunched trees (18.5 m$^3_{o.b.}$/PMH$_0$) was slightly higher than that of standing trees (18.1 m$^3_{o.b.}$/PMH$_0$). However, the tree species distribution was not similar; for example, the share of spruce was 58.2% for the bunched trees and 78.1% for the standing trees. The tree species, as well as the volumes, had a highly significant influence on the cycle productivity, whereas log presentation did not show any significant effect. The highest overall productivity was reached when harvesting Douglas fir, with 27.7 m$^3_{o.b.}$/PMH$_0$, followed by spruce with 16.5 m$^3_{o.b.}$/PMH$_0$, and broadleaved tree species with 13.8 m$^3_{o.b.}$/PMH$_0$. When distinguishing between standing and bunched trees, overall productivity was 46% lower for standing than for bunched broadleaved trees but 30% higher for standing Douglas fir. The harvesting cycle productivity of spruce was not significantly affected by whether the trees were standing or bunched at the machine operating trails, as shown in Figure 4 for all harvesting cycles.

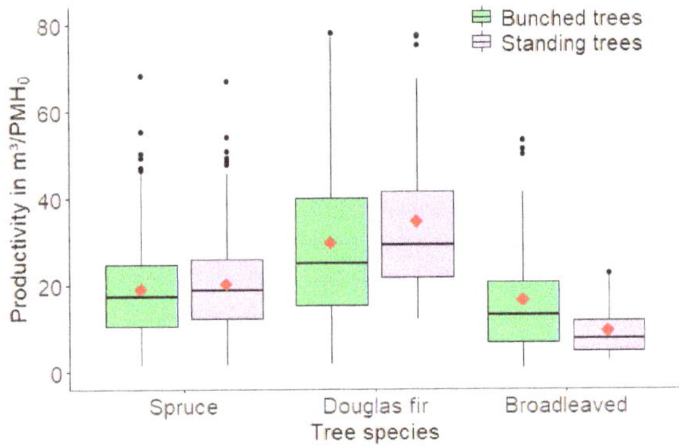

Figure 4. Productivity boxplot of the harvesting cycles in $m^3_{o.b.}$ per PMH_0 for bunched (green) and standing (purple) trees of different tree species (spruce, Douglas fir and broadleaved tree species) during harvesting operation.

The effect of piece volume on the harvester productivity was analyzed in more detail for coniferous tree species. After removing broadleaved tree species, Equation (1) was fitted using a nonlinear regression technique between productivity and merchantable volume. The expected increase of productivity with increasing volume is shown in Figure 5. While a cycle productivity of 20 m^3/PMH_0 was reached for piece volumes of 0.3 $m^3_{o.b.}$, the productivity was expected to be doubled (40 m^3/PMH_0) for piece volumes of approximately 1 m^3. The coefficient of determination R^2 (adjusted by degrees of freedom) was 51.5% (Figure 5). This means that more than half of the variation in harvester cycle productivity was explained by tree merchantable volume.

$$y = 42.51 \times V^{0.637} \quad (1)$$

where y is productivity in $m^3_{o.b.}$ per PMH_0 and V is merchantable volume in $m^3_{o.b.}$.

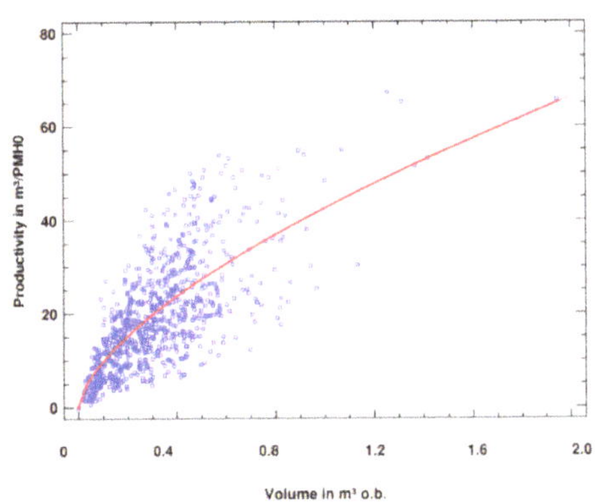

Figure 5. Harvester cycle productivity ($m^3_{o.b.}/PMH_0$) of conifers in relation to volume ($m^3_{o.b.}$).

3.4. Resulting Costs

The timber harvesting and extraction operation costs totaled 46,323 € or 69 €/$m^3_{o.b.}$ for all four working steps and involved operational units (Table 2). Costs for bunching the trees at the machine operating trail were 18% of the total costs and included the motor-manual felling and winching with the MFC.

Table 2. Time consumption and costs of the timber harvesting and extraction operations.

Operational Units	Hours (Scheduled Machine Hours (SMH))	Hourly Costs (€/SMH)	Total Costs (€)
Motor-manual felling	156	38	5928
Mini forestry crawler	35	70	2450
Harvester	92	200	18,400
Forwarder 1	90	90	8550
Forwarder 2	90	93	8835
Machine transport	18	120	2160

4. Discussion

Forest managers are currently facing new challenges: (i) changes in forest composition through both an increased number of species and a greater proportion of site-adapted broadleaved species, along with higher structural diversity in order to improve the forest's resistance and resilience towards climate change [1,2]; (ii) changes in forest certification or state regulations on the machine operating trail spacing to support intensified soil protection strategies [8,16]; and (iii) the mechanization of forest operations in inclined terrain, in order to reduce costs and increase worker safety [7]. In this study, we analyzed the performance of a timber harvesting and extraction system, addressing all the above-mentioned challenges. While the additional time consumption for the use of a tethering winch only represented 2.4% of a SMH, it should be noted that one forwarder was used for approximately 28 h as an anchor, contributing a non-negligible cost factor of 5.4% of the total costs.

Compared with a conventional fully mechanized system with 20-m spacing between machine operating trails, the system studied in this case study may involve higher costs, as 18% of the total costs were attributed to motor-manual felling and winching the trees from the midfields to the machine operating trails. Increased costs due to midfield operations were also found in other studies [11,17]. It should be noticed that highly structured forest stands with admixture of broadleaved tree species may lead to processing difficulties when fully mechanized systems are used, which will result in lower productivity and higher costs. Nevertheless, time consumption of MFCs may be improved by better work organization in order to reduce the waiting time (which accounted for nearly 20% of all work time), and thereby reduce the costs. However, increased machine operating trail spacing has positive aspects as well and is in compliance with sustainable forest operations [18]: lower traffic, increased forest production area, and better societal acceptance. Thus, the extra cost for the provision of ecosystem services can be valued as that due to the application of environmental protection regulations and societal acceptance.

In the current study, we used merchantable volumes (not full tree volumes) to quantify production performance. Thus, large differences in the volumes of the different tree species occurred. For a 56-year-old stand, spruce volumes were relatively low and of low quality, due to high rates of tree breakages from snow. On the contrary, due to pruning and early release of future crop trees, Douglas firs had—despite their younger age (35 years)—larger volumes and higher log qualities compared with the Norway spruces. Higher log quality and generally larger log size meant Douglas fir had the highest productivity values, followed by Norway spruce and broadleaved tree species. The latter are known to often complicate and prolong the processing phase due to more complex and variable stems and crown architectures [19].

Processing productivity by the harvester increased significantly for broadleaved tree species when trees were bunched at the machine operating trails. In contrast, considerable positive effects of bunching were not observed for Norway spruce or Douglas fir. Thus, such a system may be more valuable in forest stands with a large proportion of broadleaved tree species. In conifer-dominated stands, other timber harvesting and extraction methods could be more productive; for example, systems using harvesters with a 15-m boom reach. Another system could involve motor-manual felling of the trees in the midfield and winching them with an integrated winch attached to the harvesters.

However, in that case, the harvester operator would need to pay particular attention to worker safety, as forest workers and harvesters would work next to each other. Further, this would lead to potential waiting times until the trees were felled, which could hamper the productivity and thus the efficient use of the harvester. In this respect, a positive aspect of the described timber harvesting and extraction system with midfields is that the motor-manual work, including the bunching and the mechanized harvester-forwarder system, are decoupled in time. This allows more planning and logistical flexibility among the machine systems and additionally increases the safety situation.

It is important to consider that the tethered equipment was at its upper limit for traction-assistance in this study. Depending on regional regulations, work systems similar to the one described here can also be implemented with a winch-assisted set-up. There, the working equipment is indeed fully secured by the cable through an anchor or other machine unit, providing access to terrain the machine would not be able to reach on its own [20]. In terms of worker safety, it is therefore essential to respect the defined operational limits for the type of equipment applied in the designated work system.

5. Conclusions

Due to new regulations, an increased share of mixed-species forests and extended machine operating trail spacing are becoming more and more popular in southwestern Germany. The described timber harvesting and extraction system, with its good performance among broadleaved tree species, has proven its suitability for these changing operational conditions. Nevertheless, further field work and research is required to further adapt and improve upon the overall sustainability of such systems, with special attention towards occupational safety and cost reduction. In this respect, a focus should be placed on optimal log presentation towards the harvester's boom reach, at reduced winching distances and motor-manual work effort, with the aim of defining optimal operational layouts, to realize comparable harvester performance as when bunching directly at the machine operating trail.

Author Contributions: F.B. conceived and designed the study, with contributions from J.S.; F.B., P.A. and J.S. collected the data; F.B. and E.T. analyzed the data, with contributions from P.A.; F.B., E.T., S.H. and J.S. wrote the manuscript. All authors have read and agreed to the published version of the manuscript.

Funding: This study was undertaken within the framework of the project "SOLVE" (Timber harvesting and transportation systems adapted to altered forest structures due to climate change), which was funded by the German Federal Ministry of Food and Agriculture (BMEL) and the German Federal Ministry for the Environment, Nature Conservation, Building and Nuclear Safety (BMUB) in the context of the "Förderrichtlinie Waldklimafond" (Förderkennzeichen 28W-B-3-048-01). J.S. was supported by the European Social Fund and by the Ministry of Science, Research and Arts Baden-Württemberg while working at the University of Freiburg.

Acknowledgments: The authors thank Max Bierer, Christian Ludowicy and Mathieu Fortin for their support, as well as the ForstBW A.ö.R., in particular the Maschinenbetrieb Schrofel, the Forstrevier Kleines Wiesental, and the KELLER—Forstbetrieb.

Conflicts of Interest: The authors declare no conflict of interest.

References

1. Saha, S.; Bauhus, J. Trade-offs between climate change adaptation and mitigation objectives for forests in south-western Germany. *Forstarchiv* **2016**, *87*, 60–61.
2. Berendt, F.; Fortin, M.; Jaeger, D.; Schweier, J. How climate change will affect forest composition and forest operations in Baden-Württemberg—A GIS-based case study Approach. *Forests* **2017**, *8*, 298. [CrossRef]
3. Kühmaier, M.; Stampfer, K. Development of a multi-attribute spatial decision support system in selecting timber harvesting systems. *Croat. J. For. Eng.* **2010**, *31*, 75–88.
4. Cacot, E.; Maire, L.; Chakroun, M.; Peuch, D.; Montagny, X.; Perrinot, C.; Bonnemazou, M. La mécanisation du bucheronnage dans le peuplement de feuillus. In *Synthèse Réalisée Dans le Cadre du Projet Ecomef Retenu à L'appel à Projet n°10 du FUI*; Institut Technologique FCBA: Champs-sur-Marne, France, 2015.
5. Cavalli, R.; Amishev, D. Steep terrain forest operations–challenges, technology development, current implementation, and future opportunities. *Int. J. For. Eng.* **2019**, *21*, 1–7. [CrossRef]

6. Huber, C.; Stampfer, K. Efficiency of topping trees in cable yarding operations. *Croat. J. For. Eng.* **2015**, *36*, 185–194.
7. Visser, R.; Stampfer, K. Expanding ground-based harvesting onto steep terrain: A review. *Croat. J. For. Eng.* **2015**, *36*, 321–331.
8. FSC. *Deutscher FSC-Standard. Version 3-0*; FSC Deutschland: Freiburg, Germany, 2018.
9. FVA-BW. *Richtlinie zur Feinerschliessung*; Forstliche Versuchs- und Forschungsanstalt Baden-Württemberg: Freiburg, Germany, 2003.
10. Mederski, P. A comparison of harvesting productivity and costs in thinning operations with and without midfield. *For. Ecol. Manag.* **2006**, *224*, 286–296. [CrossRef]
11. Berendt, F.; Fortin, M.; Suchomel, C.; Schweier, J. Productivity, costs, and selected environmental impacts of remote-controlled mini forestry crawlers. *Forests* **2018**, *9*, 591. [CrossRef]
12. Waas, S. *Feinerschließung-Rückegassen und Rückewege*; Merkblatt der Bayerischen Landesanstalt für Wald und Forstwirtschaft, LWF: Freising, Germany, 2017; Volume 38.
13. International Organization for Standardization. *Machinery for Forestry-Winches-Dimensions, Performance and Safety Part 2: Traction Aid Winches*; International Organization for Standardization: Geneva, Switzerland, 2019; ISO/DIS 19472-2.
14. KWF. Funktiogrammvorlagen für Holzernteverfahren: "Funktiogrammvorlagen und Maschinenpiktogramme zum Download". Available online: https://www.kwf-online.de/index.php/forschungsprojekte/replan/306-funktiogramme-erstellen (accessed on 7 October 2020).
15. Björheden, R. *Forest Work Study Nomenclature. Test Edition Valid 1995–2000*; Swedish University of Agricultural Sciences Department of Operational Efficiency: Garpenberg, Sweden, 1995; ISBN 91-576-5055-1.
16. ForstBW. *Konzept zur Sicherstellung der Dauerhaften Funktionsfähigkeit von Rückegassen. Bodenschutz & Forsttechnische Befahrbarkeit, Version 1.0*; Landesbetrieb Forst Baden-Württemberg: Stuttgart, Germany, 2012.
17. Mederski, P.; Venanzi, R.; Bembenek, M.; Karaszewski, Z.; Rosińska, M.; Pilarek, Z.; Luchenti, I.; Surus, M. Designing thinning operations in 2nd age class pine stands-economic and environmental implications. *Forests* **2018**, *9*, 335. [CrossRef]
18. Marchi, E.; Chung, W.; Visser, R.; Abbas, D.; Nordfjell, T.; Mederski, P.S.; McEwan, A.; Brink, M.; Laschi, A. Sustainable Forest Operations (SFO): A new paradigm in a changing world and climate. *Sci. Total Environ.* **2018**, *634*, 1385–1397. [CrossRef]
19. Labelle, E.; Soucy, M.; Cyr, A.; Pelletier, G. Effect of tree form on the productivity of a cut-to-length harvester in a hardwood dominated stand. *Croat. J. For. Eng.* **2016**, *37*, 175–183.
20. Holzfeind, T.; Visser, R.; Chung, W.; Holzleitner, F.; Erber, G. Development and benefits of winch-assist harvesting. *Curr. For. Rep.* **2020**, *6*, 201–209. [CrossRef]

Publisher's Note: MDPI stays neutral with regard to jurisdictional claims in published maps and institutional affiliations.

sustainability MDPI

Article

The Effects of Soil Moisture on Harvesting Operations in *Populus* spp. Plantations: Specific Focus on Costs, Energy Balance and GHG Emissions

Farzam Tavankar [1], Mehrdad Nikooy [2], Francesco Latterini [3], Rachele Venanzi [4], Leonardo Bianchini [4] and Rodolfo Picchio [4,*]

[1] Department of Forestry, Khalkhal Branch, Islamic Azad University, Khalkhal 56817-31367, Iran; tavankar@aukh.ac.ir
[2] Department of Forestry, Faculty of Natural Resources, University of Guilan, Somehsara 96196-43619, Iran; nikooy@guilan.ac.ir
[3] Consiglio per la Ricerca in Agricoltura e l'Analisi dell'Economia Agraria (CREA), Centro di Ricerca Ingegneria e Trasformazioni Agroalimentari, Via della Pascolare 16, 00015 Monterotondo, Italy; francesco.latterini@crea.gov.it
[4] Department of Agricultural and Forest Sciences, University of Tuscia, Via San Camillo de Lellis, 01100 Viterbo, Italy; venanzi@unitus.it (R.V.); l.bianchini@unitus.it (L.B.)
* Correspondence: r.picchio@unitus.it

Citation: Tavankar, F.; Nikooy, M.; Latterini, F.; Venanzi, R.; Bianchini, L.; Picchio, R. The Effects of Soil Moisture on Harvesting Operations in *Populus* spp. Plantations: Specific Focus on Costs, Energy Balance and GHG Emissions. *Sustainability* **2021**, *13*, 4863. https://doi.org/10.3390/su13094863

Academic Editor: António Dinis Ferreira

Received: 19 March 2021
Accepted: 24 April 2021
Published: 26 April 2021

Publisher's Note: MDPI stays neutral with regard to jurisdictional claims in published maps and institutional affiliations.

Abstract: Background: Poplar tree plantations for wood production are part of a worldwide growing trend, especially in moist soil sites. Harvesting operations in moist sites such as poplar plantations require more study for detailed and increased knowledge on environmental and economic aspects and issues. Methods: In this study, the effects of soil moisture content (dry vs. moist) on productivity, cost, and emissions of greenhouse gases (GHG) caused by operations of different harvesting systems (chainsaw-skidder and harvester-forwarder) were evaluated in three poplar plantations (two in Italy and one in Iran). Results: The productivity (m^3 h^{-1}) of both systems in the dry sites were significantly higher (20% to 30%) than those in the moist sites. Production costs (€ m^{-3}) and GHG emissions (g m^{-3}) of both systems in the dry sites were also significantly lower than those in the moist sites. The productivity of the harvester-forwarder system was about four times higher, and its production cost was 25% to 30% lower than that of the chainsaw-skidder system, but the calculated GHG emissions by harvester-forwarder system was 50–60% higher than by the chainsaw-skidder system. Conclusions: Logging operations are to be avoided where there are conditions of high soil moisture content (>20%). The result will be higher cost-effectiveness and a reduction in the emission of pollutants.

Keywords: skidding productivity; logging cost analysis; harvesting site conditions; sustainable forest operations

1. Introduction

Poplar planting has occurred around the world for a very long time. The plantations in Iran and Italy provide an important source of wood supply. At present, in both countries there are over 100,000 ha of these monospecific plantations (50,000 ha in Iran and 66,000 ha in Italy) and they mainly consist of *Populus deltoides* and *P. euramericana* [1–3]. Although poplar plantations cannot be currently considered among the main sources of wood in both countries, their importance is rapidly increasing [4,5]. Poplar wood shows interesting features, such as uniform mechanical properties and a high percentage of juvenile wood. These make it possible to obtain several products from plantations, i.e., building and veneering material, paper pulp and wood chips for bioenergy [2,6,7]. Moreover, the poplars in both the Italian and Iranian conditions can reach a growth rate of approximately 10–30 m^3 ha^{-1} $year^{-1}$, which is substantially higher than local tree species [8–10].

Considering the growing importance of poplar plantations as a source of wood material, it is necessary to assess the technical and environmental characteristics of the harvesting operations in these plantations. Wood production from artificial stands is indeed a simplified multistage process as compared to forestry production, but proper planning of the logging operations is crucial when viewing the overall sustainability of the intervention [11]. The concept of sustainability in the forestry sector is strictly related to the paradigm of sustainable forest operations (SFO) [12]. SFO refers to the implementation of logging operations which are able to meet the requirements of all three pillars of sustainability (economy, environment and society) [13,14]. Machine productivity and operation costs are the two main factors in evaluating harvesting operations regarding the economic aspect of plantation management [15–18]. Having accurate information on the productivity of logging machines is therefore a key issue for the economic assessment of production. Work productivity evaluation is a complex issue, considering that the working performance of a given harvesting system is related to several factors, for instance, machine type, tree size, logging intensity, number of trees per hectare, terrain conditions, operator skills and planned treatment [19–23]. However, always considering the concept of SFO, assessing and optimizing work performance is not enough to obtain sustainable logging. The environmental aspect is also fundamental [24–27]. Along with soil impact and stand damage, greenhouse gases (GHG) emission related to mechanical operations is a major aspect to be evaluated so as to assess the environmental performance of a given harvesting system [28,29].

Considering the points listed above, it is a major challenge for forest managers to evaluate the manifold consequences of decisions and to estimate the economic and environmental performance of different alternatives before carrying out action.

These statements are valid for all forestry and agro-forestry interventions, but are even more important when dealing with poplar plantations, which show particular features. Poplar plantations are often located in plain or floodplain lands, which means that very often harvesting operations take place in soil conditions with high moisture content. Among all the variables, the soil moisture content during logging can significantly influence the degree of soil disturbance, with greater potential for higher soil compaction on wet/saturated soils than on dry ones [30]. Several studies have focused on the effects of different levels of soil moisture content during harvesting on soil disturbance and the physical properties of the soil [31,32]. On the other hand, the effects of different soil moisture content at the time of felling and skidding on the harvesting operations' performance has not yet been studied.

Considering these peculiarities of poplar plantations and their growing importance in both the Italian and Iranian forestry systems, the main aims of the present study were: (i) to provide a comparative analysis of different harvesting technologies for poplar plantations; (ii) to determine the influence of soil moisture on harvesting operations' performance both from an economic and an environmental point of view.

This study will make a detailed statement of what can be verified and a general statement of what cannot. More precisely, one can state that operations in specific soil conditions could produce the performance reported here under the specific conditions of this study. This knowledge is worth having, although poorly suited to any generalization, and can be used to estimate an expected forest operation's performance that may occur under different conditions with some level of approximation.

2. Materials and Methods

2.1. Study Areas

This study was carried out in three different geographical areas, two in Italy and one in Iran, investigating three different harvesting operations in poplar plantations.

The Iranian poplar plantation (IR) is located in the coastal area of the Caspian Sea in the Guilan province in northern Iran. The total area of the plantation is 60 ha, situated in flat terrain at an altitude range from 0 m to 20 m a.s.l. The average annual rainfall is from 1260 to 1340 mm and most of the precipitation occurs between the months of September and December. The average annual temperature is +15 °C, with the minimum during the winter at a few degrees below 0 °C, and the maximum at +25° during the summer. The soil type is clay loam with poor drainage. This plantation was divided into two areas of 30 ha each. The harvesting operation in 30 ha for the dry site was performed in the first half of September 2019 before the rainfall. Harvesting operations were carried out in the second half of September 2019 after rainfall in the 30 ha moist site. The soil moisture content was 14.4% in the dry site and 34.6% in the moist site. Trees were felled by chainsaw, and whole trees were extracted to roadside landings by wheeled skidder Timberjack 450C. Finally, processing operations were motor-manually performed at the landing site.

The first Italian poplar plantation (IT1) was located in the Lazio region in central Italy. The total area of the plantation is 20 ha and it is situated in flat terrain and with an altitude range from 90 m to 110 m a.s.l. The average annual rainfall is from 830 to 900 mm and most of the precipitation occurs from October to December. The average annual temperature is +14.9 °C with a minimum during the winter at 2.5 °C, and a maximum at +30.7 °C during the summer. The soil type is clay loam with a low level of organic matter, nitrogen and phosphorus and with poor drainage. This plantation was divided into two areas of 10 ha each. The harvesting operations in 10 ha on the dry site were performed in the second half of June 2018. Harvesting operations were carried out in the first half of April 2018 after rainfall in the 10 ha moist site. The soil moisture content was 12.1% in the dry site and 36.8% in the moist site. As in the Iranian site, trees were motor-manually felled by chainsaw, and whole tree extraction was carried out by a wheeled skidder Timberjack 450C. In this case too, motor-manual processing with a chainsaw was carried out at the landing site.

The second Italian poplar plantation (IT2) was located in the Veneto region in North Italy. The total area of the plantation is 20 ha and situated in flat terrain with an altitude range from 10 m to 30 m a.s.l. The average annual rainfall is from 730 to 850 mm and most of the precipitation occurs from September to November. The average annual temperature is +13.6 °C with a minimum during winter of 0.1°C, and a maximum of +29.2 °C during the summer. The soil type is clay loam texture with a low level of organic matter and with poor drainage. This plantation is divided into two areas of 10 ha each. The harvesting operations in the 10 ha dry site were performed in the first half of July 2018. Harvesting operations were carried out in the second half of September 2018 after rainfall in the 10 ha moist site. The soil moisture content was 15.0% in the dry site and 35.7% in the moist site. Trees were mechanically felled and processed by a harvester and cut to length through an extraction system by a forwarder. Technical characteristics of the machinery used in the various harvesting sites are given in Table 1. Average dendrometric characteristics of the three plantations are shown in Table 2. A preliminary analysis for dendrometric characteristics of the three different stands was done by one-way ANOVA to check for differences among the average values of the three plantations. There were no significant differences of dendrometric characteristics between the Italian sites (IT1 and IT2). However, density, basal area and standing volume of trees in the Iranian site were higher than in the Italian sites.

2.2. Data Collection and Analysis

Dendrometric data were obtained through systematic plot sampling. Grid dimension was 150 m × 150 m, the area of each circular plot was 1256 m^2 (20 m radius), and in total 20 plots in each area were established. Diameter at breast height (dbh) and height of tree species were measured by caliper and clinometer, respectively, in each plot. The volume of winched logs was calculated by Huber's formula ($V = A_m \times L$), where V is log volume (m^3), A_m is the middle point cross-sectional area of log (m^2), and L is the length of log (m).

Soil samples were collected with a steel ring (inside diameter 5 cm, length 10 cm) and immediately put in hermetic plastic bags and labeled. The wet weight of all samples was measured before transfer to the laboratory (on the same sampling day). In the laboratory, soil samples were dried in an oven at 105 °C for 24 h until reaching a constant mass to determine the soil moisture content.

Table 1. Specification of the mechanization used in the three harvesting sites.

Characteristics	Chainsaw Stihl ms880	Harvester John Deere 1470 D	Skidder Timberjack 450C	Forwarder John Deere JD1110 D
Displacement (cm^3)	122	9000	6800	4140
Power (kW)	6.4	179.7	120.0	121.0
Weight (kg)	10	19,700	10,270	17,500
Bar length (cm)	90	75	-	-
Oil tank volume (l)	0.7	290.0	150.0	300.0
Fuel tank volume (l)	1.3	470.7	159.0	150.0
Number of cylinders	1	6	6	6
Maximum traction or load (kg)	-	-	11,000	8500
Maximum operative distance (m)	-	8.6	75.0	10.5

Table 2. Average dendrometric stand and main wood characteristics before harvesting in the three poplar plantations. The wood density values (±SD) showed refer to fresh matter and recorded during the harvesting operation.

Description	IR	IT1	IT2
Plantation area (ha)	60	20	20
Tree density (stem·ha^{-1})	400	278	279
Mean DBH (cm)	38.3	42.8	40.4
Mean basal area (m^2·ha^{-1})	46.1	39.9	35.8
Mean tree height (m)	25.3	24.8	27.2
Standing volume (m^3·ha^{-1})	876.4	702.6	749.8
Wood density (kg·m^{-3})	795.8 (±11.5)	702.9 (±15.6)	721.7 (±9.4)
Wood moisture (%)	98.2 (±19.2)	95.4 (±18.5)	99.1 (±21.7)

A time-motion study was carried out to evaluate working productivity. Each working cycle was stop watched individually, separating productive time from delay time [33]. Calculated delay factor represents the quotient of delay time over net cycle time. Productivity was evaluated both on delay-free time and on actual total time, inclusive of all delays. Inclusion of delays was not capped on the basis of a maximum event duration. Scheduled Machine Hours (SMH) include all the time the machine is scheduled to work, whereas Productive Machine Hours (PMH) represent the time during which the machine actually performs work, excluding the time lost to both mechanical and non-mechanical delays.

The working group had 10 to 15 years of work experience with the machines and they were able to service and repair them.

The working cycles are reported, for the three areas, in Tables 3 and 4. Continuous time was recorded to the nearest second with a chronometer. The cycle times of the machines were divided into time elements (process steps) that were considered characteristic of this work.

Table 3. Description of felling and processing cycle elements of harvesting in the three yards.

Time Elements	IR & IT1 by Chainsaw (2 Operators)	IT2 by Harvester (1 Operator)
Moving (M)	starts when the chainsaw operator moves from the last felled tree to the next to be felled and ends when the team cleans the tree stump before the felling	starts when the harvester wheels start moving from one standing point and ends when they stop at the next standing point
Felling (F)	starts when the chainsaw operator turns on the chainsaw and performs the cut and ends with the fall of the tree	starts when the harvester head grips the stem and ends when the tree falls onto the ground
Processing (P)	starts when the chainsaw operator cuts the first branch and ends when he finishes the cross cutting of the tree	starts when the tree stem starts moving through the harvester head and ends when the harvester wheels start moving

Table 4. Description of extraction cycle elements of harvesting in the three yards.

Time Elements	IR & IT1 by Skidder (2 operators)	IT2 by Forwarder (1 Operators)
Travel unloaded (TUL)	begins when the skidder leaves the roadside landing area and ends when the skidder arrives at a suitable position (nearest distance from the logs) on the skid trail	begins when the forwarder leaves the roadside landing area and ends when the forwarder arrives at the first suitable position (nearest distance from the first logs)
Bunching—Loading (B)	begins when the skidder driver releases the cable and ends when the winching phase is finished	begins when the forwarder driver loads the first log and ends when the forwarder is fully loaded
Travel loaded (TL)	begins when the skidder starts to move and ends when the skidder arrives on roadside landing	begins when the fully loaded forwarder starts to move and ends when the forwarder arrives at the roadside landing
Landing operations (LO)	begins when the choker setter opens the load and ends when load is piled up in final position and the skidder is preparing for the next cycle	begins when the forwarder driver starts to unload the logs and ends when load is piled up in final position and the forwarder is preparing for the next cycle

Details of the harvested trees and volume in each treatment are given in Table 5.

Table 5. Harvested trees, volume and working cycles for work productivity analysis in each treatment (data reported to FU of 1 t of fresh mass showed in Table A1).

Parameter	IR Dry	IR Moist	IT1 Dry	IT1 Moist	IT2 Dry	IT2 Moist
Felled-processed trees (N)	601	625	556	528	2790	2762
Felled processed-volume (m^3)	1314.600	1367.184	1405.200	1334.940	7498.000	7423.020
Extraction cycles (N)	100	100	100	100	100	100
Extracted volume (m^3)	672.000	690.000	791.000	602.000	1350.000	1180.000

The system boundaries for the study area were set to those of the harvesting operations, from the felling to the landing site. The Functional Unit (FU) for the analyses was the cubic meter of round wood (m^3); in Appendix A the data referring to another Functional Unit (FU) are shown (1 t of fresh mass, following the data shown in Table 2). This is important in order to compare these results more readily with other studies.

Operational costs were estimated according to the Miyata method [34] as previously explained in Spinelli et al. [35]. Economic evaluation of the different machines was carried out taking into consideration different periods of use. The skidder, harvester and forwarder depreciated by 1200 SMH per year [35,36] in a depreciation period of 10 years [37]. The chainsaw for the felling depreciated by 800 SMH per year in a depreciation period of 2 years [36]. Labor cost was set at € 15 SMH^{-1} inclusive of indirect salary costs [38]. Lubricant consumption was calculated as reported by Picchio et al. [39].

Costs for insurance, repair and service were obtained by literature analysis [35], while the fuel and lubricant prices were taken by a market survey (second semester 2019) conducted upon three company products. The calculated operational cost, as reported in similar studies [35], was increased by 10% to account for overhead costs [40].

Focusing instead on the environmental aspects, an energy consumption analysis was performed, applying the Gross Energy Requirement (GER) method [41]. Indirect input (MJ kg^{-1}) of harvesting machinery was evaluated taking into consideration the average energy value of the raw materials. This is related to several parameters, i.e., quantitative presence (%), total mass of the machine (kg), overall service life of the machine (h m^{-3}) and use of the machine during harvesting. Energy consumption related to human manpower was evaluated according to what was reported in previous works [42–44], through the application of a standard value equal to 0.030 MJ min^{-1} $worker^{-1}$.

To calculate the energy balance, the energy value of poplar wood was determined as Higher Heating Value (HHV) (CEN/TS 14918), on 30 random samples, through Parr calorimeter, model 6200 [45].

Regarding pollutant emissions during logging operations, emissions related to fuel were evaluated as the sum of the emissions during combustion (Efc) and the emissions produced within the production and logistic process (Efp). For Efc assessment of fuel energy content, the emission factor of the engine and the thermal efficiency of the combustion were taken into consideration, as reported in Klvac et al. [46] and Athanassiadis [47].

Dealing with Efp assessment, fuel energy content and emission factors were obtained by [46], but HC emission factor was taken from [47].

2.3. Statistical Analysis

The first step in statistical analysis was checking for normality using the Kolmogorov-Smirnov test and for homogeneity of variance using the Levene test. Averages of dendrometric characteristics, skid trail network, and average extraction cycle time elements in each area between the two site conditions (moist and dry soil) were compared by independent t test. A regression analysis of time study data was used to check the model's capability of predicting productivity as a function of statistically significant independent variables such as distance and load size. If the data were not normally distributed, a non-parametric Spearman's rank coefficient was applied to analyze the correlation between the variables.

A major focus was placed on extraction operations investigating the relationships between time elements and dendrometric characteristics of extracted timber, and between time elements and bunching-extraction distance. This investigation was performed by non-linear regression analysis, performed by SPSS 19.0 software (New York, NY, United States).

3. Results

Results of the *t*-test showed no statistically significant differences regarding both extraction distance and mean volume per working cycle between moist and dry sites in all the three plantations. Bunching distance was statistically lower in the IT1 dry site than in the moist one, while in IR and IT2 no statistically significant difference was found for this parameter (Table 6).

Table 6. Extraction trail and corridors' average features for the three areas in the two soil moisture conditions (mean ± SD), From the *t*-test for independent samples applied, statistically significant differences ($p < 0.05$) between the average values are highlighted (underlined text) (data reported to FU of 1 t of fresh mass showed in Table A2).

	IR		IT1		IT2	
	Dry	Moist	Dry	Moist	Dry	Moist
Mean bunching distance (m)	38.6 ± 10.1	49.8 ± 9.5	<u>17.5 ± 5.2</u>	<u>56.1 ± 8.2</u>	22.7 ± 4.2	18.4 ± 2.1
Mean extraction distance (m)	146.0 ± 20.1	144.5 ± 12.2	118.2 ± 14.3	138.1 ± 10.3	152.4 ± 12.5	158.7 ± 11.2
Mean volume per working cycle (m^3)	6.72 ± 0.15	6.90 ± 0.14	7.91 ± 0.22	6.02 ± 0.18	13.5 ± 0.31	11.8 ± 0.42

Data regarding working time analysis of felling and processing operations are given in Table 7 and Figure 1. In all three plantations felling operation time (motor-manual in IR and IT1 and mechanized in IT2) did not show any statistically significant differences between dry and moist sites. There were some differences found in processing and moving time. In every study area the most time-consuming operation was processing.

The soil moisture in IT2 did not affect working productivity, with no statistically significant difference among different soil moisture conditions for every phase, and consequently also for the overall working time. In IR1 and IT1, instead, both TET (Total Effective Time) and TGT (Total Gross Time) were significantly higher in moist conditions than in dry ones, with higher DT (Delay Time) also in the moist soil. However, as previously reported, such differences are related not to the felling operations, but only to moving (both IR and IT1) and processing (only IR).

Working time analysis data in bunching and extraction are reported in Table 8 and Figure 2. The only phase which did not show an effect of soil moisture on productivity was the landing operation (LO), while all the other phases were influenced by the moisture content of the soil. This led to a significant difference in overall working times (both TET and TGT), related to the different site conditions, in all the three yards. In particular, higher soil moisture negatively affected working productivity.

Table 7. Description statistics of felling-processing operations for the harvesting sites studied referring to a single tree. From the *t*-test for independent samples applied, statistically significant differences ($p < 0.05$) between the average values are highlighted (underlined text). Statistical comparisons are between columns.

Sites	IR-Dry	IR-Moist	IT1-Dry	IT1-Moist	IT2-Dry	IT2-Moist
Time Elements	Average Value ± SD (minutes)					
M	0.36 ± 0.05	0.48 ± 0.06	0.31 ± 0.04	0.42 ± 0.03	0.53 ± 0.05	0.55 ± 0.06
F	2.85 ± 0.22	2.75 ± 0.15	3.15 ± 0.22	3.08 ± 0.31	0.56 ± 0.04	0.59 ± 0.05
P	13.91 ± 0.38	14.85 ± 0.45	14.58 ± 0.87	15.29 ± 0.65	3.05 ± 0.25	3.15 ± 0.15
DT	0.29 ± 0.04	0.41 ± 0.11	0.38 ± 0.12	0.45 ± 0.08	0.11 ± 0.03	0.13 ± 0.03
TET	17.12 ± 0.45	18.08 ± 0.81	18.04 ± 0.88	18.79 ± 0.50	4.14 ± 0.72	4.29 ± 0.44
TGT	17.41 ± 0.52	18.49 ± 0.73	18.42 ± 0.91	19.24 ± 0.71	4.25 ± 0.65	4.42 ± 0.61

M: moving, F: felling, P: processing, DT: delay time, TET: total effective time, TGT: total gross time.

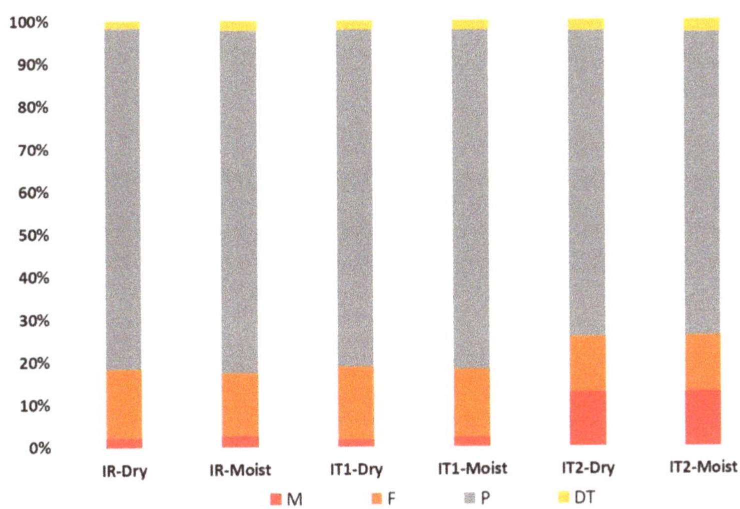

Figure 1. Working time distribution for felling-processing operations in the harvesting sites studied.

Table 8. Description statistics of bunching-extraction operations for the harvesting sites studied referring to single cycle. From the *t*-test for independent samples applied, statistically significant differences ($p < 0.05$) between the average values are highlighted (underlined text). Statistical comparisons are between columns.

Sites	IR-Dry	IR-Moist	IT1-Dry	IT1-Moist	IT2-Dry	IT2-Moist
Time Elements	Average Value ± SD (minutes)					
TUL	4.53 ± 0.44	6.86 ± 0.50	3.88 ± 0.64	5.48 ± 0.98	1.35 ± 0.50	1.89 ± 0.44
B	2.10 ± 0.80	3.40 ± 1.00	1.38 ± 0.21	3.52 ± 0.61	2.15 ± 0.44	2.39 ± 0.40
TL	4.30 ± 1.50	7.20 ± 0.50	4.11 ± 1.07	6.92 ± 1.16	3.02 ± 0.28	5.12 ± 0.31
LO	5.70 ± 2.10	6.50 ± 1.90	5.90 ± 0.92	5.80 ± 0.89	5.20 ± 1.10	5.39 ± 0.81
DT	1.00 ± 0.45	1.20 ± 0.50	1.24 ± 0.22	1.85 ± 0.18	1.02 ± 0.10	1.39 ± 0.13
TET	16.63 ± 1.80	23.96 ± 1.90	15.27 ± 1.78	21.72 ± 1.55	11.72 ± 0.97	14.79 ± 0.84
TGT	17.63 ± 1.80	25.16 ± 1.90	16.51 ± 1.89	23.57 ± 2.01	12.74 ± 1.16	16.18±1.01

TUL, travel unloaded; B, bunching; TL, travel loaded; LO, landing operations; DT, delay times; TET, total effective time; TGT, total gross time.

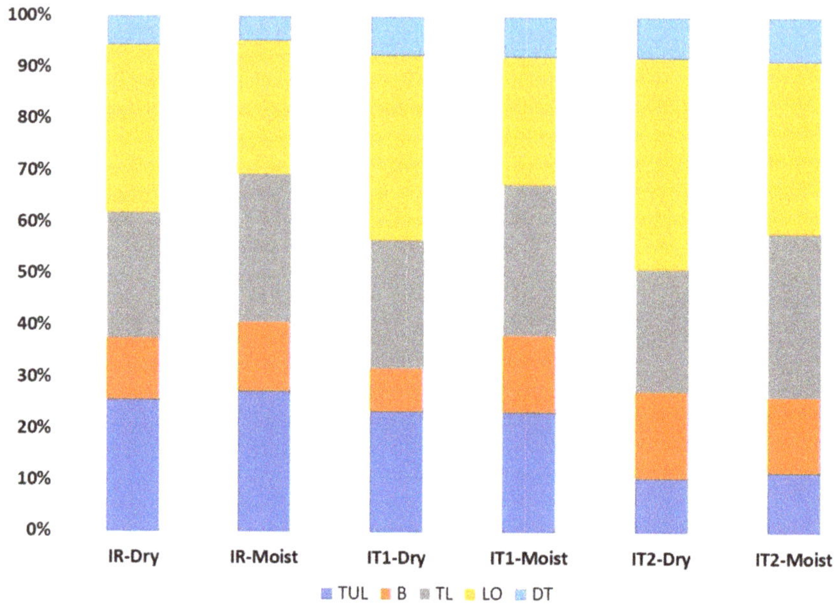

Figure 2. Working time distribution for bunching-extraction operations in the harvesting sites studied.

The analysis of the various factors influencing extraction time (Table 9) revealed that the parameter with the highest impact on time consumption was bunching–extraction distance, with R^2 values ranging from 0.6 to 0.8 for both dry and moist sites in all the three yards.

Table 9. Cycle time (Y) equations of bunching–extraction in studied sites. D: distance of bunching-extraction; and LV: load volume.

Site (Machine)	Variable	Model	Equation	R2 Adj.	p-Value
IR-Dry (Skidder)	D	Polynomial	$Y = -4 \times 10^{-5}(D)^2 + 0.0592(D) + 8.0769$	0.633	<0.001
	LV	Polynomial	$Y = 0.9524(LV)^2 + 8.2375(LV) + 28.361$	0.380	<0.001
	D and LV	Linear	$Y = 0.059(D) + 1.546(LV) - 1.523$	0.679	<0.001
IR-Moist (Skidder)	D	Polynomial	$Y = -2 \times 10^{-5}(D)^2 + 0.1343(D) + 6.2007$	0.764	<0.001
	LV	Exponential	$Y = 3.1537e^{0.2875(LV)}$	0.363	<0.001
	D and LV	Linear	$Y = 0.116(D) + 2.089(LV) - 6.076$	0.812	<0.001
IT1-Dry (Skidder)	D	Linear	$Y = 0.088(D) + 6.3003$	0.821	<0.05
	LV	Polynomial	$Y = 0.7512(LV)^2 + 4.2725(LV) + 8.030$	0.231	>0.05
	D and LV	Linear	$Y = 0.085(D) + 2.156(LV) + 0.125$	0.401	>0.05
IT1-Moist (Skidder)	D	Exponential	$Y = 9.312e^{0.0064(D)}$	0.724	<0.01
	LV	Polynomial	$Y = 0.95210(LV)^2 + 2.1225\ (LV) + 5.103$	0.412	>0.05
	D and LV	Linear	$Y = 0.109(D) + 1.816(LV) + 0.231$	0.502	> 0.05
IT2-Dry (Forwarder)	D	Polynomial	$Y = -0.0001(D)^2 + 0.078(D) + 5.103$	0.811	<0.05
	LV	Polynomial	$Y = -0.0021(LV)^2 + 0.807(LV) + 2.210$	0.452	>0.05
	D and LV	Linear	$Y = 0.080(D) + 2.086(LV) - 0.957$	0.568	> 0.05
IT2-Moist (Forwarder)	D	Polynomial	$Y = 0.0003(D)^2 + 0.004(D) + 7.7868$	0.765	<0.01
	LV	Polynomial	$Y = -0.052(LV)^2 + 1.105(LV) + 1.574$	0.431	>0.05
	D and LV	Linear	$Y = 0.128(D) + 1.974(LV) - 0.358$	0.631	>0.05

Increased bunching–extraction distance obviously led to increased bunching–extraction time; however, it is interesting to notice (Figure 3) how this effect is less evident in forwarding operations (IT2) than it is in winching operations (IR and IT1).

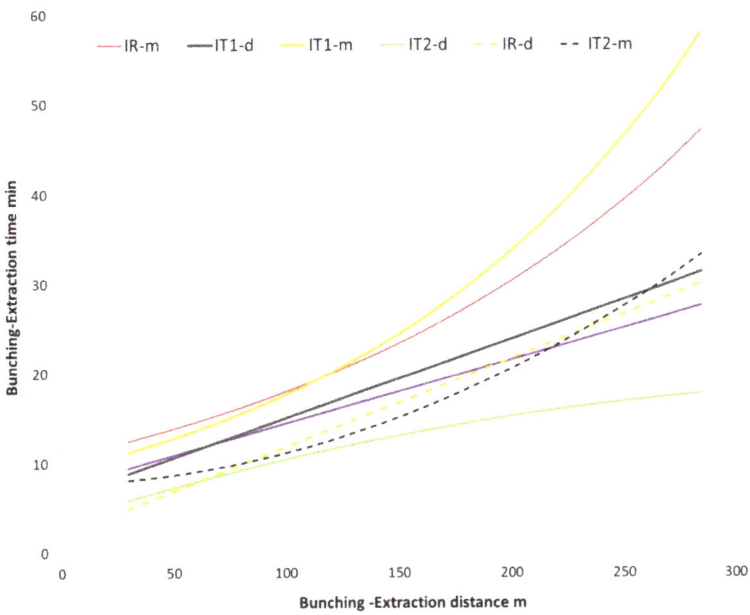

Figure 3. Graphical regression analysis referring to bunching–extraction time in relation to bunching–extraction distance in the studied areas (IR-d: Iranian dry site; IR-m: Iranian moist site; IT1-d: skidding Italian site with dry soil; IT1-m: skidding Italian site with moist soil; IT2-d: forwarding Italian site with dry soil; IT2-m: forwarding Italian site with moist soil).

Focusing on the overall harvesting system productivity (Figure 4), moist soil showed negative effects in all three plantations. In detail, SMH in dry soil conditions was 5.671 $m^3\ h^{-1}$ in IR, 6.403 $m^3\ h^{-1}$ in IT1 and 23.761 $m^3\ h^{-1}$ in IT2; while, respectively, were 12.53%, 18.68% and 16.27% lower in the moist soil. Moreover, the higher moisture content of soil also resulted in a higher percentage difference between PMH and SMH, i.e., 2.73% vs. 3.08% in IR; 3.39% vs. 4.40% in IT1; and 4.4% vs. 5.83% in IT2. Referring to the single operations, SMH in felling-processing was 7.541 vs. 7.101 $m^3\ h^{-1}$ in IR; 8.238 vs. 7.887 $m^3\ h^{-1}$ in IT1; and 37.941 vs. 36.481 $m^3\ h^{-1}$ in IT2. Bunching-extraction productivity was also negatively affected by higher soil moisture; specifically, SMH was 22.870 vs. 16.455 $m^3\ h^{-1}$ in IR; 28.746 vs. 15.325 $m^3\ h^{-1}$ in IT1 and 63.580 vs. 43.758 $m^3\ h^{-1}$ in IT2.

Focusing on harvesting costs, the results of the economic evaluation carried out within the present study are given in Tables 10 and 11.

The details of hourly costs reported in Table 10 show how the harvesting machinery applied in IT2 (harvester and forwarder) presents higher hourly costs, mostly related to the higher purchase price. However, the higher productivity of this fully mechanized harvesting system allowed IT2 to have a lower cost per m^3 of timber produced (Table 11).

Regarding the influence of soil moisture conditions on harvesting costs, it is evident that the negative influence on working performance correlated to higher moisture also led to higher harvesting costs. In detail, this was about 16%, 26% and 16% higher in the moist site than in the dry one for IR, IT1 and IT2, respectively.

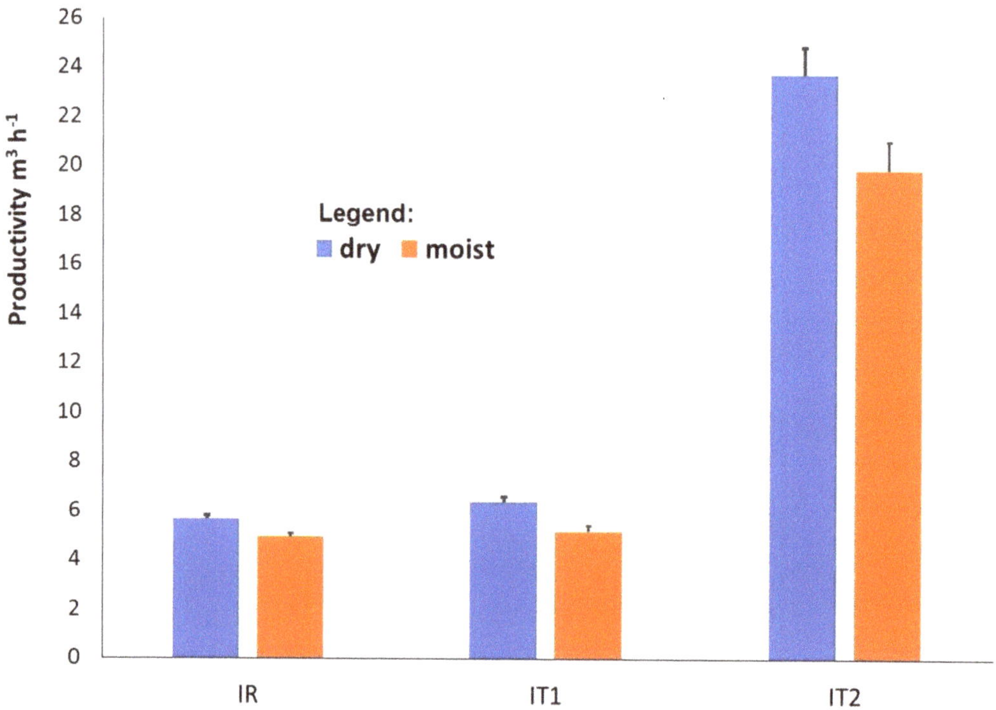

Figure 4. Average yard productivity (bars) and possible increase of performance (lines) from SMH to PMH for the six harvested sites (data reported to FU of 1 t of fresh mass showed in Figure A1).

Table 10. Summary cost assessment of mechanization used in the logging activities studied.

Description	MU	Chainsaw	Skidder	Harvester	Forwarder
Investment cost	€	1674.00	155,000.00	390,000.00	365,000.00
Service life	Years	2	10	10	10
Annual use	H	800	1000	800	800
Recovery value	€	167.40	15,500.00	39,000.00	36,500.00
Interest on capital	%	3	3	3	3
Fuel consumption	$l\ h^{-1}$	1.0	4.2	15.0	17.0
Fuel price	€ l^{-1}	2.00	0.80	0.80	0.80
Lubricant cost	% of fuel cost	20	35	35	35
Labor cost	€ h^{-1}	16.40	16.90	17.50	17.50
Crew	n°	2	2	1	1
		Fixed costs			
Depreciation	€ $year^{-1}$	753.30	13,950.00	35,100.00	32,850.00
Interest	€ $year^{-1}$	38.92	2766.75	6961.50	6515.25
Insurance and tax	€ $year^{-1}$	64.87	4611.25	11,602.50	10,858.75
Yearly fixed costs	€ $year^{-1}$	857.09	21,328.00	53,664.00	50,224.00
Hourly fixed costs	€ h^{-1}	1.07	21.33	67.08	62.78

Table 10. *Cont.*

Description	MU	Chainsaw	Skidder	Harvester	Forwarder
		Variable costs			
Fuel	€ h^{-1}	2.02	3.36	12.00	13.60
Lubricant	€ h^{-1}	0.40	1.18	4.20	4.76
Repair and maintenance	€ h^{-1}	0.94	13.95	43.88	41.06
Workers	€ h^{-1}	32.80	33.80	17.50	17.50
Hourly variable cost	€ h^{-1}	36.17	52.29	77.58	76.92
Operating cost	€ h^{-1}	37.24	73.61	144.66	139.70
Profit and overhead	%	10	10	11	12
Profit and overhead	€ h^{-1}	3.72	7.36	15.91	16.76
Total operating cost	**€ h^{-1}**	**40.96**	**80.98**	**160.57**	**156.47**

Table 11. Harvesting costs for one cubic meter of wood and percentage of costs at two main operations (felling–processing and bunching–extraction to landing) in the studied sites (data reported to FU of 1 t of fresh mass showed in Table A3).

Description	MU	IR Dry	IR Moist	IT1 Dry	IT1 Moist	IT2 Dry	IT2 Moist
Real unit cost (SMH)	€ m^{-3}	8.97	10.69	7.79	10.48	6.69	7.98
Felling-Processing percentage	%	60.5	54.0	63.8	49.6	63.2	55.2
Bunching-Extraction percentage	%	39.5	46.0	36.2	50.4	36.8	44.8
Hypothetical unit cost (PMH)	€ m^{-3}	8.68	10.33	7.47	9.94	6.39	7.54
Felling-Processing percentage	%	61.5	54.6	65.1	51.0	64.6	56.7
Bunching-Extraction percentage	%	38.5	45.4	34.9	49.0	35.4	43.3

Regarding environmental aspects, the results of the analysis of energy efficiency are given in Table 12. The highest energy input was reported for IT2, due to the complete mechanization of the overall harvesting operations. The effects of moisture on environmental performance can be observed in all of the three plantations where higher soil moisture led to lower energy efficiency, more exactly, 97.7% vs. 97.0% in IR; 98.1% vs. 96.9% in IT1 and 97.0% vs. 96.6% in IT2.

Table 12. Total energy inputs and balance in the studied harvesting yards (data reported to FU of 1 t of fresh mass showed in Table A4).

Description	M.U.	Energetic Output	Direct Input	Indirect Input	Human Labor Input	Total Inputs	Output/Inputs Ratio	System Efficiency
IR-d	MJ m^{-3}	11,658	257.62	4.22	0.70	262.54	44.4	97.7%
	GJ ha^{-1}	10,217	225.78	3.70	0.61	230.09		
IR-m	MJ m^{-3}	11,658	338.67	5.83	0.80	345.30	33.8	97.0%
	GJ ha^{-1}	10,217	296.81	5.11	0.70	302.72		
IT1-d	MJ m^{-3}	11,658	212.05	3.37	0.62	216.04	54.0	98.1%
	GJ ha^{-1}	8191	148.99	2.37	0.44	151.79		
IT1-m	MJ m^{-3}	11,658	352.75	6.24	0.76	359.76	32.4	96.9%
	GJ ha^{-1}	8191	247.84	4.39	0.53	252.76		
IT2-d	MJ m^{-3}	11,658	337.20	7.80	0.06	345.06	33.8	97.0%
	GJ ha^{-1}	8741	252.83	5.85	0.05	258.73		
IT2-m	MJ m^{-3}	11,658	386.36	9.18	0.08	395.62	29.5	96.6%
	GJ ha^{-1}	8741	289.69	6.89	0.06	296.63		

In IR and IT1 a major part of the energy input is related to bunching–extraction operations in both moist and dry soil conditions. Instead, in IT2, felling operations via harvester were the reason for the highest portion of energy input in this yard, in both soil conditions (Figure 5). Interestingly, moist soil led to an increased portion of energy input related to bunching–extraction in all three yards (77.2% vs. 81.6% in IR; 74.6% vs. 84.0% in IT1 and 25.8% vs. 32.7% in IT2), showing how this operation was most influenced by soil moisture when regarding environmental issues.

As shown in Table 13 and Figures 6 and 7, soil moisture also showed negative effects regarding pollutant emissions in the three different yards, with increasing emissions for all the investigated parameters in moist soil conditions. What is more, mechanized felling via harvester (IT2) led to higher emissions in comparison to motor-manual felling (IR and IT1).

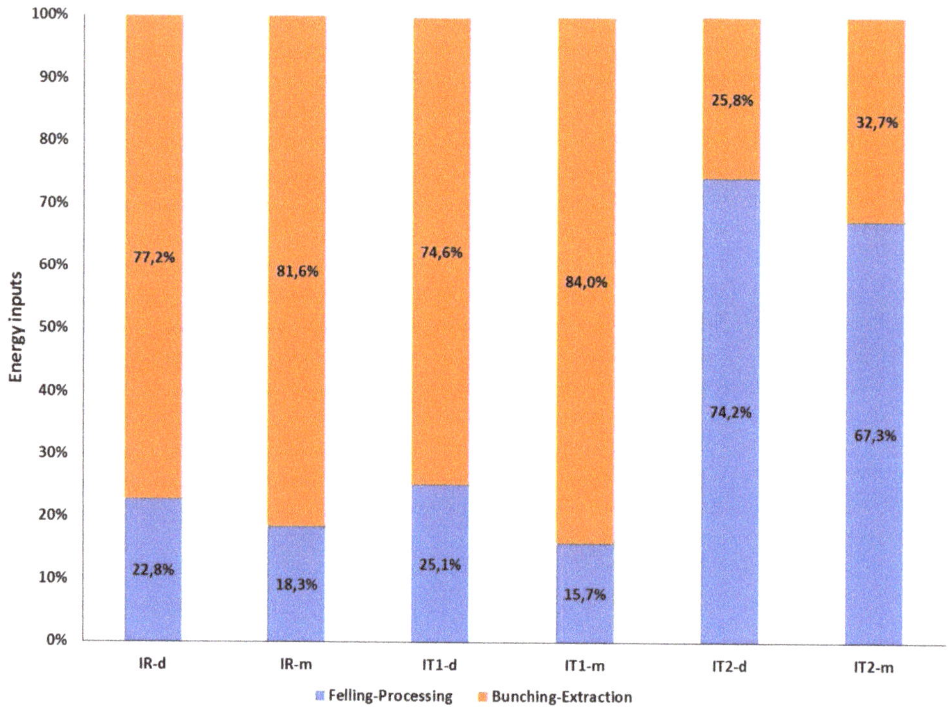

Figure 5. Energy inputs percentage for each harvesting operation assessed in the studied sites.

Table 13. Total emission assessed in the studied harvesting yards (data reported to FU of 1 t of fresh mass showed in Table A5).

Harvesting Sites	CO_2	CO	HC	N_{ox}	PM_{10}
			g m^{-3}		
IR-d	1650.72	26.20	0.42	25.16	3.77
IR-m	1699.04	27.31	0.46	26.60	4.08
IT1-d	1602.18	24.08	0.35	24.20	3.35
IT1-m	1627.43	25.19	0.41	25.04	3.65
IT2-d	2651.56	48.35	0.74	41.85	6.51
IT2-m	2789.36	55.08	0.78	47.91	7.73

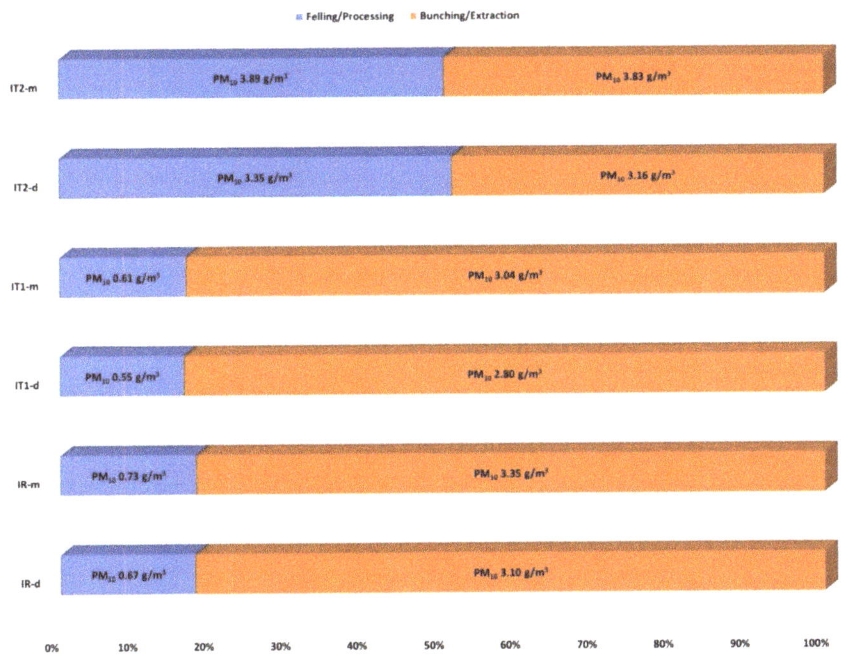

Figure 6. Percentage distribution of total PM10 emission in the harvesting sites studied, data shown for single operation (data reported to FU of 1 t of fresh mass showed in Figure A2).

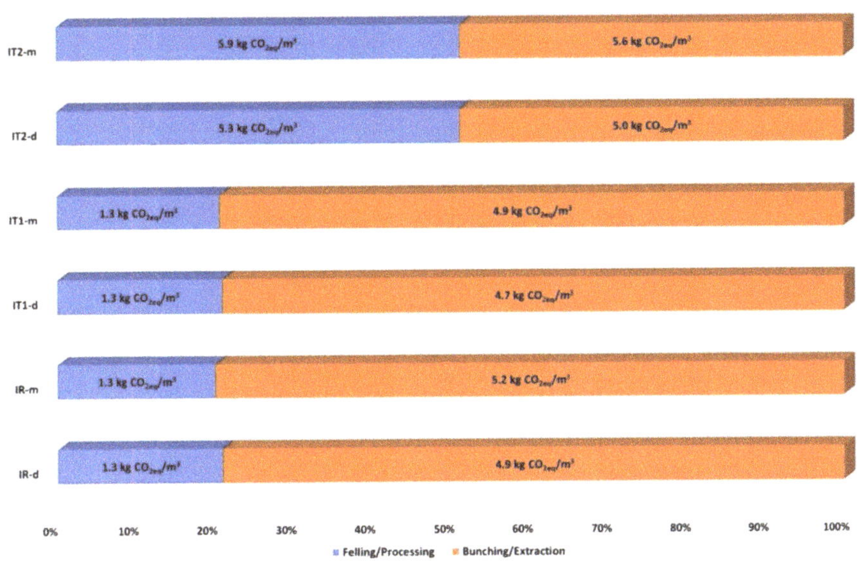

Figure 7. Percentage distribution of GHG emission in the harvesting sites studied, data shown for single operation and reported in CO_2 equivalent (data reported to FU of 1 t of fresh mass showed in Table A3).

4. Discussion

4.1. Comparison of Harvesting Systems Performance

Several studies on work productivity and cost analysis in poplar plantations are available in the literature, even if most of these deal with Short Rotation Coppice (SRC) plants for bioenergy production [48]. Indeed, few studies have focused on productivity analysis in poplar plantations for timber production. In a poplar plantation located in Serbia, Danilovic et al. [49] reported a productivity for mechanized felling–processing via harvester of 30.3 to 34.7 m^3 h^{-1}, depending on working method and stem dimension. In the same year, Spinelli et al. [50] carried out an extensive productivity and cost analysis in a 25 year old poplar plantation in Italy, reporting an average work productivity (SMH) for motor-manual felling and processing via chainsaw of 6.3 m^3 h^{-1} and a value of 21.1 m^3 h^{-1} for the same operation performed via harvester.

The productivity values found in the present study are higher than reported in the above cited studies, both for motor-manual and mechanized felling processing. This difference is more pronounced in comparison to Spinelli et al. [50]. Such a gap concerning mechanized felling–processing can be partially related to the lower average dbh of the stems in the previous study (around 30 cm vs. 40.4 cm). However, the major difference between the present studies and the literature, which can explain the higher productivity found in the present analysis, is the lower percentage of delay times. In particular, delay percentage ranges from 1.7% in IR to 2.9% in IT2, while the average delay for motor-manual felling-processing in Spinelli et al. [50] was 29.6%, decreasing to 13.0% in mechanized operations, while in Danilovic et al. [49] delay accounted for 28.5 of the working time.

Concerning bunching–extraction operations, no study on winching and forwarding in poplar plantations for timber production were found in the literature. However, there are several studies on bunching and extraction via TimberJack 450 cable skidder, which analyzed work productivity in different forest stands. Lotfalian et al. [51] found a winching productivity of 20.2 m^3 h^{-1} in beech high stand thinning with an average extraction distance of 289 m, while Mousavi [52] reported a bunching–extraction productivity of about 11 m^3 h^{-1} in beech selection cutting with an average skidding distance of 439 m. Nikooy et al. [18] reported instead a lower value of 5.2 m^3 h^{-1} productivity of Timberjack 450C in timber extraction of path cutting in a pine plantation. Such lower productivity is probably related to the lower dimension of trees considering that skidding distance was comparable to those in both IR and IT1 [18]. Therefore, as a general trend, bunching-extraction productivity in the present study showed higher values than previous works reported in the literature for the same machinery in different stands and silvicultural interventions. This difference is related to both the type of forest intervention (clear cut) and to the flat terrain of IR and IT1, which facilitated logging operations.

Regarding forwarding operations, also in this case it is possible to make a comparison regarding work productivity only between different stands, considering the lack of studies on poplar plantations for timber production. The forwarder is the most commonly applied machinery in CTL (Cut to Length) harvesting operations, and it has been widely applied in artificial plantations, mostly of softwood species [53]. This machinery can reach a very high working productivity [54,55], even if a proper planning of the intervention is needed to reduce the impact which can occur considering the average dimension of a forwarder [13]. Comparing the findings of the present study with other forestry interventions in artificial plantations, it is evident that the substantial average dimension of stems, the short extraction distance and the flat terrain features in IT2 led to higher work productivity. In detail, Puttock et al. [56] reported a SMH productivity of 11.2 m^3 h^{-1} in a poplar-dominated mixed-wood stand in Southern Ontario during thinning interventions; Eriksson and Lindroos [57] showed PMH productivity for forwarding in clear cutting in pine and spruce stands of 21.4 m^3 h^{-1}. Another study performed in Romania reported a SMH productivity of 15.35 m^3 h^{-1} in a clear cut of spruce stand, with an average slope of 10% and an average extraction distance of 479 m [58]. In another study recently carried out in Poland, Magagnotti et al. [11] reported a forwarding productivity of 24.4 m^3 h^{-1} SMH for a poplar plantation.

Focusing on harvesting costs, it is possible to notice how felling and processing operations accounted for the major part of these in all the yards in both soil moisture conditions, as reported by previous literature [38,59]. Felling and processing costs, with values ranging from 4.22 € m^{-3} (IT2 dry) to 5.77 € m^{-3} (IR moist), were in line with the literature findings for several Italian poplar plantations for high value timber production, notwithstanding higher work productivity. Spinelli et al. [50] reported a unit cost of about € 5 m^{-3} for both motor-manual and mechanized felling–processing. This can be explained by the higher purchase costs of the machinery applied in the studied plantations. Skidding and forwarding costs are also in line with the literature, for similar harvesting systems but in different kinds of stands. In the present study the lowest cost for extraction was shown by IT2 dry at € 2.46 m^{-3}, while the highest cost was related to winching in IT1 moist (€ 5.28 m^{-3}). Regarding winching operations through cable skidder, Jourgholami and Majnounian [33] reported 6.15 € m^{-3} as the cost of Timberjack 450C in timber extraction on a pine plantation, while the findings of Lotfalian et al. [51] showed 5.15 USD m^{-3}. Focusing on forwarding, extraction costs were assessed by Cabral et al. [60] at about 1.95 € m^{-3}, while about 7.5 € m^{-3} were reported by Kaleja et al. [61], and about 9.2 € m^{-3} by Magagnotti et al. [11].

Concerning environmental impact, a comparison was carried out between the findings of the present study and other similar studies, regarding both high and medium level of mechanization. In both cases, energy inputs and energy balance in the investigated poplar plantations were substantially higher [37,62].

System efficiency values were high in all three yards in both the soil conditions (ranging from 96.6% to 98.1%), and thus were in line with previous literature findings in other forest interventions [16,62,63].

GHG emissions, mostly regarding CO_2, were lower than in previous literature findings, which reported a range between 3 and 33 kg CO_2eq, while the values of the present work ranged from 1.6 to 2.8 kg CO_2eq [5,16,24,46,64].

4.2. Influence of Soil Moisture on Harvesting Performance

There is a considerable amount of literature regarding working productivity evaluation, but a major part of the studies focused on the influence on working performance of parameters such as terrain features, working distance, age, species composition, labor skills, etc. However, not much attention has been directed towards soil moisture conditions and productivity. Studies were, however, conducted on soil impact related to logging activities [65–68].

In all of the three yards, higher soil moisture led to lower work productivity, thus to higher harvesting costs, in accordance to what was reported in the few studies dealing with this topic [69,70]. High soil moisture negatively affected both motor-manual and mechanized operations, resulting in higher working times, except for felling (both with chainsaw and harvester) and processing (only with harvester). Longer working time in higher soil moisture conditions for mechanized operations (bunching –extraction) is related to the lower driving speed which the machinery was able to achieve with moist soil. The high level of moisture of the terrain caused reduced tire grip and the operators had to reduce the working speed for safety reasons. Interestingly, this did not happen for felling–processing operations by harvester, which were not affected by soil moisture regarding working time. Soil moisture also negatively affected motor-manual operations, specifically, moving (IR and IT1) and processing (only IR). This is equally related to worker safety issues, with operators that had to be more cautious during the logging activities, due to the fact that high moisture in the soil made the trail slippery, and therefore prone to accidents.

High soil moisture showed negative effects on the environmental performance of logging activities in the studied poplar plantations. There were two reasons for these negative effects: the longer working time, which required more time in which motors were running, and the higher torque needed to move the machinery in moist soil, considering the lower grip and higher attrition. Thus, there were higher emissions of pollutants.

5. Conclusions

Although poplar plantations are important sources of timber in both Iran and Italy, few studies have focused on work productivity evaluation under different aspects related to sustainability in such kinds of stand. Moreover, these plantations are often located in plain or floodplain lands, therefore harvesting operations can occur in soil conditions with a high moisture content.

The aims of this study were: (i) to evaluate different harvesting systems in poplar plantations and (ii) to evaluate the influence of soil moisture on economic and environmental performance of logging operations.

In order to assess different harvesting systems in poplar plantations, from what was analyzed it is possible to state that a fully mechanized harvesting system (harvesting forwarder) is the most productive and economically sustainable with respect to semi-mechanical harvesting/processing and skidding extraction. However, in terms of energy balance and emissions, it is possible to state exactly the opposite, that the best harvesting system was semi-mechanical harvesting/processing and skidding extraction. These are aspects to be carefully considered during operations planning, but they must also be analyzed in terms of greater efficiency of the mechanization used, seeking to bring mechanical technologies that are increasingly efficient also in environmental terms to the forestry sector.

In order to assess the influence of soil moisture on economic and environmental performance of the logging operations, the findings revealed that high moisture content led to lower work productivity in all of the three investigated plantations, with detrimental effects on harvesting costs, which were found to be higher in moist soil conditions in all three yards. Moreover, environmental features related to pollutant emissions were higher in moist soil conditions, as a consequence of the longer time and the major torque required for the machinery to perform the logging activities.

It can be concluded from these findings that it is advisable to avoid logging operations in conditions of high soil moisture (>20%), to decrease the impact on the soil, to create higher cost-effectiveness, and to reduce the emissions of pollutants.

Author Contributions: Conceptualization, F.T., M.N. and R.P.; Data curation, F.T., M.N., F.L., R.V. and R.P.; Formal analysis, F.T., M.N., R.V. and R.P.; Investigation, F.T. and R.P.; Methodology, F.T., M.N., R.V. and R.P.; Supervision, F.T., R.V. and R.P.; Validation, F.T., M.N., F.L., R.V., L.B. and R.P.; Writing—original draft, F.T., M.N., F.L., L.B. and R.P.; Writing—review & editing, F.T., F.L., R.V., L.B. and R.P. All authors have read and agreed to the published version of the manuscript.

Funding: This research received no external funding.

Institutional Review Board Statement: Not applicable.

Informed Consent Statement: Not applicable.

Data Availability Statement: The data presented in this study could be available on request from the corresponding author.

Acknowledgments: This research was in part carried out within the framework of the MIUR (Italian Ministry for Education, University and Research) initiative "Departments of Excellence" (Law 232/2016), WP3 and WP 4, which financed the Department of Agriculture and Forest Science at the University of Tuscia.

Conflicts of Interest: The authors declare no conflict of interest.

Appendix A

The data assessed in the Appendix A are referred to another Functional Unit (FU) respect to how reported in the main text. The FU in this case was 1 t of fresh mass (following the data showed in Table 2). This was important in order to give more possibility to compare these results with other studies.

Table A1. Harvested trees, fresh mass and working cycles for work productivity analysis in each treatment.

Parameter	IR Dry	IR Moist	IT1 Dry	IT1 Moist	IT2 Dry	IT2 Moist
Felled-processed trees (N)	601	625	556	528	2790	2762
Felled processed-wood (t)	1046.159	1088.005	987.715	938.329	5411.307	5357.194
Extraction cycles (N)	100	100	100	100	100	100
Extracted wood (t)	534.778	549.102	555.994	423.146	974.295	851.606

Table A2. Extraction trail and corridors' average features for the three areas in the two soil moisture conditions (mean ± SD). From the *t*-test for independent samples applied, statistically significant differences ($p < 0.05$) between the average values are highlighted (underlined text).

	IR		IT1		IT2	
	Dry	Moist	Dry	Moist	Dry	Moist
Mean bunching distance (m)	38.6 ± 10.1	49.8 ± 9.5	17.5 ± 5.2	56.1 ± 8.2	22.7 ± 4.2	18.4 ± 2.1
Mean extraction distance (m)	146.0 ± 20.1	144.5 ± 12.2	118.2 ± 14.3	138.1 ± 10.3	152.4 ± 12.5	158.7 ± 11.2
Mean mass per working cycle (t)	5.35 ± 0.12	5.49 ± 0.11	5.56 ± 0.15	4.23 ± 0.13	9.74 ± 0.22	8.52 ± 0.30

Table A3. Harvesting costs for one wood, fresh tons and percentage of costs at two main operations (felling–processing and bunching–extraction to landing) in the studied sites.

Description	MU	IR Dry	IR Moist	IT1 Dry	IT1 Moist	IT2 Dry	IT2 Moist
Real unit cost (SMH)	€ t^{-1}	11.27	13.43	9.79	13.17	8.41	10.03
Felling–Processing percentage	%	60.5	54.0	63.8	49.6	63.2	55.2
Bunching–Extraction percentage	%	39.5	46.0	36.2	50.4	36.8	44.8
Hypothetical unit cost (PMH)	€ t^{-1}	10.91	12.98	9.39	12.49	8.03	9.47
Felling–Processing percentage	%	61.5	54.6	65.1	51.0	64.6	56.7
Bunching–Extraction percentage	%	38.5	45.4	34.9	49.0	35.4	43.3

Table A4. Total energy inputs and balance in the studied harvesting yards, referring to surface unit and to fresh mass.

Description	M.U.	Energetic Output	Direct Input	Indirect Input	Human Labor Input	Total Inputs	Output/Inputs Ratio	System Efficiency
IR-d	MJ t^{-1}	14,649	323.72	5.30	0.88	329.91	44.4	97.7%
	GJ ha^{-1}	10,217	225.78	3.70	0.61	230.09		
IR-m	MJ t^{-1}	14,649	425.57	7.33	1.01	433.90	33.8	97.0%
	GJ ha^{-1}	10,217	296.81	5.11	0.70	302.72		
IT1-d	MJ t^{-1}	16,585	301.68	4.79	0.88	307.36	54.0	98.1%
	GJ ha^{-1}	8191	148.99	2.37	0.44	151.79		
IT1-m	MJ t^{-1}	16,585	501.85	8.88	1.08	511.82	32.4	96.9%
	GJ ha^{-1}	8191	247.84	4.39	0.53	252.76		
IT2-d	MJ t^{-1}	16,153	467.23	10.81	0.08	478.12	33.8	97.0%
	GJ ha^{-1}	8741	252.83	5.85	0.05	258.73		
IT2-m	MJ t^{-1}	16,153	535.35	12.72	0.11	548.18	29.5	96.6%
	GJ ha^{-1}	8741	289.69	6.89	0.06	296.63		

Table A5. Total emission assessed in the studied harvesting yards, referring to fresh mass.

Harvesting Sites	CO_2	CO	HC	N_{ox}	PM_{10}
			$g\ t^{-1}$		
IR-d	2074.29	32.92	0.53	31.62	4.74
IR-m	2135.01	34.32	0.58	33.43	4.77
IT1-d	2279.39	34.26	0.50	34.43	4.77
IT1-m	2315.31	35.84	0.58	35.62	5.19
IT2-d	3674.05	66.99	1.03	57.99	9.02
IT2-m	3864.99	76.32	1.08	66.38	10.71

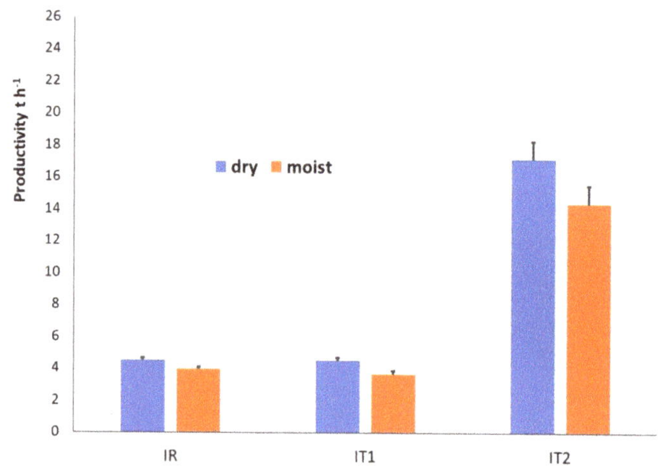

Figure A1. Average yard productivity (bars) and possible increase of performance (lines) from SMH to PMH for the six harvested sites.

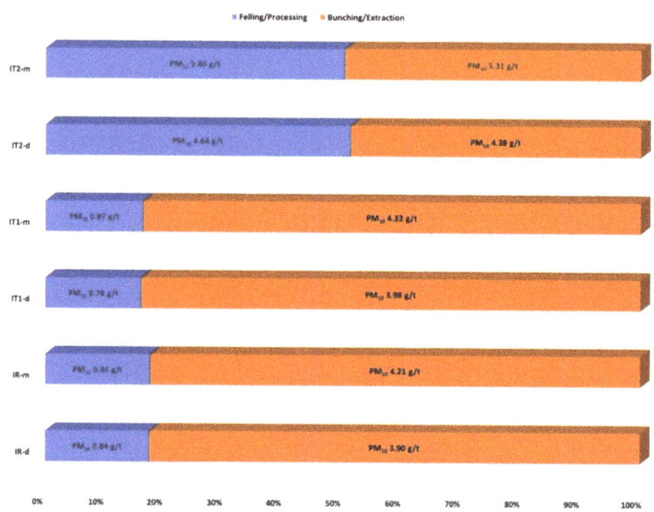

Figure A2. Percentage distribution of total PM10 emission in the harvesting sites studied, data shown for single operation, referring to fresh mass.

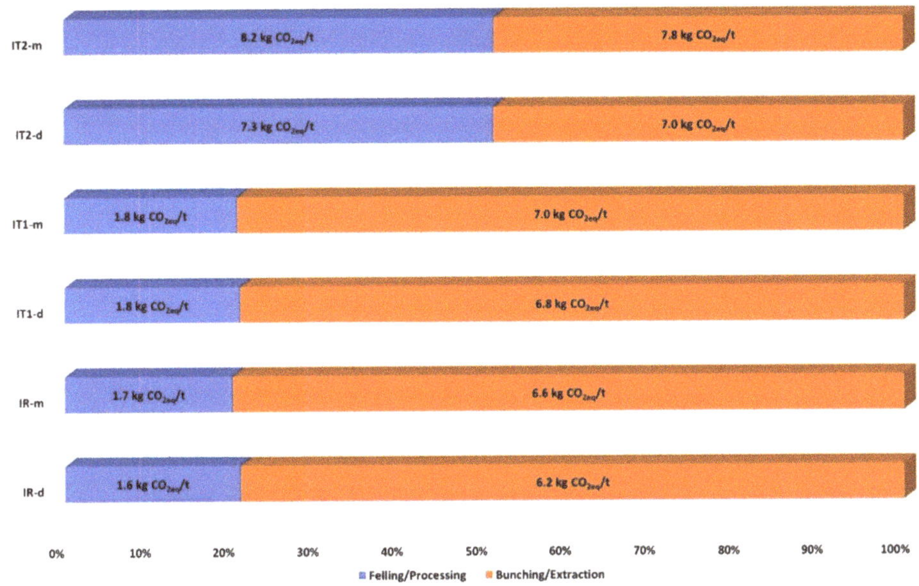

Figure A3. Percentage distribution of GHG emission in the harvesting sites studied, data shown for single operation and reported in CO_2 equivalent, data referring to fresh mass.

References

1. Azizi, M.; Faezipour, M.M.; Bayatkashkooli, A.; Taheri, F. Quantitative study of poplar plantations in three Iranian provinces. *For. Sci. Pract.* **2013**, *15*, 363–369. [CrossRef]
2. FAO. Poplars and other fast-growing trees—Renewable resources for future green economies, 2016 Synthesis of Country Progress Reports. In Proceedings of the 25th Session of the International Poplar Commission, Berlin, Germany, 13–16 September 2016; Working Paper IPC/15. Forestry Policy and Resources Division, FAO: Rome, Italy, 2016. Available online: http://www.fao.org/forestry/ipc2016/en/ (accessed on 1 June 2020).
3. Italian Inventory of Forests and Carbon Pools. 2005. Available online: https://www.sian.it/inventarioforestale/jsp/documentazione.jsp (accessed on 1 June 2020).
4. Anaraki, K.A.; Lashgarara, F.; Kiadaliri, H. Effect of socio-economic factors on development of poplar plantation in Guilan province (case study: Somesara). *Iran. J. For. Poplar Res.* **2012**, *20*, 346–356.
5. González-García, S.; Bacenetti, J.; Murphy, R.J.; Fiala, M. Present and future environmental impact of poplar cultivation in the Po Valley (Italy) under different crop management systems. *J. Clean. Prod.* **2012**, *26*, 56–66. [CrossRef]
6. Davison, J.; Riggs, W. Hybrid poplar production 1998–2003 in Eureka and Churchill Counties. *Univ. Nev. Coop. Ext. Fact Sheet* **2004**, 5–25.
7. Castro, G.L.; Fragnelli, G. New technologies and alternative uses for poplar wood. *Boletín Inf. Cideu* **2006**, *2*, 27–36.
8. Lashkarbolouki, E.; Modirrahmati, A.R.; Rahmani, R.; Kahneh, E.; Koopar, S.A.M. Phenology and growth characteristics of seven clones of Populus deltoides in Astaneh Ashrafie, Guilan. *Iran. J. For. Poplar Res.* **2010**, *18*, 527–538.
9. Colangelo, M.; Camarero, J.J.; Borghetti, M.; Gazol, A.; Gentilesca, T.; Ripullone, F. Size matters a lot: Drought-affected Italian oaks are smaller and show lower growth prior to tree death. *Front. Plant Sci.* **2017**, *8*, 135. [CrossRef]
10. Motta, R.; Nola, P. Growth trends and dynamics in sub-alpine forest stands in the Varaita Valley (Piedmont, Italy) and their relationships with human activities and global change. *J. Veg. Sci.* **2001**, *12*, 219–230. [CrossRef]
11. Magagnotti, N.; Spinelli, R.; Kärhä, K.; Mederski, P.S. Multi-tree cut-to-length harvesting of short-rotation poplar plantations. *Eur. J. For. Res.* **2021**, *140*, 345–354. [CrossRef]
12. Marchi, E.; Chung, W.; Visser, R.; Abbas, D.; Nordfjell, T.; Mederski, P.S.; McEwan, A.; Brink, M.; Laschi, A. Sustainable Forest Operations (SFO): A new paradigm in a changing world and climate. *Sci. Total Environ.* **2018**, *634*, 1385–1397. [CrossRef]
13. Picchio, R.; Latterini, F.; Mederski, P.S.; Tocci, D.; Venanzi, R.; Stefanoni, W.; Pari, L. Applications of GIS-Based Software to Improve the Sustainability of A Forwarding Operation in Central Italy. *Sustainability* **2020**, *12*, 5716. [CrossRef]
14. Schweier, J.; Blagojević, B.; Venanzi, R.; Latterini, F.; Picchio, R. Sustainability assessment of alternative strip clear cutting operations for wood chip production in renaturalization management of pine stands. *Energies* **2019**, *12*, 3306. [CrossRef]

15. Spinelli, R.; Lombardini, C.; Marchi, E.; Aminti, G. A low-investment technology for the simplified processing of energy wood from coppice forests. *Eur. J. For. Res.* **2019**, *138*, 31–41. [CrossRef]
16. Picchio, R.; Venanzi, R.; Di Marzio, N.; Tocci, D.; Tavankar, F. A Comparative Analysis of Two Cable Yarder Technologies Performing Thinning Operations on a 33 Year Old Pine Plantation: A Potential Source of Wood for Energy. *Energies* **2020**, *13*, 5376. [CrossRef]
17. Spinelli, R.; Magagnotti, N.; Lombardini, C.; Mihelič, M. A Low-Investment Option for the Integrated Semi-Mechanized Harvesting of Small-Scale, Short-Rotation Poplar Plant. *Small-Scale For.* **2021**, *20*, 59–72. [CrossRef]
18. Nikooy, M.; Esmailnezhad, A.; Naghdi, R. Productivity and cost analysis of skidding with Timberjack 450C in forest plantations in Shafaroud watershed, Iran. *J. For. Sci.* **2013**, *59*, 261–266. [CrossRef]
19. Wang, J.; Long, C.; McNeel, J.; Baumgras, J. Productivity and cost of manual felling and cable skidding in central Appalachian hardwood forests. *For. Prod. J.* **2004**, *54*, 45–51.
20. Naghdi, R.; Rafatniaa, N.; Sobhani, H.; Jalali, G.H.; Hosseini, S.M. A survey of the efficiency of timber jack 450C wheeled skidder in sheared forests. *J. Nat. Resour.* **2005**, *57*, 657–687.
21. Demir, M. Investigation of timber harvesting mechanization progress in Turkey. *Afr. J. Biotechnol.* **2010**, *9*, 1628–1634.
22. Picchio, R.; Pignatti, G.; Marchi, E.; Latterini, F.; Benanchi, M.; Foderi, C.; Venanzi, R.; Verani, S. The Application of Two Approaches Using GIS Technology Implementation in Forest Road Network Planning in an Italian Mountain Setting. *Forests* **2018**, *9*, 277. [CrossRef]
23. Picchio, R.; Latterini, F.; Mederski, P.S.; Venanzi, R.; Karaszewski, Z.; Bembenek, M.; Croce, M. Comparing Accuracy of Three Methods Based on the GIS Environment for Determining Winching Areas. *Electronics* **2019**, *8*, 53. [CrossRef]
24. González-García, S.; Krowas, I.; Becker, G.; Feijoo, G.; Moreira, M.T. Cradle-to-gate life cycle inventory and environmental performance of Douglas-fir roundwood production in Germany. *J. Clean. Prod.* **2013**, *54*, 244–252. [CrossRef]
25. Lima, K.S.; Meira Castro, A.C.; Santos Baptista, J.; Silva, U. Wood-Logging Process Management in Eastern Amazonia (Brazil). *Sustainability* **2020**, *12*, 7571. [CrossRef]
26. Sohrabi, H.; Jourgholami, M.; Jafari, M.; Shabanian, N.; Venanzi, R.; Tavankar, F.; Picchio, R. Soil recovery assessment after timber harvesting based on the sustainable forest operation (SFO) perspective in iranian temperate forests. *Sustainability* **2020**, *12*, 2874. [CrossRef]
27. Ozkaya, G.; Erdin, C. Evaluation of sustainable forest and air quality management and the current situation in europe through operation research methods. *Sustainability* **2020**, *12*, 10588. [CrossRef]
28. Lontsi, R.T.; Corre, M.D.; Iddris, N.A.; Veldkamp, E. Soil greenhouse gas fluxes following conventional selective and reduced-impact logging in a Congo Basin rainforest. *Biogeochemistry* **2020**, *151*, 153–170. [CrossRef]
29. Gusti, M.; Di Fulvio, F.; Biber, P.; Korosuo, A.; Forsell, N. The Effect of alternative forest management models on the forest harvest and emissions as compared to the forest reference level. *Forests* **2020**, *11*, 794. [CrossRef]
30. Naghdi, R.; Solgi, A. Effects of skidder passes and slope on soil disturbance in two soil water contents. *Croat. J. For. Eng.* **2014**, *35*, 73–80.
31. Startsev, A.D.; McNabb, D.H. Effects of skidding on forest soil infiltration in west-central Alberta. *Can. J. Soil Sci.* **2000**, *80*, 617–624. [CrossRef]
32. Cambi, M.; Grigolato, S.; Neri, F.; Picchio, R.; Marchi, E. Effects of forwarder operation on soil physical characteristics: A case study in the Italian alps. *Croat. J. For. Eng. J. Theory Appl. For. Eng.* **2016**, *37*, 233–239.
33. Jourgholami, M.; Majnounian, B. Productivity and cost of wheeled skidder in Hyrcanian Forest. *Int. J. Nat. Eng. Sci.* **2008**, *2*, 99–103.
34. Miyata, E.S. *Determining Fixed and Operating Costs of Logging Equipment*; General Technical Report NC-55; Department of Agriculture, Forest Service, North Central Forest Experiment Station: St. Paul, MN, USA, 1980.
35. Spinelli, R.; Magagnotti, N.; Lombardini, C. Performance, capability and costs of small-scale cable yarding technology. *Small-Scale For.* **2010**, *9*, 123–135. [CrossRef]
36. Picchio, R.; Spina, R.; Maesano, M.; Carbone, F.; Lo Monaco, A.; Marchi, E. Stumpage value in the short wood system for the conversion into high forest of a oak coppice. *For. Stud. China* **2011**, *13*, 252–262. [CrossRef]
37. Picchio, R.; Sirna, A.; Sperandio, G.; Spina, R.; Verani, S. Mechanized harvesting of eucalypt coppice for biomass production using high mechanization level. *Croat. J. For. Eng.* **2012**, *33*, 15–24.
38. Bodaghi, A.I.; Nikooy, M.; Naghdi, R.; Venanzi, R.; Latterini, F.; Tavankar, F.; Picchio, R. Ground-based extraction on salvage logging in two high forests: A productivity and cost analysis. *Forests* **2018**, *9*, 729. [CrossRef]
39. Picchio, R.; Maesano, M.; Savelli, S.; Marchi, E. Productivity and energy balance in the conversion into high forest system of a Quercus cerris L. coppice in Central Italy. *Croat. J. For. Eng.* **2009**, *1*, 15–26.
40. Hartsough, B. Economics of harvesting to maintain high structural diversity and resulting damage to residual trees. *West. J. Appl. For.* **2003**, *18*, 133–142. [CrossRef]
41. Picchio, R.; Magagnotti, N.; Sirna, A.; Spinelli, R. Improved winching technique to reduce logging damage. *Ecol. Eng.* **2012**, *47*, 83–86. [CrossRef]
42. Christie, C.J.-A. Improving the energy and fluid balance of workers involved in harvesting tasks. *Occup. Ergon.* **2010**, *9*, 119–126. [CrossRef]
43. Christie, C.J.-A. Relationship between energy intake and expenditure during harvesting tasks. *Occup. Ergon.* **2008**, *8*, 1–10.

44. Balimunsi, H.; Grigolato, S.; Picchio, R.; Nyombi, K.; Cavalli, R. Productivity and energy balance of forest plantation harvesting in Uganda. *For. Stud. China* **2012**, *14*, 276–282. [CrossRef]
45. Canagaratna, S.G.; Witt, J. Calculation of temperature rise in calorimetry. *J. Chem. Educ.* **1988**, *65*, 126. [CrossRef]
46. Klvac, R.; Fischer, R.; Skoupy, A. Energy use and emissions from the medium distance cableway system operation phase. *Croat. J. For. Eng* **2012**, *33*, 79–88.
47. Athanassiadis, D. Energy consumption and exhaust emissions in mechanized timber harvesting operations in Sweden. *Sci. Total Environ.* **2000**, *255*, 135–143. [CrossRef]
48. Vanbeveren, S.P.P.; Spinelli, R.; Eisenbies, M.; Schweier, J.; Mola-Yudego, B.; Magagnotti, N.; Acuna, M.; Dimitriou, I.; Ceulemans, R. Mechanised harvesting of short-rotation coppices. *Renew. Sustain. Energy Rev.* **2017**, *76*, 90–104. [CrossRef]
49. Danilović, M.; Tomašević, I.; Gačić, D. Efficiency of John Deere 1470D ECOIII harvester in poplar plantations. *Croat. J. For. Eng.* **2011**, *32*, 533–548.
50. Spinelli, R.; Magagnotti, N.; Sperandio, G.; Cielo, P.; Verani, S.; Zanuttini, R. Cost and productivity of harvesting high-value hybrid poplar plantations in Italy. *For. Prod. J.* **2011**, *61*, 64–70. [CrossRef]
51. Lotfalian, M.; Moafi, M.; Foumani, B.S.; Akbari, R.A. Time study and skidding capacity of the wheeled skidder Timberjack 450C. *Int. J. Sci. Technol. Educ. Res.* **2011**, *2*, 120–124.
52. Mousavi, R. Effect of log length on productivity and cost of Timberjack 450C skidder in the Hyrcanian forest in Iran. *J. For. Sci.* **2012**, *58*, 473–482. [CrossRef]
53. Mederski, P.S.; Borz, S.A.; Đuka, A.; Lazdiņš, A. Challenges in forestry and forest engineering—Case studies from four countries in East Europe. *Croat. J. For. Eng.* **2020**, *42*, 117–134. [CrossRef]
54. Gagliardi, K.; Ackerman, S.; Ackerman, P. Multi-product forwarder-based timber extraction: Time consumption and productivity analysis of two forwarder models over multiple products and extraction distances. *Croat. J. For. Eng.* **2020**, *41*, 231–242. [CrossRef]
55. Danilović, M.; Stojnić, D.; Karić, S.; Sučević, M. Transport of technical roundwood by forwarder and tractor assembly from poplar plantations. *Nov. Meh. Šumarstva Časopis Teor. Praksu Šumarskoga Inženjerstva* **2014**, *35*, 11–22.
56. Puttock, D.; Spinelli, R.; Hartsough, B.R. Operational Trials of Cut-To-Length Harvesting of Poplar in a Mixed Wood Stand. *Int. J. For. Eng.* **2005**, *16*, 39–49. [CrossRef]
57. Eriksson, M.; Lindroos, O. Productivity of harvesters and forwarders in CTL operations in northern Sweden based on large follow-up datasets. *Int. J. For. Eng.* **2014**, *25*, 179–200. [CrossRef]
58. Apăfăian, A.I.; Proto, A.R.; Borz, S.A. Performance of a mid-sized harvester-forwarder system in integrated harvesting of sawmill, pulpwood and firewood. *Ann. For. Res.* **2017**, *60*, 227–241. [CrossRef]
59. Williams, C.; Ackerman, P. Cost-productivity analysis of South African pine sawtimber mechanised cut-to-length harvesting. *South. For.* **2016**, *78*, 267–274. [CrossRef]
60. Cabral, O.M.D.J.V.; Lopes, E.D.S.; Rodrigues, C.K.; Filho, A.F. Impact of distance between strip roads on productivity and costs of a forwarder in commercial thinning of pinus taeda stands. *Croat. J. For. Eng.* **2020**, *41*, 243–249. [CrossRef]
61. Kaleja, S.; Spalva, G.; Stola, J. Productivity and cost of logbear F4000 forwarder in thinning depending on driving conditions. *Eng. Rural Dev.* **2018**, *17*, 1458–1463. [CrossRef]
62. Baldini, S.; Picchio, R.; Savelli, S. Analisi energetica nelle utilizzazioni di un ceduo di eucalipto con una meccanizzazione leggera. *J. Agric. Eng* **2007**, *3*, 49–56.
63. Vusić, D.; Šušnjar, M.; Marchi, E.; Spina, R.; Zečić, T.; Picchio, R. Skidding operations in thinning and shelterwood cut of mixed stands–Work productivity, energy inputs and emissions. *Ecol. Eng.* **2013**, *61*, 216–223. [CrossRef]
64. Valente, C.; Spinelli, R.; Hillring, B.G. LCA of environmental and socio-economic impacts related to wood energy production in alpine conditions: Valle di Fiemme (Italy). *J. Clean. Prod.* **2011**, *19*, 1931–1938. [CrossRef]
65. Bembenek, M.; Tsioras, P.A.; Karaszewski, Z.; Zawieja, B.; Bakinowska, E.; Mederski, P.S. Effect of day or night and cumulative shift time on the frequency of tree damage during CTL harvesting in various stand conditions. *Forests* **2020**, *11*, 743. [CrossRef]
66. Picchio, R.; Mederski, P.S.; Tavankar, F. How and How Much, Do Harvesting Activities Affect Forest Soil, Regeneration and Stands? *Curr. For. Rep.* **2020**, *6*, 115–128. [CrossRef]
67. Cambi, M.; Certini, G.; Neri, F.; Marchi, E. The impact of heavy traffic on forest soils: A review. *For. Ecol. Manag.* **2015**, *338*, 124–138. [CrossRef]
68. Naghdi, R.; Solgi, A.; Zenner, E.K.; Tsioras, P.A.; Nikooy, M. Soil disturbance caused by ground-based skidding at different soil moisture conditions in Northern Iran. *Int. J. For. Eng.* **2016**, *27*, 169–178. [CrossRef]
69. Grzywiński, W.; Turowski, R.; Naskrent, B.; Jelonek, T.; Tomczak, A. The impact of season on productivity and time consumption in timber harvesting from young alder stands in lowland poland. *Forests* **2020**, *11*, 1081. [CrossRef]
70. Chmielewski, S.; Porter, B. Valorization of the selected methods of harvesting and skidding of the pine stands in pre-final cut. *Tech. Roln. Ogrod. Les* **2012**, *3*, 23–26.

sustainability

MDPI

Article

Cost Analysis of a Novel Method for Ecological Compensation—A Study of the Translocation of Dead Wood

Ola Lindroos [1,*], Malin Söderlind [2], Joel Jensen [1] and Joakim Hjältén [3]

1 Department of Forest Biomaterials and Technology, Faculty of Forest Sciences, Swedish University of Agricultural Sciences, SE-90183 Umeå, Sweden; joel.jensen@slu.se
2 Norra Skog, Box 4076, SE-90403 Umeå, Sweden; malin.soderlind@norra.se
3 Department of Wildlife, Fish and Environmental Studies, Faculty of Forest Sciences, Swedish University of Agricultural Sciences, SE-90183 Umeå, Sweden; joakim.hjalten@slu.se
* Correspondence: ola.lindroos@slu.se

Abstract: Translocation of dead wood is a novel method for ecological compensation and restoration that could, potentially, provide a new important tool for biodiversity conservation. With this method, substrates that normally have long delivery times are instantly created in a compensation area, and ideally many of the associated dead wood dwelling organisms are translocated together with the substrates. However, to a large extent, there is a lack of knowledge about the cost efficiency of different methods of ecological compensation. Therefore, the costs for different parts of a translocation process and its dependency on some influencing factors were studied. The observed cost was 465 SEK per translocated log for the actual compensation measure, with an additional 349 SEK/log for work to enable evaluation of the translocation's ecological results. Based on time studies, models were developed to predict required work time and costs for different transportation distances and load sizes. Those models indicated that short extraction and insertion distances for logs should be prioritized over road transportation distances to minimize costs. They also highlighted a trade-off between costs and time until a given ecological value is reached in the compensation area. The methodology used can contribute to more cost-efficient operations and, by doing so, increase the use of ecological compensation and the benefits from a given input.

Keywords: restoration; no-net-loss; biodiversity conservation; wood living species; mining; forwarder; forest operations; cost-efficiency; boreal forest; Sweden

Citation: Lindroos, O.; Söderlind, M.; Jensen, J.; Hjältén, J. Cost Analysis of a Novel Method for Ecological Compensation—A Study of the Translocation of Dead Wood. *Sustainability* **2021**, *13*, 6075. https://doi.org/10.3390/su13116075

Academic Editors: Rachele Venanzi, Janine Schweier and Rodolfo Picchio

Received: 16 April 2021
Accepted: 25 May 2021
Published: 28 May 2021

1. Introduction

Anthropogenic disturbance has altered ecosystems worldwide, resulting in habitat loss and species extinctions over a wide range of biomes [1,2]. Forest ecosystems are no exception, and their exploitation has led to changes in ecosystem structures and processes, and to biodiversity loss [2,3]. Species associated with dead wood (saproxylic species) are especially vulnerable.

To conserve biodiversity and continue economic development simultaneously is a major challenge as human society depends on functional ecosystems in numerous ways [4]. Economic growth and biodiversity conservation are often perceived to be incompatible. There is continually increasing pressure on corporations by consumers and stakeholders to be environmentally conscious with greater focus being directed towards alternative approaches in adapting to this [5]. One such approach is the relatively recent concept of ecological compensation (biodiversity offsetting), which is based on the principle that those who destroy or damage natural values are to compensate for this by the creation or protection of natural values at a different/substitute location [6]. Thus, ecological compensation potentially provides an approach that links biodiversity conservation and human development associated with economic growth. Although legislation mandating biodiversity offsets exist in many countries at present, biodiversity offsets are still under development.

Biodiversity offsets are the last step in the mitigation hierarchy, with the overarching goal to achieve no net loss of biodiversity [6]. These actions can involve protection of areas that are otherwise at risk of exploitation, through ecological restoration or other positive management interventions and, in some circumstances, the re-creation of habitat that has been lost. Restoration of degraded habitat is often used for ecological compensation and, although our knowledge of the effects of different restoration methods on biodiversity has improved in recent years [7–10], it often suffers from the innate problem that even if substrates or habitat for species that we want to favor are provided, those species may not be able to migrate to the restored areas [7,9]. Furthermore, the delivery time of some types of substrates is long, e.g., large diameter dead wood takes a very long time to develop and reach late decompositions stages, and it could take centuries after restoration before these kind of substrates are available for specialist species. One way to circumvent this problem is the translocation of substrates and associated species from the impact area to compensation areas.

Translocation means that habitats and substrates with long delivery times are instantly created in a compensation area. In addition, in the best-case scenario, many of the organisms associated with these high-quality substrates are translocated together with the substrates rather than having to migrate there. Thus, fulfilment of the "fields of dreams" hypothesis would be unnecessary for successful compensation and/or restoration [11].

Due to the concept of ecological compensation being a novel one, there is little in the way of relevant research on methods and outcomes of ecological compensation. However, there have been assessments of the ecological functionality and methods for evaluation of the compensation measures (e.g., [12]), and also research into the social and economic effects on local societies (e.g., [13]). However, there has been limited focus on the costs of carrying out the actual compensation measures. When such costs have been investigated, it has often been in terms of the total costs for compensation projects carried out, with little focus on comparing alternatives in order to find and develop cost-efficient practices (e.g., [14]). Even when the ecological compensation constitutes a minor part of large-scale projects, such as the construction of roads and establishment of mines, cost-efficiency is, nevertheless, instrumental for increasing both the use of ecological compensation and increasing the benefits from a given economic input.

Evaluation of cost-efficiency is a central part of research in, for example, forest operations. Based on methods for cost assessment, such research focuses on evaluations of, and improvements to, work carried out, and on predicting the outcomes of planned operations. The aim is to estimate the cost of the work required to produce the desired outcome of the operation. To do so, the cost assessment focuses on two main parts: the cost per unit of time, and the time required per produced unit. The cost per time unit is based on the labor costs, operational costs and investment costs related to the operation [15]. The time required per produced unit is often called time consumption, and often there is a focus on the inverse of the time consumption—the productivity (produced units per time). The time consumption is dependent on the conditions in which the operations are carried out and is measured by dividing the work into distinct parts (work elements) in order to isolate influencing factors better. By doing so, relationships between time consumption and the influencing factors are easier to distinguish [16]. There is a wealth of research on time consumption and productivity of conventional forest operations, with well-known general relationships. For instance, the required time consumption per produced unit increases with transport distance and decreases with the number of units that can be handled at a given time (e.g., per load) [17–20]. Cost-based evaluations of conventional operations are common [21], and there have also been evaluations of how ecological considerations influence the costs of operations [22]. However, cost-based evaluations of ecological compensation are scarce.

The largest copper mine in Sweden is the Aitik mine, founded in 1968 by the mining company Boliden AB. The ore extraction process at the mine produces 50,000 tons of tailings daily. These are subsequently transported to a sand magazine. Recently, the Boliden AB was granted a permit to increase the sand storage area in the Aitik mine by 376 ha. The affected forest land had high to very-high natural value and The Land and Environmental Court of Appeal decided that offsets areas should be created to compensate for the impact. In addition, as part of the compensation measures it was decided that substantial amounts of dead wood and associated species of insect, wood fungi, lichens, bryophytes and lichens living in or on the dead wood should be translocated from the impact to a 397 ha compensation area 5 km (in a straight line) west-southwest from the affected area.

The aim of this study was to assess the cost of translocating dead wood from an affected area to the compensation area. More specifically, the focus was on assessing the costs for different parts of the translocation process and their dependency on some influential factors (i.e., transportation distances and load sizes). Moreover, being part of a scientific project to evaluate the ecological outcome of the translocation, the study also provided input on some of the costs associated with carrying out such an evaluation.

This project is unique in its usage of translocation of natural values as a primary means of ecological compensation in boreal forests. The results presented here are thus highly significant for future endeavors involving ecological compensation (in a boreal setting) and may be valuable to other companies facing similar challenges.

2. Materials and Methods

The study was conducted during the creation of off-set areas required to compensate for the impact of increasing the sand storage area in the Aitik mine. Hence, it was an observational study (and not an experimental study), with only limited possibilities to interfere with the work in order to collect desired data.

2.1. Work Phases

The translocation of dead wood consisted of seven phases, from the identification of substrates (phases 1 and 2) to the translocating work (phases 3–7) (Figure 1). All phases, except the log marking, were necessary for the actual translocation work. However, for the area identification, felling and the insertion phases, there was extra work carried out in order to facilitate the planned scientific evaluation of the translocation.

Figure 1. Progression of the work phases in the translocation project.

The machine operators who carried out the work in the phases Extraction, Road transport and Insertion were given instructions to drive as normal, but to handle the logs with additional care since they could be more fragile than newly harvested and fresh logs.

2.2. Area Description

The impact and compensation areas are located in the north boreal vegetation zone [23] (Ahti et al. 1968). The area had previously been under silvicultural management, predominantly subjected to selective felling, but had not been managed over recent decades. The forests were dominated by conifers (Scots pine (*Pinus sylvestris*) and Norway spruce (*Picea abies*)), and with scattered broadleaves (mainly Downy birch (*Betula pubescens*)).

The affected area covered 376 hectares, of which 167 hectares consisted of forests of high or very high conservation value as defined by Swedish Standards Institute [24]. The area included 21 red-listed species and forest structures important for biodiversity (e.g., very large old trees, snags and logs of unusual dimensions).

The compensation area covered 397 hectares, of which 192 hectares was productive forest (mean annual increment of at least 1 m^3 per ha and year) with high conservation value, 113 hectares of productive forest with low conservation value but with no forest of very-high conservation value. The remaining areas were non-productive forest, mires or open water [25].

To allow for a scientific assessment of the outcome of the translocation of the dead wood for biodiversity, the whole compensation area was systematically inventoried prior to translocation of the dead wood. Thirty circular plots were randomly distributed in areas with productive forest. The translocated dead wood was placed in the plots to enable the planned scientific evaluation. The evaluation focused on assessing ecological effects due to the quality and quantity of dead wood. Hence, each plot received either 48, 16 or zero translocated logs, with 10 replications of each log concentration, giving a total of 30 experimental plots. For each concentration there were instructions on the number of logs of the different types to be placed in each plot. Plots were 50 m in diameter and separated from each other by at least 150 m. Detailed information on the scientific evaluation can be found in [26].

The inventory work and establishment of the plots were carried out by two nature value consultants, who were also involved in work phases 2, 4 and 7.

2.3. Substrate Identification

The impact area was systematically inventoried by the nature value consultants for standing and lying trees, and the meeting of criteria for any of the eight log assortments identified based on tree species (Scots pine or Norway spruce), posture (standing or downed) and decomposition stage (fresh, early or intermediate) as suitable for translocation (Table 1, examples in Figure 2). Trees meeting the criteria were marked.

Table 1. Log assortments that were translocated, and the mean log volume in each assortment, from sampling carried out for this study.

Species	Posture of Tree	Decomposition Stage	Translocated Logs	Sampled Logs		Log Volume (m^3)	
			N	*n*	(%)	Mean	SD
Spruce	Downed	Early decomposition	80	16	20	0.13	0.05
		Intermediate decomposition	80	27	34	0.15	0.05
	Standing	Kelo type (intermediate decomp.)	80	26	33	0.18	0.08
		Fresh	80	17	21	0.33	0.12
Pine	Downed	Early decomposition	20	11	55	0.15	0.07
		Intermediate decomposition	66	15	23	0.23	0.12
	Standing	Kelo type (intermediate decomp.)	94	23	24	0.27	0.13
		Fresh	140	20	14	0.47	0.16
		All pooled	640	155	24	0.24	0.15

2.4. Tree Felling and Log Marking

The translocation work began on 2 October 2017, starting with felling of standing trees, cutting felled and lying trees to logs and marking the logs. The nature value consultants chose which of the marked trees to create logs from, and to which dimensions. The objective was to find logs with diameters greater than 15 cm and lengths between 3 and 5 m. Some trees marked in the previous phase were excluded from translocation as they were too decayed or too thin. Only dead wood substrates that were considered feasible to move without breaking were selected for the translocation. The number of relocation logs created per tree was not recorded, but the number of logs was just slightly higher than the number of trees marked during the substrate identification phase (Table 4).

The felling and cross-cutting work was carried out by two chainsaw operators, each with at least 25 years of experience in motor-manual forestry operations. Both used Husqvarna 550XPG chainsaws with 25.72 cm (18 inch) blades.

The logs were carefully handled so as not to reduce their value to the project e.g., by not disturbing the attached fruiting bodies of wood fungi. The logs were inventoried for red-listed or indicator species, tagged for dead wood assortment, photographed, and their upper side marked with paint so that they could be placed in the same position after translocation. Finally, they were marked with a unique ID. A total of 671 logs were created. Due to a limited number of lying dead pine trees, with only 20 early and 66 intermediary decomposition logs found instead of the desired 80 in each class, they were supplemented with logs from the pine fresh wood and pine kelo style classes, respectively (see N in Table 1).

Figure 2. Examples of different classes of translocated logs. Downed intermediate decomposition logs of pine(**A**) and spruce (**B**), fresh standing pine (**C**) and standing pine of kelo type (**D**). Photo: Nordlund Konsult AB.

2.5. Extraction

All 671 marked logs were extracted from the affected area to the roadside, using a conventional forwarder (Table 2). The size of the forwarder and its impact on the ground was not considered when the machine was being chosen, since the area was to be converted. Logs were arranged by assortment type at two separate landings. In order to protect the translocation logs, the bottom of the pile was made from a layer of conventional roundwood. The extraction was carried out by a machine operator with 18 years of experience of forwarding work.

2.6. Road Transport

The extracted logs were transported using a conventional timber truck (Table 2) from the affected area to three separate landings at the compensation area. The first load (4 logs per assortments, 32 logs in total) was transported to the first landing, loads 2 and 3 were transported to the second landing (28 per log assortment, 224 logs in total) and loads 4, 5 and 6 were transported to the third landing (48 per log assortment, 384 logs in total). The logs were loaded in order of decreasing durability, with fresh wood being placed at the bottom, then kelo type trees, early lying dead wood and, finally, intermediate lying dead wood. In total, 640 of the 671 logs were transported, with the numbers per assortments reported in Table 1.

The road transport was carried out by two machine operators with at least 18 years of experience with self-loading timber trucks.

Table 2. Specifications for the machines used for the work phases Extraction, Road Transport and Insertion.

Feature	Extraction	Road Transport	Insertion
Machine type	Forwarder	Self-loading timber truck	Forwarder
Make and model	Komatsu 865	Scania R440	Terri ATD
Manufacturing year	2013	2012	2001
Work time (hours)	10,800	-	4000
Driven distance (km)	-	850,000	-
Laden mass (tonnes [a])	18	19.6	3
Width (m)	3.2	2.6	1.8
Length (m)	10	22	6.5–7.1
Load capacity (m^3sob)	15	50	4
Propulsion	Wheels with bogie band	Wheels	Tracks
Crane model	CRF 11c	FTG V13	Mowi 2046
Crane reach (m)	12	8	4.6

[a] metric tonnes (1 tonne = 1000 kg).

2.7. Insertion

The insertion of logs to the compensation area was carried out using a small tracked forwarder (Table 2) under the supervision of nature value consultants. The size of the forwarder was chosen to minimize the impact on the ground, and so that it could navigate between the trees in the stand. A total of 640 logs were inserted in 20 plots, with 2 logs of each log assortment placed at each of 10 (16 logs per plot) of the 20 plots, and 6 logs of each log assortment at each of the remaining 10 plots (26 logs per plots). Within plots, log assortments were randomly distributed.

The insertion was carried out by two machine operators with 3 and 5 years of part-time experience, respectively, with small forwarders.

2.8. Cost Assessment

Costs were collected in the national currency Swedish krona (SEK), which at the time of the study had an exchange rate of ca 10.0 SEK per Euro and 8.6 SEK per USD.

The total cost for the compensation project was calculated by totaling the total costs for each work phase. The total cost for each work phase was calculated by multiplying the times required to carry out the work involved in the phase by the hourly cost for the respective work.

Cost per translocated log was calculated for each work phase by dividing the total cost for the work phase by the number of translocated logs according to:

$$C_x = \frac{\sum (T_i \times c_i)}{O} \quad (1)$$

where C is the cost per translocated log for work phase x, T is the time required for work i in the work phase, and c is the hourly cost for work i. O is the number of translocated logs (irrespective of the number of objects handled in work phase x).

For the phases Extraction, Road Transport and Insertion, the work was assessed in more detail by use of time studies in order to enable the creation of models for cost estimations under various work conditions, i.e., with time requirements other than the observed total time requirements in this study.

2.9. Time Studies

The work-time required for translocation, divided into the different work phases, was collected from the self-reported work-time of the operators. For the three last phases (Extraction, Road transport and Insertion) detailed time studies conducted on site were carried out on a number of loads.

For the studied loads, total values of load size, time consumption and distance driven were recorded. Driving speeds were derived from distance and time recordings. The time consumption per observed load in the phases was recorded, split over six work elements (Table 3) to isolate influential variables. Extraction was the first work phase to be time studied and suffered from initial technical challenges to record the driving distances. Hence, between 9–11 of the 26 loads were properly recorded with driving distances at work element level (see "*n*" under Transport distance in Table 6).

Table 3. Definition of work elements used in this study.

Work Element	Definition
Loading	From the first log being gripped to the last log being put in the loadspace
Driving full	From the last log being put in the loadspace to the first log being gripped for unloading
Unloading	From the first log being gripped for unloading to the last log being put on the ground
Driving empty	From the last log being put on the ground to the first log being gripped for loading
Miscellaneous	Work-related activities that fit into none of the above work elements
Delay	Non-work-related disturbances (phone calls, machine breakdowns, etc.)

The work element Loading was observed at even higher resolution for the Road Transport and Insertion phases, by recording loading time per log assortment. In addition to time, distances driven, load sizes (number of logs) and number of crane cycles (loading/unloading) were also recorded.

During extraction, loading times per log assortment were not recorded, since the logs were loaded in the order they were found in the forest, which was not organized by log assortment. Furthermore, it was difficult to determine when the work element Loading was complete, due to the variation in load capacity usage and the indistinct loading area (i.e., loading both when driving from and to the landing).

All time studies were carried out by the same person. During Extraction and Road Transport, the person was located in the machine cabins and, for the Insertion, the person followed the forwarder on foot. The time consumption by the time studies work was recorded in units of a second.

2.10. Calculations

Speed was calculated as the distance driven divided by the time consumption for the work element and/or load (round trips) of interest. Productivity, here defined as the output (in terms of logs or m^3) per time unit, was calculated as the load size divided by the time consumption for the work element and/or load (round trips) of interest of the respective work element.

Between 14 and 55% of the logs in the eight translocated log assortments were sampled to give volume estimations (Table 1). Individual log volumes were calculated assuming a cylinder:

$$(V = \frac{\pi * d^2}{4} \times l) \quad (2)$$

where V is the volume, d is the log diameter over bark at half-length and l is the full length of the log. Hence, an even tapering of the logs was assumed. The volume unit used was solid cubic meter of wood over bark (m^3).

Load sizes during Extraction, Road Transport and Insertion were calculated as the total volume of logs in the load, by multiplying the load's number of logs of each log assortment with the mean log volume for each log assortment (i.e., from Table 1). Utilized load capacity was calculated by dividing load size by the machine's load capacity (Table 2).

2.11. Statistical Analysis

The impact of transport distance and load size during the work phases Extraction, Road transport and Insertion, on the time consumption of work elements Loading, Driving Full, Unloading and Driving empty, was assessed using linear regression. Data preparation was carried out using Microsoft Excel, whereas all the regression analyses were carried out using Minitab 17 (Mintab Inc., State College, PA, USA) with the critical significance level set to 5%.

3. Results

The results are here presented in three levels of detail. First, the work phase costs are reported, based on the costs invoiced. Second, detailed analysis of the work carried out with some work phases is reported, along with the analysis of the relationships between work time required and work conditions. Third, those relationships are used to model estimated costs under other conditions, in order to highlight cost-driving factors as well as to demonstrate the possibility of using the analysis to estimate costs for future projects.

3.1. Costs

The total costs of the seven work phases involved in the compensation project was 520,800 SEK (Table 4). When distributing the total costs over the number of translocated logs, the cost per log was 813.8 SEK. The insertion of logs to the plots in the compensation area was the most expensive work phase, accounting for around 29% of the costs. About two thirds of the insertion cost was associated with the physical translocation of logs, while one third pertained to the work of the nature value consultants. The cheapest work phase was the Road Transport, which accounted for 5% of the costs.

Table 4. Time consumption, hourly cost, total cost, number of objects (logs/trees), cost per translocated log (n = 640) and the work phases' proportion of the cost.

Work Phase	Work Time including Actions Driven by		Time Consumption (hours)	Hourly Cost (SEK/hour)	Total Cost (SEK)	Number of Objects	Cost per Translocated Log (SEK/log)	Proportion of Cost (%)
	Compensation	Evaluation						
Area identification	Yes	Yes	60	525	31,500	-	49.2	6
Substrate identification	Yes	No	90	525	47,250	600	73.8	9
Felling	Yes	Yes	240	400	96,000	671	150.0	19
Log marking	No	Yes	240	525	126,000	671	196.9	24
Extraction	Yes	No	47.5	900	42,750	671	66.8	8
Road transport	Yes	No	27	900	24,300	640	38.0	5
Insertion total			240	637.5	153,000	640	239.1	29
Insertion of logs	*Yes*	*Yes*	*120*	*750*	*90,000*	*640*	*140.6*	*17*
Natural value consultant(s) at insertion	*No*	*Yes*	*120*	*525*	*63,000*	*640*	*98.4*	*12*
Total			884.5		520,800	640	813.8	100

The cost for work directly related to the planned scientific evaluation of the ecological compensation project was made up of all log marking, all of the nature value consultants' contribution at insertion, 20 h (33%) of the time consumption for area identification, 6 h (5%) of the time for insertion of logs (i.e., 20% of the unloading time, see time required below) and 40 h (20%) of the tree felling. When totaling those times and the related hourly cost, the cost related to carrying out the scientific evaluation was 348.8 SEK per log and the cost for the actual compensation work was 465.0 SEK/log. Hence, the included operational preparations for the scientific evaluation of the project required an additional 75% of financial resources, in addition to the resources needed for the compensation work.

3.2. Work Analysis

On average, 18 logs were put on each load during extraction, which was almost twice as many as when inserting the logs (Table 5). However, in terms of utilization of the different vehicles' payload capacities, the conditions were reversed. In fact, only one third of the payload capacity was, on average, used during extraction, whereas more than two thirds were used during insertion. This was due to a much smaller payload capacity on the forwarder used for insertion (Table 2).

Table 5. Load sizes and logs per crane cycle for the work phases Extraction, Road Transport and Insertion.

Feature	Extraction			Road Transport			Insertion		
	n	Mean	SD	*n*	Mean	SD	*n*	Mean	SD
Load size (Number of logs)	26	18.4	6.1	6	106.7	48.75	17	10.4	3.8
Load size (m^3)	26	5.1	2.3	6	28.6	11.6	17	2.8	0.6
Utilized load capacity (%)	26	34	15	6	57	23	17	69	16
Logs per crane cycle at loading (*n*)	26	0.98	0.06	6	2.14	0.43	17	1.02	0.09
Logs per crane cycle at unloading (*n*)	25	1.02	0.09	6	2.38	0.29	17	1.00	0.00

Logs were generally loaded and unloaded individually during the extraction and insertion, whereas slightly more than two logs were handled at a time when loading and unloading during road transport (Table 5).

On average, the total distance driven during Extraction was almost 1.6 km per load, 49 km during Road Transport, and 2.3 km during Insertion (Table 6). However, the distances varied between loads. For Extraction and Road Transport, the loading work required the highest amount of work time, whereas it was the driving with and without load, that required most work time for Insertion. The first three loads (to landings 1 and 2) of the Road transport loads were hauled to insertion landings 1 and 2, at about 13 km driving distance from the extraction landing. Loads 4–6 were hauled to landing 3, about 33 km from the extraction landing. Thus, over all loads, the distances reported had high variation (high SD in Table 6), but the speeds were rather consistent.

Table 6. Time consumption, transport distance and speed per load for the four work phases, distributed over work elements.

Work Phase	Work Element	Time Consumption per Load (h)			Transport Distance per Load (km)			Speed per Load (km/h)		
		n	Mean	SD	*n*	Mean	SD	*n*	Mean	SD
Extraction	Driving empty	26	0.15	0.10	11	0.32	0.19	11	2.35	0.34
	Loading	26	0.38	0.20	10	0.39	0.27	10	1.20	0.35
	Driving full	26	0.12	0.10	9	0.20	0.16	9	2.15	0.44
	Unloading	26	0.14	0.06	10	0.07	0.05	10	0.49	0.30
	Miscellaneous	26	0.03	0.04	-	-	-	-	-	-
	Delay	26	0.05	0.10	-	-	-	-	-	-
	All pooled [1]	26	0.87	0.45	24	1.57	0.94	-	-	-
Road transport	Driving empty	6	0.63	0.32	6	25.3	13.48	6	40.0	2.78
	Loading	6	0.83	0.40	6	0.71	0.83	6	0.77	0.69
	Driving full	6	0.59	0.27	6	22.9	10.88	6	38.5	1.05
	Unloading	6	0.60	0.35	6	0.22	0.10	6	0.52	0.52
	Miscellaneous	6	0.23	0.17	-	-	-	-	-	-
	Delay	6	0.28	0.26	-	-	-	-	-	-
	All pooled [1]	6	3.17	1.03	6	49.08	24.62	-	-	-
Insertion	Driving empty	17	0.21	0.09	17	0.96	0.27	17	4.82	1.44
	Loading	17	0.15	0.05	17	0.06	0.06	17	0.44	0.41
	Driving full	17	0.26	0.09	17	0.93	0.21	17	3.90	1.51
	Unloading	17	0.29	0.12	17	0.30	0.24	17	0.98	0.56
	Miscellaneous	17	0.11	0.10	-	-		-	-	-
	Delay	17	0.12	0.28	-	-		-	-	-
	All pooled [1]	17	1.14	0.33	17	2.25	0.46	-	-	-

[1] For most studied loads, total values were recorded but not always at a work element level.

On average, the driving speed with and without a load was about 2 km/h during Extraction, whereas it was about twice as high during Insertion (Table 6). The average driving speed during Road Transport was almost 40 km/h. In all three work phases, the driving speed was slightly higher when driving without payload than with payload.

In total, the mean productivities were 27.6 extracted logs per hour (SD 21.9), 50.2 road transported logs per hour (SD 19.3) and 11.5 logs inserted per hour (SD 3.3) This corresponds to a volume-based productivity of 7.6 (SD 8.1), 9.2 (SD 3.9) and 2.8 (SD 0.6) m^3 per hour.

The analysis of the observed work element relationships with external factors yielded significant relationships between distance and the time required for driving with and without load (Table 7). Moreover, the time required for loading as well as unloading depended significantly on the load size (i.e., number of logs per load). For Extraction, the distance driven during loading was found to be the strongest predictor of time consumption per load (Loading, time per load (h) = 0.1041 × distance (km), $n = 10$, $p < 0.001$, R^2-adj = 89.7). However, given the features of the loading work, the loading is dependent on the number of logs loaded, and the distance driven during loading is directly dependent on the logs' dispersion. Indeed, the loading distance driven was strongly dependent on the load size (distance (km) = 0.0263 × load size (number of logs), $n = 10$, $p < 0.001$, R^2-adj = 83.6). Following this logic, and due to the rather small sample size, load size was used for the models in Table 7.

Table 7. Models to predict time consumption per load as a function of transport distance or load size, based on regression analysis of observations in Table 6. In the case where no significant models could be found for a work element, the time consumption was predicted using the mean value from the observations.

Work Phase	Work Element	Time Consumption per Load [(a)]	p-Value	R^2-adj (%)	n
Extraction	Driving empty	0.4431 a *	<0.001	98.8	11
	Loading	0.02057 b	<0.001	86.6	26
	Driving full	0.4399 a	<0.001	98.4	9
	Unloading	0.007667 b	<0.001	93.7	25
	Miscellaneous	0.03	-	-	26
	Delay	0.05	-	-	26
Road transport	Driving empty	0.02462 a	<0.001	99.5	6
	Loading	0.007812 b	<0.001	98.4	6
	Driving full	0.02574 a	<0.001	99.9	6
	Unloading	0.005843 b	<0.001	97.1	6
	Miscellaneous	0.23	-	-	6
	Delay	0.28	-	-	6
Insertion	Driving empty	0.2262 a	<0.001	93.6	17
	Loading	0.0612 + 0.00847 b	0.004	40.38	17
	Driving full	0.2770 a	<0.001	92.3	17
	Unloading	0.02830 b	<0.001	97.7	17
	Miscellaneous	0.11	-	-	17
	Delay	0.12	-	-	17

[(a)] a = Transport distance in km, b = load size in number of logs. When no significant relationships with a or b were found, the mean values from Table 6 were used. * The coefficient value for a corresponds to a speed, which can be calculated by dividing 1 by the coefficient (e.g., the model's speed for driving empty during extraction = 1/0.4431 = 2.3 km/h).

For Insertion, the loading time model yielded a relatively low level of explained variation. This was because two different work modes were used: one where several assortments (6 loads) were loaded, and one where fewer assortments (11 loads) were loaded. When analyzing just the load with fewer assortments (i.e., with less driving between log piles during loading), the model explained a much higher level of the variation (Loading time per load (h) = 0.0434 + 0.00832 × load size (number of logs), $n = 11$, $p < 0.001$, R^2-adj = 85.7). However, both models resulted in very similar time predictions.

3.3. Modelling of Costs and Cost Sensitivity

Using the models described in Table 7, it is possible to analyze the effect of distance and load size on the time consumption for Extraction, Road Transport and Insertion. By combining this with the hourly cost for the work, the effect on costs can also be analyzed.

For instance, the total time required per load for the translocation related Insertion work can be estimated by adding the six work element models thus:

$$\text{Total insertion time} = 0.2262a_1 + 0.0612 + 0.00847b + 0.2770a_2 + 0.8 \times 0.02830b + 0.11 + 0.12 \quad (3)$$

where a_1 and a_2 are the distances (in km) expected to be driven when driving without and with load, respectively, and b is the number of logs expected in the load. Since 20% of the observed unloading time during Insertion was considered attributable to the work related to carrying out the scientific evaluation, the model for predicting the unloading time is multiplied by 0.8 (i.e., (100% − 20%)/100).

The cost per load is calculated by multiplying the hourly cost for insertion work (from Table 5) with Equation (3), and the cost per log is calculated by dividing the cost per load by *b*. It should be noted that this modeling assumes that a large number of logs are translocated and thus results in a large number of loads. The fewer the number of loads transported, the more influential the threshold effects of utilized load capacity will be. For instance, when considering translocating only a few logs, there will be a substantial difference in costs per log if a last trip with a single log in the load is required.

By applying this methodology and assuming that distances with, and without, a load are the same (for simplicity), the general effects of distance can be explored (Figure 3A). The distance-independent work per load is high for the Road transport, as indicated by the long time required for the distance of 0 km. However, when distributing the time based on load size, the Road Transport has the lowest time required for distance-independent work of all three work phases (Figure 3B). Insertion starts at the highest amount of time required per log, but extraction exceeds the time required at an expected driving distance of 1.5 km. However, since the insertion forwarder has a lower hourly cost than the extraction forwarder, the intersection is at 0.8 km when looking at the cost per log (Figure 3C). In this example, it costs 29.4, 16.7 and 45.2 SEK/log for the distance-independent work for the Extraction, Road Transport and Insertion phases, respectively. On top of this, there is an additional 44.2, 0.4 and 22.5 SEK/log in transportation costs for each kilometer.

As indicated by the models (Table 7), load size also affects the time consumption and thus costs, since it influences the loading and unloading work elements. Based on similar examples as those in Figure 3, it can be seen that the expected time per load increases most for each extra log in the load for Insertion (Figure 4A). When being analyzed as time per log, it can be seen that the load size effect is accentuated the smaller the loads are (Figure 4B). A similar pattern is also present for the cost per log (Figure 4C), and the effect is due to how load size-independent time/costs are distributed over fewer logs the smaller the load is. In contrast, the load size effect flattens out at a certain number of logs in the load, and this happens at a lower load size when the log-independent time requirements per load (and costs) are low. For 1 log per load, the time required is expected to be 0.37, 1.73 and 0.78 h/log for the Extraction, Road Transport and Insertion, respectively, under the assumptions shown in Figure 4. Correspondingly, the costs would be 336, 1559 and 581 SEK/log.

By applying different load sizes and hourly costs, the effects can be explored further than here, and the costs for possible alternatives can be evaluated.

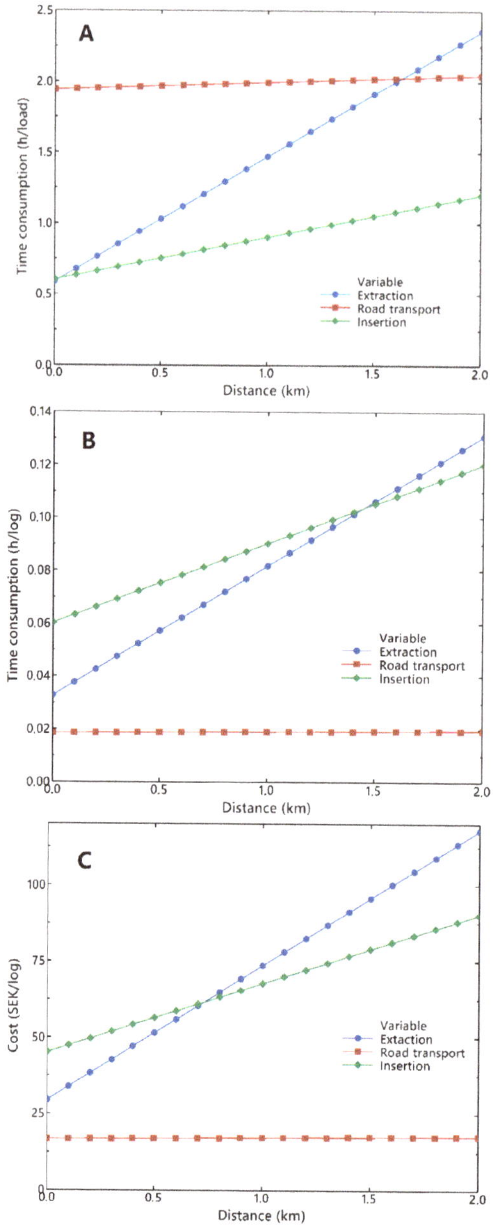

Figure 3. Predicted time consumption (**A**) per load and (**B**) per log, and (**C**) predicted cost per log for the actual translocation work as a function of expected driving distance, when assuming that the distances driven with and without load are identical. Hence, 1 km in the figure gives a total distance of 2 km for driving with and without load. Load sizes are assumed to 18, 105 and 10 logs for the Extraction, Road Transport and Insertion, respectively. Correspondingly, the hourly costs are assumed to be 900, 900 and 750 SEK, respectively.

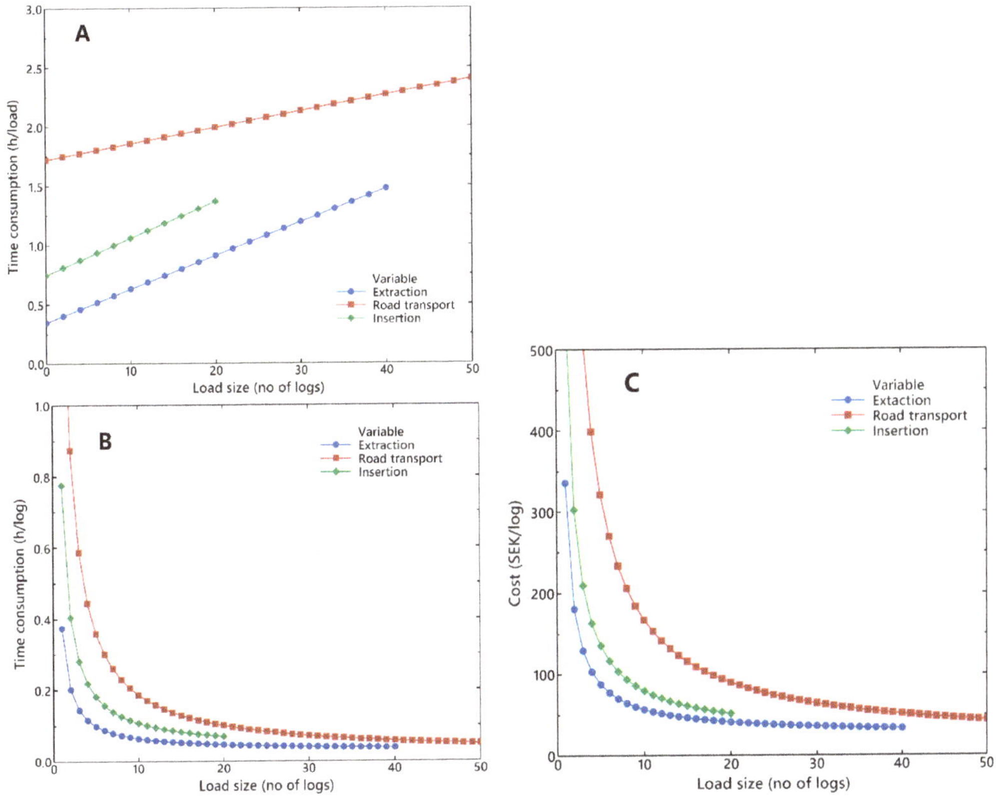

Figure 4. Predicted time consumption (**A**) per load and (**B**) per log, and (**C**) predicted cost per log as a function of expected load size. To mirror approximately the maximal load capacity of the study's vehicles, load sizes have been limited to 40 and 20 logs for Extraction and Insertion, respectively. Driving distances are assumed to be 0.3, 24 and 0.9 km for driving with and without load, respectively, for the Extraction, Road Transport and Insertion, respectively. Correspondingly, the hourly costs are assumed to be 900, 900 and 750 SEK, respectively.

4. Discussion

Translocation of dead wood and associated organisms is a novel method for ecological compensation and restoration that could potentially provide a new important tool for biodiversity conservation. For example, by using this method, habitats/substrates with normally very long delivery times (e.g., large diameter dead wood in late stages of decomposition) are instantly created in a compensation area and, in the best-case scenario, many of the organisms are translocated together with the substrates. Hence, the time to achieve high ecological value in the compensation area is likely to be substantially shortened by the translocation. However, neither the outcome for biodiversity conservation nor the cost for translocation of dead wood have yet been properly evaluated. This study is, to the best of our knowledge, the first attempt to assess the cost of the translocation process. Our analyses revealed big differences in cost for different phases in the translocation process and thus identified phases where efforts should be made to reduce costs and improve efficiency.

The observed cost was 813.8 SEK per translocated log, of which 42% was for work related to facilitating the planned scientific evaluation of the translocation's ecological results. Thus, when omitting work related to the scientific evaluation and including only the work related to the actual translocation plan, the cost was 465 SEK/log.

Due to the different nature of translocation work compared to normal forest operations, it is difficult to compare the observed time consumptions and costs to previous research. However, it can be noted that the driving speeds are in line with previous research on extraction [18,27] and road transport [28,29]. There is a lack of research about the forwarder used for the insertion (Terri ATD), but it can be noted that it was driven at a considerably higher speed than during extraction with conventional forwarders. Hence, the small load volume of the insertion forwarder is, to some extent, compensated by its higher speed. In relation to the forwarder used in the extraction, the insertion forwarder's work becomes more competitive the longer the distances driven (Figure 3B,C).

In relation to costs, there is a limited amount of research to compare with. However, the cost of conventional logging can serve to put the costs in perspective. The average costs for harvesting trees and extracting the logs to roadside in Sweden are approximately 95 SEK/m^3 for final felling, with average tree volumes of 0.23 m^3 under bark and with 44% of the costs being attributable to the extraction [30]. So, assuming the insertion corresponds to two extraction costs, conventional harvesting, extraction and insertion would be 137 SEK per m^3. Conventional Swedish road transport costs are approximately 86 SEK per m^3 in Northern Sweden [30]. So, in total, conventional felling, extraction, road transport and insertion could be estimated to cost 223 SEK per m^3. With a mean log volume of 0.24 m^3 over bark as in this study, this corresponds to approximately 54 SEK/log, whereas it was observed to be almost seven times higher (363 SEK/log) for the corresponding work phases of actual translocation work (i.e., with the costs of the scientific evaluation excluded). There are substantial shortcomings in the comparison, but it, nevertheless, clearly highlights the considerably higher costs related to translocation of ecologically valuable, and sensitive, logs compared to fresh logs for industrial uses.

In this observational study, total time consumptions and costs were based on self-reported data from the contractors executing the translocation work. The self-reported data was also provided to the customer for reimbursement purposes, and thereby motivating correctness in relation to agreements and to maintain the business relationship. Thus, the self-reported costs were the actual costs invoiced for executing the studied work. The conducted time studies of some work phases gave input on how the invoiced costs would change under different conditions. Additional time studies could provide similar information for also the other work phases. To estimate the actual time requirements and the actual hourly costs (and not invoiced) would naturally also be of interest, but would require a larger study set-up and also data for estimating hourly costs of contractors.

Although the full cost of the project to expand the sand magazine is not known, the compensation costs can be expected to be relatively small in comparison. Compensation costs are, nevertheless, one of many costs in projects that often require large investments, so cost efficiency can contribute to at least two beneficial effects. The first is by making ecological restoration more likely to be used when costs can be kept low. The second is by providing increased benefits for a given cost, when the most cost-efficient measures can be chosen.

There are several possible ways to influence the costs, by choices related to the planning and execution of the operations, and by the qualitative requirements of the result of the operation.

The results showed that there is an increased cost with increased driving distances for all three work phases in which logs are transported (Figure 3B). The cost increase was substantial for both extraction and insertion. Hence, the closer to roads that the logs to be extracted can be collected, the lower the costs will be. Correspondingly, insertion costs will be lower the closer to the road the logs can be placed in the compensation area. In contrast, road transport contributed very little to overall cost and the cost increased very little with distance. Thus, an increase in road transport distance will have relatively little effect on overall cost, so translocation over larger distances is possible without substantial impact on overall cost. In fact, a doubling of the road distance driven would only have resulted in a 4% increase in costs per translocated log (0.40 SEK/km and log × 50 km/456 SEK/log).

For future translocation projects, this knowledge should be beneficial in allowing the prioritization of short extraction and insertion distances over road transportation distances when trying to minimize costs.

Another well-known time and cost driving factor that was not studied is the effect of log concentration. This can, briefly, be described as the more logs that are located in one place when being loaded or unloaded, the faster and thus cheaper the work [31]. Hence, if logs can be collected within a limited area during each round trip with the forwarder, the cheaper it will be. Similarly, it will be more efficient if the loading and unloading of trucks and the insertion forwarder can be carried out in concentrated areas. In fact, the organization of logs at the roadside landings was one important area of improvement highlighted by the operators in the study, in order to minimize the need for relocating the vehicles when loading/unloading different log types.

The operations described in this study were carried out for the first time by the people involved. It can be expected that repeating the operation will be quicker as these people gain more experience, and hence, the costs will decrease. A key way of improving extraction may be to use a different type of grapple than the conventional one used, preferably one that is small, straight and pointed to grip one log at a time without damaging the logs being loaded or the adjacent logs. Better planning and marking of log positions and log types, as well as clearer extraction trails, have been suggested by the operators to reduce the time for extraction, but it would also create extra work during extraction preparation. Furthermore, it would be desirable to only work with as many logs as are required for the relocation. In this study, 671 logs were produced during felling and subsequently extracted, but only 640 were road transported and inserted. The excess could be seen as a buffer to ensure that the desired number of logs of different qualities were available for the Road Transport and Insertion phases but, ideally, such an excess would not be necessary.

A rough estimation for the new improved planning and execution is that it could result in a total cost of 397 SEK/log, corresponding to a cost reduction of 15%. The estimation is the result of removing the time estimated to be associated with the 31 excess logs, a 30% decrease in time consumption during loading for both extraction and road transport (corresponding to better organization of roadside landings), and a subsequent 9% decrease in time consumption during extraction, road transport and insertion (corresponding to a general improvement in work efficiency due to experience).

The qualitative requirements of the operation greatly affected the cost. This was partly manifested by the high cost for the insertion, in part due to the choice of using a smaller forwarder to reduce the risk of damage to the ground in the compensation area. Thus, if ground damage had not been an issue, the use of a forwarder with a higher payload and thereby better cost-efficiency might have reduced costs. However, the quality of the translocated logs affected the cost even more, as indicated by the low load capacity utilizations (Table 5). If payload capacities were fully met, as with conventional logging, the costs would be substantially decreased. The challenge, however, is that the more decomposed and therefore fragile the logs are, the greater ecological value they might have as more rare- and threatened-species are associated with the late stages of decomposition [32,33]. Hence, there is a trade-off between costs and the qualitative properties of the translocated logs. Increasing payload usage by loading more logs with high ecological values carries with it an increased risk that log structures and species living on the logs would be severely damaged when piling them on top of each other. In fact, the most valuable logs in the impact area were not translocated since it was not considered feasible to move them due to their fragility [26]. Another way to utilize the payload capacity would be to translocate less ecologically valuable logs that can survive being transported in full loads. Yet another way would be to construct loads such that the most fragile logs were loaded on top with the most solid logs at the bottom of a load.

The level of decomposition of the logs could also be expected to influence the loading and unloading work. The more decomposed the logs are, the more carefully they need to be handled. The size of the logs is also a determining factor on the loading and unloading times. During road transport, the observed difference in mean loading times between the fastest log assortment (spruce logs, early decomposition) and the slowest (pine logs, intermediate decomposition) was 13.3 s per log. The same log categories were loaded fastest and slowest during insertion, with a mean difference of 23.4 s per log. It was not possible to evaluate whether the observed differences were statistically significant due to the aggregated data collection method, neither was it possible to evaluate possible reasons (such as differences in log sizes between classes).

When considering the choices related to qualitative results of the translocation, there is a trade-off between costs and time until a given ecological value is reached in the compensation area. If it is acceptable to wait for the ecological values to develop, economy of scale can be used in order to cut costs. Conversely, the time required to achieve high ecological values in the compensation area can be reduced by accepting higher costs.

To know the acceptable timeline for ecological values to develop, there is a need for knowledge about the speed at which such values develop under different treatments. The work carried out in this translocation project included facilitation of such scientific evaluation, which focused on evaluation of the effect of differences in concentration of dead wood. Hence, this study covered some of the costs of knowledge production and found that the operational work to carry out the scientific evaluation required an additional 75% of financial resources in addition to the resources needed for the actual compensation work. The operational work to lay out the experiment is only one part of an evaluation project. Thus, the costs related the researchers' design of the experiment, the inventories required to follow up the results of the translocation, the data analysis and publication of results should be totaled to give the full cost of the scientific evaluation. A modest estimation is that those costs not accounted for in this study are of a magnitude 10–100 times higher than the observed cost related to carrying out the scientific evaluation.

To the best of our knowledge, this study is unusual in that it added a cost-efficient dimension to ecological compensation. The data from a single observational study should naturally be handled with care when being used in other contexts, and there is a need for additional studies to verify and refine the findings. However, the general results are in line with similar studies from conventional forest operations, indicating that similar cost-driving aspects are in action. Hence, the methodology seems applicable to operations within the field of ecological compensation, and could be applied to enable cost-benefit evaluations of possible alternatives. This would enable efficient use of scarce resources, both when it comes to the strict delivery of the compensation (i.e., to spend as few resources as possible on a decided compensation action) as well as to the ecological quality of different actions (i.e., to choose the action that gives a desired benefit-to-cost level).

Author Contributions: M.S. and O.L. conceived and designed the study, with contributions from J.H.; M.S. collected the data; M.S. and O.L. analyzed the data, with contributions from J.J. and J.H., O.L., M.S., J.J. and J.H. wrote the manuscript. All authors have read and agreed to the published version of the manuscript.

Funding: This research was funded by SLU, Boliden AB and the Swedish Research Council on Environment, Agricultural Sciences and Spatial Planning (Formas), grant number 2019-00923.

Acknowledgments: This paper was based on data from the Master's thesis of Söderlind [34]. The authors thank Boliden AB and Sveaskog AB for allowing access to the impact area and help with practicalities, Nordlund Konsult for invaluable help with planning and carrying out the translocations and the machine operators who volunteered to participate in the study.

Conflicts of Interest: The authors declare no conflict of interest. The funders had no role in the design of the study, in the collection, analyses, or interpretation of data, in the writing of the manuscript, or in the decision to publish the results.

References

1. Butchart, S.H.M.; Walpole, M.; Collen, B.; van Strien, A.; Scharlemann, J.P.W.; Almond, R.E.A.; Baillie, J.E.M.; Bomhard, B.; Brown, C.; Bruno, J.; et al. Global Biodiversity: Indicators of Recent Declines. *Science* **2010**, *328*, 1164–1168. [CrossRef] [PubMed]
2. FAO. *Global Forest Resources Assessment 2010*; FAO: Rome, Italy, 2010; pp. 1–378.
3. Ceballos, G.; Ehrlich, P.; Barnosky, A.; García, A.; Pringle, R.; Palmer, T. Accelerated modern human-induced species losses: Entering the sixth mass extinction. *Sci. Adv.* **2015**, *1*, e1400253. [CrossRef]
4. Lubchenco, J. Entering the century of the environment: A new social contract for science. *Science* **1998**, *279*, 491–497. [CrossRef]
5. Verrier, B.; Rose, B.; Caillaud, E.; Remita, H. Combining organizational performance with sustainable development issues: The Lean and Green project benchmarking repository. *J. Clean. Prod.* **2014**, *85*, 83–93. [CrossRef]
6. BBOP. *Biodiversity Offset Design Handbook-Updated*; Forest Trends: Washington, DC, USA, 2012.
7. Bell, D.; Hjältén, J.; Nilsson, C.; Jorgensen, D.; Johansson, T. Forest restoration to attract a putative umbrella species, the white-backed woodpecker, benefited saproxylic beetles. *Ecosphere* **2015**, *6*, 1–14. [CrossRef]
8. Hjältén, J.; Hägglund, R.; Johansson, T.; Roberge, J.-M.; Dynesius, M.; Olsson, J. Forest restoration by burning and gap cutting yield distinct immediate effects on saproxylic beetles. *Biodiver. Conserv.* **2017**, *26*, 1623–1640. [CrossRef]
9. Hyvärinen, E.; Kouki, J.; Martikainen, P. Prescribed fires and retention trees help to conserve beetle diversity in managed boreal forests despite their transient negative effects on some beetle groups. *Insect Conserv. Divers.* **2009**, *2*, 93–105. [CrossRef]
10. Versluijs, M.; Eggers, S.; Hjältén, J.; Löfroth, T.; Roberge, J.-M. Ecological restoration in boreal forest modifies the structure of bird assemblages. *For. Ecol. Manag.* **2017**, *401*, 75–88. [CrossRef]
11. Palmer, M.A.; Ambrose, R.F.; Poff, N.L. Ecological theory and community restoration ecology. *Restor. Ecol.* **1997**, *5*, 291–300. [CrossRef]
12. Bull, J.W.; Suttle, K.B.; Gordon, A.; Singh, N.J.; Milner-Gulland, E.J. Biodiversity offsets in theory and practice. *Oryx* **2013**, *47*, 369–380. [CrossRef]
13. Bidaud, C.; Schreckenberg, K.; Jones, J.P. The local costs of biodiversity offsets: Comparing standards, policy and practice. *Land Use Policy* **2018**, *77*, 43–50. [CrossRef]
14. Cuperus, R.; Bakermans, M.M.; De Haes, H.A.U.; Canters, K.J. Ecological compensation in Dutch highway planning. *Environ. Manag.* **2001**, *27*, 75–89. [CrossRef] [PubMed]
15. Ackerman, P.; Belbo, H.; Eliasson, L.; de Jong, A.; Lazdins, A.; Lyons, J. The COST model for calculation of forest operations costs. *Int. J. For. Eng.* **2014**, *25*, 75–81. [CrossRef]
16. Magagnotti, N.; Spinelli, R.; Acuna, M.; Bigot, M.; Guerra, S.; Hartsough, B.; Kanzian, C.; Kärhä, K.; Lindroos, O.; Roux, S.; et al. *Good Practice Guidelines for Biomass Production Studies*; COST Action FP-0902, WG 2 Operations Research and Measurement Methodologies; CNR IVALSA: Sesto Fiorentino, Italy, 2012.
17. Eriksson, M.; Lindroos, O. Productivity of harvesters and forwarders in CTL operations in Northern Sweden based on large follow-up datasets. *Int. J. For. Eng.* **2014**, *25*, 179–200. [CrossRef]
18. Manner, J.; Nordfjell, T.; Lindroos, O. Automatic load level follow-up of forwarders' fuel and time consumption. *Int. J. For. Eng.* **2016**, *27*, 151–160. [CrossRef]
19. Cabral, O.M.D.J.V.; Lopes, E.D.S.; Rodrigues, C.K. Impact of distance between strip roads on productivity and costs of a forwarder in commercial thinning of Pinus Taeda Stands. *Croat. J. For. Eng.* **2020**, *41*, 243–249. [CrossRef]
20. Santos, D.W.F.D.N.; Magalhães Valente, D.S.; Fernandes, H.C.; Souza, A.P.D.; Cecon, P.R. Modeling technical, economic and environmental parameters of a forwarder in a Eucalyptus forest. *Int. J. For. Eng.* **2020**, *31*, 197–204.
21. Gagliardi, K.; Ackerman, S.; Ackerman, P. Multi-product forwarder-based timber extraction: Time consumption and productivity analysis of two forwarder models over multiple products and extraction distances. *Croat. J. For. Eng.* **2020**, *41*, 231–242. [CrossRef]
22. Eliasson, L.; Grönlund, Ö.; Lundström, H.; Sonesson, J. Harvester and forwarder productivity and net revenues in patch cutting. *Int. J. For. Eng.* **2021**, *32*, 3–10. [CrossRef]
23. Ahti, T.; Hämet-Ahti, L.; Jalas, J. Vegetation zones and their sections in northwestern Europe. In *Annales Botanici Fennici*; Societas Zoologica Botanica Fennica Vanamo: Helsinki, Finland, 1968; pp. 169–211.
24. Swedish Standards Institute. *Geographic Information—Data on Forest and Forest Management*; SS 637009; Swedish Standards Institute: Stockholm, Sweden, 2016. (In Swedish)
25. Forsgren, A.; Forsberg, A.; Lindmark Burck, S.; Granér, A.; Josefsson, T.; Nordin, J.; Nordlund, E.; Nordlund, M. *Kompensationsplan: Utökning sandmagasin Aitik. [Compensation plan: Increased sand magazine Aitik]*; Boliden Mineral AB: Gällivare, Sweden, 2016. (In Swedish)
26. Tranberg, O.; Hekkala, A.-M.; Löfroth, T.; Jönsson, M.; Sjögren, J.; Lindroos, O.; Hjältén, J. Translocation of deadwood in biodiversity offsetting, a novel way to compensate for habitat loss due to land exploitation. (Manuscript).
27. Manner, J.; Palmroth, L.; Nordfjell, T.; Lindroos, O. Load level forwarding work element analysis based on automatic follow-up data. *Silva Fenn.* **2016**, *50*, 1546. [CrossRef]
28. Nurminen, T.; Heinonen, J. Characteristics and Time Consumption of Timber Trucking in Finland. *Silva Fenn.* **2007**, *41*, 471–487. [CrossRef]
29. Svenson, G.; Fjeld, D. The impact of road geometry, surface roughness and truck weight on operating speed of logging trucks. *Scand. J. For. Res.* **2017**, *32*, 515–527. [CrossRef]

30. Eliasson, L. *Skogsbrukets kostnader och intäkter 2018. (Costs and Revenues in Forestry)*; Webresultat Nr 35; Skogforsk: Uppsala, Sweden, 2019; p. 2. (In Swedish)
31. Manner, J.; Nordfjell, T.; Lindroos, O. Effects of the number of assortments and log concentration on time consumption for forwarding. *Silva Fenn.* **2013**, *47*, 1030. [CrossRef]
32. Stokland, J.N.; Siitonen, J.; Jonsson, B.G. *Biodiversity in Dead Wood*; Cambridge University Press: Cambridge, UK, 2012.
33. Junninen, K.; Komonen, A. Conservation ecology of boreal polypores: A review. *Biol. Conserv.* **2011**, *144*, 11–20. [CrossRef]
34. Söderlind, M. *Cost Analysis of Ecological Compensation—A Study of the Relocation of Dead Wood at the Aitik Mine*; Rapport 12, Dept. of Forest Biomaterials and Technology, Swedish University of Agricultural Sciences: Umeå, Sweden, 2018.

MDPI
St. Alban-Anlage 66
4052 Basel
Switzerland
Tel. +41 61 683 77 34
Fax +41 61 302 89 18
www.mdpi.com

Sustainability Editorial Office
E-mail: sustainability@mdpi.com
www.mdpi.com/journal/sustainability

www.ingramcontent.com/pod-product-compliance
Lightning Source LLC
Chambersburg PA
CBHW061442210926
51850CB00008B/1590
9783036549774